Bioconjugation Protocols

METHODS IN MOLECULAR BIOLOGY™

John M. Walker, SERIES EDITOR

METHODS IN MOLECULAR BIOLOGY™

Bioconjugation Protocols

Strategies and Methods

Edited by

Christof M. Niemeyer

Fachbereich Chemie der Universität Dortmund
Dortmund, Germany

HUMANA PRESS ✳ TOTOWA, NEW JERSEY

© 2004 Humana Press Inc.
999 Riverview Drive, Suite 208
Totowa, New Jersey 07512

www.humanapress.com

This publication is printed on acid-free paper. ∞
ANSI Z39.48-1984 (American Standards Institute)

Permanence of Paper for Printed Library Materials.
Cover illustration: Fig. 1 from Chapter 18, "Generation and Characterization of Ras Lipoproteins Based on Chemical Coupling," by Melanie Wagner and Jürgen Kuhlmann.

Production Editor: Jessica Jannicelli
Cover design by Patricia F. Cleary.

For additional copies, pricing for bulk purchases, and/or information about other Humana titles, contact Humana at the above address or at any of the following numbers: Tel.: 973-256-1699; Fax: 973-256-8341; E-mail: humana@humanapr.com; or visit our Website: www.humanapress.com

Printed in the United States of America. 10 9 8 7 6 5 4 3 2 1

e-ISBN: 1-59259-813-7

Library of Congress Cataloging in Publication Data

Bioconjugation protocols : strategies and methods / edited by
Christof M. Niemeyer.
    p. ; cm. -- (Methods in molecular biology ; 283)
  Includes bibliographical references and index.
  ISBN 1-58829-098-0 (alk. paper)
  1. Bioconjugates--Laboratory manuals.
  [DNLM: 1. Immunohistochemistry--methods. 2. Adjuvants,
Immunologic. 3. Gene Transfer Techniques. 4. Molecular
Probes--chemical synthesis. QW 25 B6147 2004] I. Niemeyer,
Christof M. II. Series: Methods in molecular biology (Clifton, N.J.) ; v. 283.
  QP517.B49B565 2004
  612.1'111--dc22
                        2004004802

Preface

There are a number of outstanding volumes that provide a comprehensive overview of bioconjugation techniques. However, many of the conventional approaches to the synthesis of chemically modified protein conjugates lack efficient means to control the stoichiometry of conjugation, as well as the specific site of attachment of the conjugated moiety. Moreover, the recent developments in microarray technologies as well as in nanobiotechnology—a novel field of research rapidly evolving at the crossroads of physics, chemistry, biotechnology, and materials science—call for a summary of modern bioconjugation strategies to overcome the limitations of the classical approaches. *Bioconjugation Protocols: Methods and Strategies* is intended to provide an update of many of the classic techniques and also to introduce and summarize newer approaches that go beyond the pure biomedical applications of bioconjugation. The purpose of *Bioconjugation Protocols: Methods and Strategies* is therefore to provide instruction and inspiration for all those scientists confronting the challenges of semisynthesizing functional biomolecular reagents for a wide variety of applications ranging from novel biomedical diagnostics, to therapeutics, to biomaterials.

Part I contains seven protocols for the preparation of protein conjugates. The use of noncovalent conjugation mediated by the versatile streptavidin–biotin system, described in Chapters 1 and 2, enables the synthesis of therapeutic enzyme–antibody conjugates. Streptavidin itself is the target for covalent bioconjugation with thermoresponsive polymers (Chapter 3), and Chapter 4 describes the effective covalent conjugation of biomolecules with polyethylene glycols. Covalent techniques are also used to produce stoichiometrically defined bispecific antibodies (Chapter 5) and the synthesis of immunoconjugates with a defined regioselective chemical modification is reported in Chapter 6. In Chapter 7, an enzymatic procedure to conjugate proteins with small molecules is presented.

In Part II, various approaches to the synthesis of nucleic acid conjugates are illuminated within seven protocols. Chapters 8 and 9 focus on the fluorescent labeling of nucleic acids for microarray analyses and single-molecule detection, respectively, whereas in Chapter 10 a method is described for the sequence-specific enzymatic labeling of DNA by means of methyltransferases. Chapter 11 deals with the *in situ* labeling of DNA amplicons during PCR, which is useful for the ultrasensitive immunological detection of protein antigens. Chapter 12 describes the synthesis of stoichiometrically well-defined DNA oligonucleotide–streptavidin conjugates, useful as biomolecular adapters in microarray techniques and the nanosciences. Chapters 13 and 14 report on the synthesis of nucleic acid peptide and protein conjugates using either DNA or peptide nucleic acids, respectively.

Part III is focused on approaches toward semisynthetic glycosyl- and lipid-conjugates of proteins and their implications for medicinal chemistry. Chapter 15 gives an overview on protein lipidation by means of synthetic chemistry, and concrete examples of this young field of research—such as the synthesis of lipidated peptides and the semisynthesis of Rab and Ras lipoproteins—are presented in Chapters 16–18. The conjugation of peptides to saccharide moieties is outlined in the next two chapters. Chapter 19 describes the conjugation of glycopeptide thioesters to recombinant protein fragments, and Chapter 20 outlines the enzymatic condensation of glycosylated peptides.

Part IV contains three protocols on the biofunctionalization of inorganic surfaces. The on-chip synthesis of peptide nucleic acids to generate microarrays for the high-throughput analysis of RNA and DNA is described in Chapter 21. The biofunctionalization of gold nanoparticles is reported in Chapter 22, allowing one to readily prepare facile hybrid reagents useful in a variety of bioanalytical assays. Finally, Chapter 23 discusses the production of probes for atomic force microscopy, taking advantage of biofunctionalized carbon nanotubes.

The collection of protocols in this volume clearly emphasizes a novel trend in bioconjugation chemistry, which is the interplay of advanced organic synthesis, molecular biology, and materials science. In particular, the solid-phase synthesis of peptides and nucleic acids is combined with the chemo- and regioselective ligation of expressed protein fragments and the biofunctionalization of solid substrates. Conversely, an increasing number of approaches takes advantage of the high specificity of enzyme catalysts to produce well-defined bioconjugates.

Bioconjugation Protocols: Methods and Strategies is intended to provide information and inspiration to all levels of scientists from novices to those professionally engaged in the field of bioconjugate chemistry. I would like to thank all the contributing authors for providing manuscripts, John Walker for editorial guidance, and the staff of Humana Press for professional production of this volume. I thank my coworkers for their enthusiastic contributions to our research and writing projects. Finally, my thanks to my family, especially to Trixi, Sarah, and Amelie, who are the light of my life.

Christof M. Niemeyer

Contents

Contributors

MICHAEL ADLER • *Chimera Biotec GmbH, Dortmund, Germany*
KIRILL ALEXANDROV • *Max-Planck-Institut für Molekulare Physiologie, Dortmund, Germany*
ANDREA BAUER • *Functional Genome Analysis, Deutsches Krebsforschungszentrum, Heidelberg, Germany*
MICHAEL BAUM • *Febit AG, Mannheim, Germany*
VERENA BEIER • *Functional Genome Analysis, Deutsches Krebsforschungszentrum, Heidelberg, Germany*
SUSANNE BRAKMANN • *Applied Molecular Evolution, Institute for Zoology, University of Leipzig, Leipzig, Germany*
OLE BRANDT • *Functional Genome Analysis, Deutsches Krebsforschungszentrum, Heidelberg, Germany*
BUELENT CEYHAN • *Biologisch-Chemische Mikrostrukturtechnik, Fachbereich Chemie der Universität Dortmund, Dortmund, Germany*
DAVID R. COREY • *Departments of Pharmacology and Biochemistry, University of Texas Southwestern Medical Center at Dallas, Dallas, TX*
ZHONGLI DING • *Department of Bioengineering, University of Washington, Seattle, WA*
THOMAS DUREK • *Max-Planck-Institut für Molekulare Physiologie, Dortmund, Germany*
THOMAS DZIUBLA • *Institute for Environmental Medicine, University of Pennsylvania School of Medicine, Philadelphia, PA*
TATIANA GIORGI • *Biologisch-Chemische Mikrostrukturtechnik, Fachbereich Chemie der Universität Dortmund, Dortmund, Germany*
ROGER S. GOODY • *Max-Planck-Institut für Molekulare Physiologie, Dortmund, Germany*
ROBERT F. GRAZIANO • *Medarex Inc., Bloomsbury, NJ*
PAUL GUPTILL • *Medarex, Inc., Bloomsbury, NJ*
POMPI HAZARIKA • *Biologisch-Chemische Mikrostrukturtechnik, Fachbereich Chemie der Universität Dortmund, Dortmund, Germany*
INES HEINEMANN • *Max-Planck-Institut für Molekulare Physiologie, Dortmund, Germany*
ALLAN S. HOFFMAN • *Department of Bioengineering, University of Washington, Seattle, WA*
JÖRG D. HOHEISEL • *Functional Genome Analysis, Deutsches Krebsforschungszentrum, Heidelberg, Germany*

ANETTE JACOB • *Functional Genome Analysis, Deutsches Krebsforschungszentrum, Heidelberg, Germany*

KUNIHIRO KAIHATSU • *Departments of Pharmacology and Biochemistry, University of Texas Southwestern Medical Center at Dallas, Dallas TX*

JÜRGEN KUHLMANN • *Max-Planck-Institut für Molekulare Physiologie, Dortmund, Germany*

FLORIAN KUKOLKA • *Biologisch-Chemische Mikrostrukturtechnik, Fachbereich Chemie der Universität Dortmund, Dortmund, Germany*

MARINA LOVRINOVIC • *Biologisch-Chemische Mikrostrukturtechnik, Fachbereich Chemie der Universität Dortmund, Dortmund, Germany*

MARKUS MEUSEL • *MACHEREY-NAGEL GmbH & Co. KG, Düren, Germany*

MARGHERITA MORPURGO • *Pharmaceutical Sciences Department, University of Padova, Padova, Italy*

JUAN-CARLOS MURCIANO • *Institute for Environmental Medicine, University of Pennsylvania Medical Center, Philadelphia, PA*

SILVIA MURO • *Institute for Environmental Medicine, University of Pennsylvania Medical Center, Philadelphia, PA*

VLADIMIR R. MUZYKANTOV • *Institute for Environmental Medicine, University of Pennsylvania Medical Center, Philadelphia, PA*

CHRISTOF M. NIEMEYER • *Biologisch-Chemische Mikrostrukturtechnik, Fachbereich Chemie der Universität Dortmund, Dortmund, Germany*

ALEXANDER PESCHLOW • *The University Chemical Laboratory, Cambridge University, Cambridge, UK*

GORAN PLJEVALJČIĆ • *Department of Molecular Biology, The Scripps Research Institute, La Jolla, CA*

MARTINA REIBNER • *Biologisch-Chemische Mikrostrukturtechnik, Fachbereich Chemie der Universität Dortmund, Dortmund, Germany*

FALK SCHMIDT • *Institut für Organische Chemie de RWTH Aachen, Aachen, Germany*

VLADIMIR V. SHUVAEV • *Institute for Environmental Medicine, University of Pennsylvania School of Medicine, Philadelphia, PA*

PATRICK S. STAYTON • *Department of Bioengineering, University of Washington, Seattle, WA*

ACHIM STEPHAN • *Functional Genome Analysis, Deutsches Krebsforschungszentrum, Heidelberg, Germany*

THOMAS J. TOLBERT • *Department of Chemistry, Indiana University, Bloomington, IN*

FRANCESCO M. VERONESE • *Pharmaceutical Sciences Department, University of Padova, Padova, Italy*

CARL-WILHELM VOGEL • *Cancer Research Center of Hawaii, University of Hawaii, Honolulu, HI*

MARTIN VÖLKERT • *Max-Planck-Institut für Molekulare Physiologie, Dortmund, Germany*

RON WACKER • *Chimera Biotec GmbH, Dortmund, Germany*

MELANIE WAGNER • *Max-Planck-Institut für Molekulare Physiologie, Dortmund, Germany*

HERBERT WALDMANN • *Max-Planck-Institut für Molekulare Physiologie, Dortmund, Germany*

ELMAR WEINHOLD • *Institut für Organische Chemie der RWTH Aachen, Aachen, Germany*

RAINER WIEWRODT • *Department of Medicine, University of Pennsylvania School of Medicine, Philadelphia, PA*

CHI-HUEY WONG • *Department of Chemistry, The Scripps Research Institute, La Jolla, CA*

ADAM T. WOOLLEY • *Department of Chemistry and Biochemistry, Brigham Young University, Provo, UT*

I

ANTIBODY AND ENZYME CONJUGATES

1

Streptavidin–Biotin Crosslinking of Therapeutic Enzymes With Carrier Antibodies

Nanoconjugates for Protection Against Endothelial Oxidative Stress

Vladimir V. Shuvaev, Thomas Dziubla, Rainer Wiewrodt, and Vladimir R. Muzykantov

Summary

The streptavidin–biotin system may be used to synthesize immunoconjugates for targeted delivery of drugs, including therapeutic enzymes. The size of antibody–enzyme conjugates, which is controlled by the extent of biotinylation and molar ratio between the conjugate components, represents an important parameter that in some cases dictates subcellular addressing of drugs. This chapter describes the methodology of formation and characterization of polymeric immunoconjugates in the nanoscale range. A theoretical model of streptavidin conjugation based on general principles of polymer chemistry is considered. Factors that influence size and functional characterization of resulting polymer conjugates, as well as advantages and limitations of this approach, are described in detail. The protocols describe the formation of immunoconjugates possessing an antioxidant enzyme, catalase, directed to endothelial cells by anti-platelet endothelial cell adhesion molecule antibodies. However because of the modular nature of the streptavidin–biotin crosslinker system, the techniques herein can be easily adapted for the preparation of nanoscale immunoconjugates delivering other protein drugs to diverse cellular antigens.

Key Words: Immunoconjugates; vascular immunotargeting; polymerization; nanoscale carrier; catalase; streptavidin; biotin; dynamic light scattering; drug delivery.

1. Introduction

Targeted drug delivery, as attained by conjugating therapeutic enzymes with affinity carrier antibodies, promises a significant improvement over the current therapeutic means and, therefore, has remained the focus of intense research

From: *Methods in Molecular Biology, vol. 283: Bioconjugation Protocols: Strategies and Methods*
Edited by: C. M. Niemeyer © Humana Press Inc., Totowa, NJ

for several decades. For example, endothelial cells lining the luminal surface of the vasculature represent an important target for delivery of antithrombotic, anti-inflammatory, antioxidant agents and genetic materials. Cell adhesion molecules (e.g., platelet endothelial cell adhesion molecule [PECAM] and intercellular adhesion molecule [ICAM]) represent very attractive endothelial determinants for vascular immunotargeting, for example, in the context of inflammation. Some drugs require intracellular uptake. Recent studies revealed that although endothelial cells do not internalize monomeric antibodies against PECAM and ICAM, one can facilitate intracellular delivery of therapeutic cargoes by controlling size of the anti-PECAM and anti-ICAM immunoconjugates in the nanoscale range *(1–3)*.

The biotin–streptavidin system can be used to synthesize nanoscale therapeutic immunoconjugates, providing an interesting alternative to other commonly pursued intravenous drug targeting strategies, such as liposomes and polymeric nanocarriers *(4–7)*. These immunoconjugates are typically characterized by (1) their high drug incorporation efficiency, (2) high drug to carrier ratio, (3) a wide tunable range of particles sizes with the same or similar composition, and (4) a relatively rigid and biodegradable structure. In optimal conditions, the degree of drug inclusion is so high that the level of free drug becomes negligible and a separation step may often be omitted. Several reporter and therapeutic enzymes conjugated with anti-PECAM and anti-ICAM have been successfully delivered in therapeutic levels to pulmonary endothelium *(1,2,8–11)*.

This chapter describes the methodology and detail protocols for the generation of nanoscale immunoconjugates using the polymeric form of the streptavidin–biotin system in addition to methods to control their size and shows examples of targeted delivery of an antioxidant therapeutic enzyme, catalase, to vascular endothelium in cell culture.

The distinguishing feature of the streptavidin–biotin system is the extraordinary affinity ($K_d = 10^{-15}$ M) of this noncovalent interaction. It may be compared only with systems involving liganded metal ions either as partial covalent bonds or chelates. This extremely specific almost irreversible reaction is widely used in biology and medicine *(12)*. If a biotin derivative is covalently linked to proteins, these biotinylated proteins will bind to streptavidin and form a conjugate. These conjugates can be categorized into two types depending on protein biotinylation level: oligomeric and polymeric conjugates. Oligomeric conjugates, which are readily used in many labeling techniques, occur when the protein contains less than two biotin residues per protein (**Fig. 1A**). However, when the average biotinylation level of the proteins (e.g., biotinylated antibody and enzyme) is equal to or exceeds 2, polymer structures can be formed (**Fig. 1B,C**). Because the linkage occurs through the paired coupling of spe-

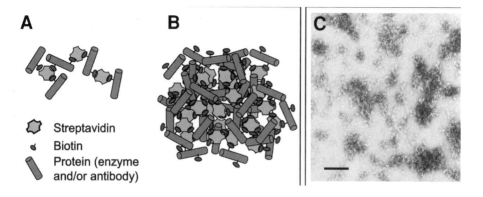

Fig. 1. Scheme of protein conjugation with streptavidin–biotin system. Proteins with less than 2 biotin/molecule form an oligomeric structure (**A**). Higher biotinylated proteins can switch polymerization reactions in the presence of streptavidin as a crosslinker with formation of polymeric structure (**B**). Polymer size depends on reactive conditions, probably a result of the high rate of streptavidin–biotin reaction and formation of internal core inaccessible for free copolymers. (**C**) Electron micrograph of negatively stained immunoconjugates. Conjugates were placed on grids precoated with thin carbon films, and negative staining was performed with uranyl acetate. Images were taken from representative areas at an original magnification of ×50,000 and enlarged to ×440,000. Scale bar = 100 nm.

cific subunits isolated on separate molecular species, it is convenient to relate these conjugates to the classic step (condensation) polymerization chemistry *(13)*. In this circumstance, the modified Carathor's equation applies to this reaction scheme.

$$X_n = \frac{1+r}{1+r-frp}$$

where X_n is the average number of monomer residues (both streptavidin and protein) per conjugate, r is the ratio of streptavidin molecules to protein molecules, f is the number of proteins that can bind to streptavidin, and p is the extent of reaction (number of available linkage sites that are actually linked). From this equation, it is noted that polymerization occurs only when the denominator approaches zero. Hence, it is possible to approximately know *a priori* what protein:streptavidin ratio, r, will provide the maximum conjugate size for a given biotinylation level. Also, because of the high sensitivity of the extent of reaction on X_n, small changes in polymerization procedures can have a large impact upon the final size of the conjugate.

2. Materials

2.1. Equipment

1. Dynamic light scattering apparatus 90Plus Particle Sizer (Brookhaven Instruments Corp., Holtsville, NY) or similar apparatus.
2. UV-VIS spectrophotometer.
3. Microplate reader.
4. Gamma-counter.
5. Fluorescent microscope.

2.2. Reagents and Proteins

1. Succinimidyl-6-(biotinamido) hexanoate (NHS-LC-Biotin; Pierce, Rockford, IL).
2. 2-(4'-hydroxyazobenzene) benzoic acid (HABA; Pierce).
3. O-Phenylenediamine (OPD, in tablets of 60 mg; Sigma, St. Louis, MO).
4. $Na^{51}CrO_4$ (Perkin Elmer, Boston, MA).
5. Dimethylformamide (DMF).
6. Hydrogen peroxide (H_2O_2).
7. Glycerol.
8. Catalase from bovine liver (Calbiochem, CA).
9. Streptavidin from *Streptomyces avidinii* (Calbiochem).
10. Avidin (Pierce).
11. Horseradish peroxidase (HRP).
12. Monoclonal anti-PECAM antibody (clone 62 was generously provided by Dr. Nakada; Centocor, Malvern, PA).
13. Mouse IgG (Calbiochem, San Diego, CA).

2.3. Buffers, Media, and Cells

1. Phosphate-buffered saline (PBS): 0.1 *M* sodium phosphate, 0.15 *M* sodium chloride, pH 7.2.
2. HABA stock solution: 10 m*M* HABA in 10 m*M* NaOH. Add 24.2 mg to 10 mL of 10 m*M* NaOH. The solution may be stored at 4°C.
3. HABA/avidin working solution: dissolve 10 mg of avidin in 19.4 mL PBS and add 600 µL of 10 m*M* HABA stock solution.
4. Cell culture medium: M199 medium (Gibco, Grand Island, NY), 10% fetal calf serum (Gibco) supplemented with 100 µg/mL heparin (Sigma), 2 m*M* L-glutamine (Gibco), 15 µg/mL endothelial cell growth supplement (Upstate, Lake Placid, NY), 100 U/mL penicillin, and 100 µg/mL streptomycin.
5. RPMI 1640 medium without phenol red (Gibco).
6. Human umbilical vein endothelial cells (HUVEC; Clonetics, San Diego, CA).

3. Methods

3.1. Biotinylation of IgG Antibodies and Catalase

1. Dissolve IgG (anti-PECAM antibody or any other mouse IgG) and catalase in 0.1 *M* PBS to concentrations of 3.5 and 5.0 mg/mL, respectively. Considering

that molecular masses of IgG and catalase are 150 and 240 kDa, respectively, their molar concentrations are 23.3 μM and 20.8 μM, respectively (*see* **Notes 1** and **2**).

2. Prepare fresh 0.1 M NHS-LC-biotin in DMF, that is, dissolve 4.5 mg NHS-LC-biotin in 100 μL of anhydrous DMF. Keep solutions of proteins and NHS-LC-biotin on ice.
3. Add appropriate volume of 0.1 M NHS-LC-biotin to protein solution to have 5-, 10-, and 15-fold biotin:protein molar excess using the following equation:

$$V_{NHS-Biotin} = kV_{protein}\left(\frac{C_{protein}}{C_{NHS-Biotin}}\right)$$

where $V_{NHS-biotin}$ and $V_{protein}$ are the volumes of NHS-LC-biotin and protein (antibody or catalase), respectively in μL; $C_{NHS-biotin}$ and $C_{protein}$ are the molar concentrations of NHS-LC-biotin and protein (antibody or catalase), respectively in mM; k is molar excess of biotin label. Thus, add 1.2, 2.3, and 3.5 μL of 0.1 M NHS-LC-biotin to 1 mL of antibody solution and 1.0, 2.1, and 3.1 μL of 0.1 M NHS-LC-biotin to 1 mL of catalase solution. Vortex the samples.

4. Incubate the samples on ice for 2 h.
5. Remove unbound biotin derivatives by dialysis against 1.0 L of PBS with three changes.
6. Measure protein concentration in catalase and antibody preparations by A_{280} absorbance using following coefficients: A(0.1%) 1.04 for catalase and 1.7 for IgG. Bradford assay (Bio-Rad) or other protein assays may be used as well.
7. Split antibody preparation in Eppendorf tubes 100 μL/tube and store at –80°C (or –20°C) because IgG at 4°C can easily aggregate and even partial aggregation of IgG may significantly affect further conjugation. Catalase may be stored in PBS at 4°C during several months without significant loss of its activity.

3.2. Estimation of Protein Biotinylation Level

1. Add 450 μL of HABA/avidin working solution into plastic spectrophotometric cuvet and measure absorbance at 500 nm Å. Add 50 μL of sample, mix it in cuvet and measure A'_{500}. Biotin competes with HABA for same binding sites on avidin and releases HABA in free solution that in turn decreases the absorbance of the dye. Because the reaction may require 2–5 min to be complete, check the absorbance several times and take into calculation only the value after absorbance was stabilized for at least 15 s (*see* **Notes 3–5**).
2. Calculate molar concentration of biotin using following equation:

$$[biotin, \mu M] = \frac{(D' \times A^{\circ} - A') \times D'' \times 10^6}{\varepsilon_{HABA}}$$

where D' is dilution coefficient for HABA/avidin working solution $D' = 0.9$; D'' is dilution coefficient of the sample $D'' = 50\ \mu L/500\ \mu L = 10$; 10^6 is a coefficient to express biotin concentration in μM; ε_{HABA} is molar extinction coefficient of HABA bound to avidin at 500 nm that equals 34000 AU $\times M^{-1}cm^{-1}$.

Considering all known parameters the equation may be easily transformed into a simple formula:

$$[biotin, \mu M] = (0.9 \times \overset{\circ}{A} - A') \times 294$$

3. Calculate protein biotinylation using following equation:

$$\text{Protein Biotinylation} = \frac{[biotin, \mu M]}{[Protein, \mu M]}$$

3.3. Conjugation

Immunoconjugates may be prepared by a one-step or two-step procedure. In both cases, the molar ratio between biotinylated protein that should be specifically delivered (i.e., catalase) and biotinylated antibody that targets specific antigen on cell surface (i.e., anti-PECAM) is kept constant. As a rule, the catalase:anti-PECAM ratio is 1:1 mol/mol. In contrast, an optimal concentration of streptavidin (with respect to the desired conjugation size) varies and should be determined for each preparation of biotinylated ligands. In the one-step procedure, biotinylated proteins (enzyme and antibody) are premixed and then streptavidin is added to conjugate them. In two-step procedure, biotinylated enzyme is first conjugated with streptavidin and then antibody is added to form the larger secondary conjugate (*see* **Note 6**).

3.3.1. One-Step Procedure

3.3.1.1. STREPTAVIDIN TITRATION

1. Prepare 110 μL of catalase/antibody mixture with the molar ratio 1:1 in PBS. For that, mix 58 μL of 5.0 mg/mL catalase and 52 μL of 3.5 mg/mL anti-PECAM. All components for conjugation are to be kept on ice.
2. Split the mixture into 5 aliquots of 20 μL each in 1.5-mL transparent Eppendorf tubes.
3. Add 10.0 mg/mL streptavidin solution in PBS to have final molar ratio streptavidin:(catalase + antibody) 0.5, 1.0, 1.5, 2.0, and 2.5 (i.e., add 1.3, 2.6, 4.0, 5.3, and 6.6 μL of streptavidin, respectively). The conjugation should be performed while vortexing. Continuous and regular mixing is critical for correct conjugation.
4. Measure the mean effective diameter of the obtained conjugates by dynamic light scattering (DLS). Add 180 μL of PBS to each conjugate sample, mix it well, and transfer the diluted sample into NMR tube for the ₊nalysis on DLS apparatus 90Plus Particle Sizer. Count rate should be from 100 kcps to 1 Mcps. Run the sample for at least 3 min and determine effective diameter (*see* **Note 7**).
5. Plot the effective diameter of conjugates as a function of streptavidin/(catalase + antibody) molar ratio. Make additional points if necessary. The standard streptavidin titration curve is bell shaped, similar to the classical antigen–antibody precipitation titration curves. Higher biotinylated component(s) produces larger conjugates (*see* **Subheading 3.5.1.**).

3.3.1.2. Preparation of Conjugate Stock

1. Chose optimal streptavidin:(catalase + antibody) molar ratio that gives you the required size of conjugate (*see* **Subheading 3.3.1.1., step 5**).
2. Prepare 100 µL of catalase + antibody mixture with the molar ratio 1:1 in PBS. Mix 53 µL of 5.0 mg/mL catalase and 47 µL of 3.5 mg/mL anti-PECAM.
3. Calculate the volume of 10 mg/mL streptavidin that should be added to reach a specific molar ratio. For example, if you found that the optimal molar ratio is 2, add 26.4 µL of 10.0 mg/mL streptavidin.
4. Add 5 µL of the conjugate preparation to 195 µL of PBS. Transfer the diluted sample into NMR tube and measure the effective diameter of obtained conjugates by DLS. Upscaling may change the size of conjugate. If this occurs, then adjustment in volume of streptavidin should be done to correct optimal streptavidin:protein molar ratio.

3.3.2. Two-Step Procedure

3.3.2.1. Titration by Streptavidin

1. Transfer 10.6-µL aliquots of 5.0 mg/mL biotinylated catalase into five 0.5-mL Eppendorf tubes.
2. In the first conjugation step, add 1.3, 2.6, 4.0, 5.3, and 6.6 µL of 10 mg/mL streptavidin per tube. Addition of streptavidin solution should be as fast as possible while the sample is at constant vortexing. Keep vortexing for several seconds after streptavidin was added. Spin it down briefly.
3. Incubate 5 min on ice.
4. In the second conjugation step, add 9.4 µL of 3.5 mg/mL anti-PECAM to all five samples in a similar way as in first conjugation step. Final streptavidin:(catalase + antibody) molar ratios are 0.5, 1.0, 1.5, 2.0, and 2.5, respectively.
5. Measure the effective diameter of obtained conjugates by DLS (*see* **Subheading 3.3.1.1., step 4**, and **Note 7** for details).
6. Plot the effective diameter of conjugates as a function of streptavidin:(catalase + antibody) molar ratio. Make additional points if necessary. Higher biotinylated component(s) produces larger conjugates (*see* **Subheading 3.5.1.**, for example).

3.3.2.2. Preparation of Conjugate Stock

1. Chose optimal streptavidin:(catalase + antibody) molar ratio that gives you required size of the conjugate (*see* plot obtained in **Subheading 3.3.2.1., step 6**).
2. Transfer 53 µL of 5.0 mg/mL biotinylated catalase into 1.5-mL transparent Eppendorf tubes.
3. In the first conjugation step, add specified quantity of streptavidin. For example, if you found that the final streptavidin:(catalase + antibody) molar ratio has to be 2, add 26.4 µL of 10 mg/mL streptavidin. Addition of streptavidin solution should be as fast as possible while the sample is at constant vortexing. Keep vortexing for several seconds after streptavidin was added, spin it down and incubate sample during 5 min on ice.

4. In the second conjugation step, add 47 μL of 3.5 mg/mL anti-PECAM to primary conjugate in a similar way as in first conjugation step.
5. Mix 5 μL of the final conjugate preparation with 195 μL of PBS. Transfer the diluted sample into NMR tube and measure an effective diameter of obtained conjugates by DLS. Upscaling may slightly change the size of conjugate compared with results of streptavidin titration using small volumes. In this case adjustment in volume of streptavidin should be done to correct optimal streptavidin:protein molar ratio.

3.3.3. Storage of Conjugate

Because conjugates tend to aggregate, keeping them at 4°C for longer than several hours is not recommended. Freezing is also not recommended for the same reason of material aggregation after thawing. To store conjugate for further use, add glycerol to 50% and keep the preparation at –20°C. Under these conditions, no significant changes in conjugate size, catalase enzymatic activity, and antibody binding occur for at least 1 wk (*see* **Note 8**).

3.4. Characterization of Conjugates In Vitro

To be therapeutically functional, immunoconjugates should preserve both its activities: enzymatic activity of catalase and antibody binding to cell antigen. Because the conjugation process may affect both protein components, functional activity of the conjugates must be tested in vitro before more expensive and challenging in vivo studies. For example, protection assay against H_2O_2-induced injury of endothelial cell culture reveals functional activity of the catalase conjugate.

3.4.1. Catalase Activity

1. Prepare 10X stock solution of assay buffer: 50 mM sodium phosphate, pH 7.0.
2. Prepare working solution: dilute 75 μL of 3% H_2O_2 in 10 mL 1X assay buffer.
3. Take 1 mL of working solution and add catalase (as free enzyme or conjugate) to final concentration of 0.1–1.0 μg of catalase/mL.
4. Place the sample immediately in a quartz cuvet into UV-VIS spectrophotometer and follow the kinetics of H_2O_2 degradation at 242 nm.
5. Measure the slope of the curve ΔA/min using initial linear fragment and calculate catalase activity as follows:

$$\text{Catalase activity, U/mg} = 23.0(\Delta A/\text{min})/\text{mg of catalase}$$

3.4.2. Protection Against Hydrogen Peroxide Cytotoxicity

1. Pretreat a 24-well plate with 0.5 mL/well of 1% gelatin for 1 h, remove the solution ,and allow it to dry out for 1 h. Plate HUVEC (4th passage) in the plate at a cell density of 50,000–100,000 cell/well in cell culture medium. Grow cells for

3–4 d until confluent culture. One day before the experiment, replace the medium with a fresh one containing 200,000 cpm/mL of [^{51}Cr] as Na^{51}CrO$_4$.

2. The next day, wash out free [^{51}Cr] with fresh cell culture medium and add 0.25 mL/ well of catalase/anti-PECAM immunoconjugates (*see* **Subheading 3.3.1.2.**) diluted with the medium at a concentration of 5–10 µg of catalase/well. Incubate cells for 1 h at 37°C. Wash out unbound conjugates first with fresh cell culture medium and then with phenol red-free RPMI medium.

3. Induce oxidative stress by addition of 5 m*M* H$_2$O$_2$ in RPMI (i.e., 257 µL/50 mL of RPMI) and incubate cells at 37°C for 5 h.

4. Place an aliquot of 20 µL of supernatant into 96-well low-binding plate at 0, 15, 30, 45, and 60 min for the H$_2$O$_2$ degradation assay. In the meantime, prepare calibration curve by placing 0, 5, 10, 15, and 20 µL of 5 m*M* H$_2$O$_2$ in duplicates and adjust the volume with RPMI.

5. For the H$_2$O$_2$ degradation assay, prepare fresh OPD/HRP working solution: dissolve one 60-mg tablet of OPD in 17.5 mL of PBS on rotating platform or orbital shaker and add 100 µL of 1 mg/mL HRP. Add 180 µL of OPD/HRP working solution to sample-containing wells on 96-well plate. Incubate the plate on ELISA shaker for 15 min. Stop the reaction by addition of 50 µL/well of 50% H$_2$SO$_4$. Read the absorbance in microplate reader at 490 nm. Calculate the H$_2$O$_2$ concentration in the samples using calibration curve.

6. To detect of [^{51}Cr] release, after a 5-h incubation take 100 µL of cell culture supernatant into tubes for gamma-counter. Pool the rest of supernatant and cell lysate with 0.5 mL of 1% Triton X-100, 1.0 *M* NaOH. Measure radioactivity in the samples and calculate the % of released [^{51}Cr].

3.5. Results

3.5.1. Conjugate Preparation

Catalase was biotinylated to a level of 3.25 biotin/catalase monomer (or approx 12–13 biotin/catalase tetramer) and monoclonal anti-PECAM antibody was biotinylated to 3.5 biotin/IgG. Both proteins were mixed at a molar ratio 1:1, and streptavidin was added to form conjugates. The titration curve of the conjugation is shown in **Fig. 2A** as a dependence of conjugate effective diameter determined by DLS on molar ratio streptavidin:biotinylated proteins. The curve demonstrates a continuous increase of the conjugate size at relative excess of biotinylated proteins (**Fig. 2A**, right shoulder of the curve). The maximum is reached at equimolar ratio between accessible biotin-binding sites on streptavidin and available biotins on proteins. Thus the position of the maximum will depend on effective number and flexibility of biotins on the proteins, size and structure of the proteins and mixing conditions. Further increase of streptavidin concentration results in relative excess of streptavidin and decreasing of conjugate size (**Fig 2A**, left shoulder). Noteworthy, the left shoulder has a plateau supposedly because the

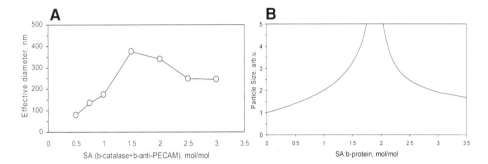

Fig. 2. Titration of biotinylated proteins (catalase and anti-PECAM antibody) with streptavidin. **A,** Catalase was mixed with antibody, and streptavidin was added as a bulk to conjugate both proteins. Size was measured by DLS. **B,** Modeling of conjugation reaction. Parametric addition of Carathor's equation with streptavidin as the limiting reagent (streptavidin:protein < 1.0) and protein as the limiting reagent (streptavidin:protein > 1). The extent of reaction was selected such that these two equations converged to a maximum value. Degree of polymerization was related to particle size through the r^2 relationship of conjugate molecular weight to radius of gyration.

reaction between biotin and streptavidin is so fast that a rate of mixing of two these components is always a limiting step.

Based on the Carathor's equation, the relative concentration of the two reagents (streptavidin and biotinylated protein) is one of the key determinants in the ultimate number of proteins per conjugate. As such, by varying the ratio of streptavidin to biotinylated protein, it is theoretically possible to control the size of particle. If we use the average biotinylation of catalase and antibody at a converging extent of reaction, a theoretical titration curve with a maximum at 1.95 streptavidin:proteins molar ratio is obtained (**Fig. 2B**). This is very close to the actual maximum obtained in experiment (compare with **Fig. 2A**). Deviations from theory are most likely the result of the presence of proteins with two distinct biotinylation levels (i.e., catalase and antibody). Also, the extent of reaction is dependent upon size of conjugate, which is not considered in the model presented. Although this model is limiting in its ability to account for varying accessible functionality with extent of reaction, it still demonstrates the sensitivity of size on reaction conditions.

The size of resulting conjugate depends on biotinylation level of protein(s). Biotinylated catalase was mixed with antibody biotinylated at different extents (**Fig. 3A**). Higher biotinylated antibody formed larger conjugates. The rate of streptavidin addition is another important parameter that affects the size of conjugate. Slow addition of streptavidin leads to increased size of forming conjugates compared to instant mixing (**Fig. 3B**). This dependence of particle

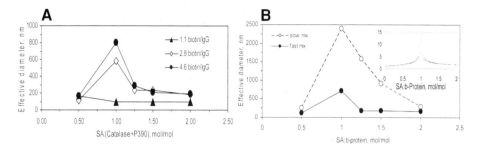

Fig. 3. Effects of biotinylation level and mixing conditions on size of immunoconjugate. (**A**) Catalase at a level of biotinylation of 12 biotin/tetramer was mixed with antibody of indicated biotinylation at a molar ratio 1:1. Streptavidin was added as a bulk. (**B**) The rate of streptavidin addition affects the size of conjugate. Streptavidin was added as a bulk (closed circles) or at slow rate of about 5 μL/min. Insert shows theoretical modeling of this effect. An increase in the extent of reaction results in an increase in the maximum of the titration curve. It is hypothesized that by adding reactants slowing, the likelihood of steric shielding of binding sites is reduced. This results in an overall increase in the extent of reaction and the conjugate size.

size on mixing conditions can also be accounted for by the extent of reaction in the Carathor's. Because the reaction rate proceeds nearly instantaneously, mixing conditions will greatly affect the extent of reaction. As such, experiment agrees well with theory that the increase in reaction mixing results in a decrease in extent of reaction, and therefore results in smaller maximum particle sizes (**Fig. 3B**, insert).

3.5.2. Conjugate Characterization

We prepared catalase/anti-PECAM immunoconjugates for further characterization. The conjugates were analyzed by high-performance liquid chromatography gel filtration. We could detect only trace amounts of free catalase in conjugate preparation, whereas practically all streptavidin and antibody were apparently included in conjugates (**Fig. 4A**). Thus use of the conjugates does not require additional step of conjugate separation from free component. Conjugation only slightly decreased activity of catalase. Its activity in the conjugate was measured to be 80% of initial catalase activity in free solution. Furthermore, the binding of the conjugates was visualized by immunofluorescence microscopy using fluorescein isothiocyanate-labeled antimouse IgG antibody. The conjugates readily bound to cultured human endothelial cells (**Fig. 4B**). Interestingly, they are mostly localized on cell-cell borders in accordance to PECAM distribution in confluent culture (*14*).

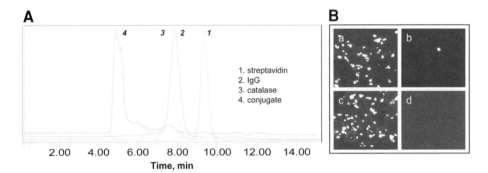

Fig. 4. Characterization of immunoconjugates. (**A**) High-performance liquid chromatography analysis of catalase/anti-PECAM immunoconjugates on SW-300 gel-filtration column (Waters, MA). The conjugation was performed by the one-step procedure. The immunoconjugate and its individual components were injected in phosphate buffer. Normalized chromatograms are shown. (**B**) Binding of the immunoconjugates to HUVECs. Catalase/anti-PECAM (a and c) or catalase/nonimmune IgG (b and d) 300-nm immunoconjugates at a concentration of 5 µg of catalase/well were incubated with confluent cell culture. Cells were fixed without (a and b) or with (c and d) after permeabilization and conjugates were stained using fluorescein isothiocyanate-labeled anti mouse IgG. Samples were analyzed by fluorescence microscopy.

The conjugates were used for protecting endothelial cells against oxidative stress (**Fig. 5**). HUVECs were preincubated with catalase/anti-PECAM antibody at different doses of catalase (0.1–5.0 µg of catalase/well as indicated) and protective properties of the bound conjugates were analyzed by H_2O_2 degradation assay, [^{51}Cr] release, and visually by phase-contrast microscopy. Enzymatic activity of the bound conjugates estimated by H_2O_2 degradation assay showed dose-dependent response up to 5 µg of catalase/well (**Fig. 5A**). However, only doses of 1.0 and 5.0 µg of catalase/well were protective as detected by [^{51}Cr] release (**Fig. 5B**). Phase-contrast microscopy also demonstrated that those doses protected cells against oxidative stress compared to cells untreated with the conjugates (**Fig. 5C**).

4. Notes

4.1. Biotinylation

1. It is important to remember that biotinylation depends on initial concentration of protein. The level of biotinylation is increased at a higher concentration of protein even at same NHS-LC-biotin:protein molar ratio.
2. Biotinylation efficiency and its effects on protein activities vary significantly from protein to protein. In case of catalase, biotinylation does not change the catalytic activity at up to a level of 4–5 biotin/catalase monomer.

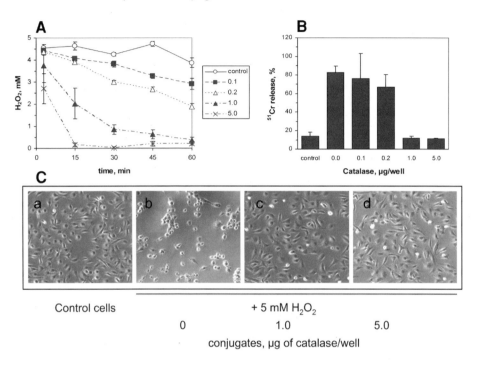

Fig. 5. Cell protection against H_2O_2-induced oxidative stress by catalase/anti-PECAM immunoconjugates. Cells were treated with the immunoconjugates as described in **Subheading 3.4.2.** at several concentrations. (**A**) Degradation of H_2O_2 by bound catalase-containing conjugates. Initial concentrations of immunoconjugates used for cell treatment are indicated in micrograms of catalase/well. (**B**) Cell death as a result of severe oxidative stress was analyzed by release of [^{51}Cr]. Cells were incubated with 5 mM H_2O_2 for 5 h. Release of [^{51}Cr] in control cells shows level of passive diffusion whereas practically complete [^{51}Cr] release in the absence of immunoconjugates demonstrates significant cell death. (**C**) Phase-contrast microscopy of control and treated cells.

4.2. Estimation of Protein Biotinylation Level

3. HABA is not readily soluble in 10 mM NaOH and requires 10–20 min of intense vortexing.

4. HABA will change its color from yellow to amber because the dye instantly interacts with avidin. HABA/avidin working solution may be stored during 2–4 wk at 4°C. Absorbance of the solution at 500 nm should be 0.9–1.3 AU (we recommend to adjust it to 0.95–1.05 AU with HABA appropriately diluted in PBS to keep constant concentration of HABA).

5. Absorbance of working solution after addition of biotin should be no less than 0.35–0.4. Otherwise the sample of biotinylated protein will have to be diluted.

4.3. Conjugation

6. A number of factors are important in conjugation and may affect a size of particles:

   a. Streptavidin:protein ratio is the most important parameter. A titration curve is required for each new preparation of biotinylated catalase or antibody.

   b. The optimal biotinylation level should be estimated experimentally for every protein. To produce 150- to 400-nm conjugates, proteins have to be biotinylated to a level of 3–4 biotin/catalase monomer or IgG. Under-biotinylated protein may form too small particles and does not reach desired size. It may be rebiotinylated. Overbiotinylated protein will form precipitates and titration curve does not show a visible maximum. Such proteins cannot be used for conjugation.

   c. Reaction between streptavidin and biotin is so fast that mixing conditions are able to change size of formed conjugates. Instant addition of streptavidin is recommended because it is easier to control. However, you can prepare larger conjugates if you inject streptavidin slower. It can be useful tool for preparation conjugates of different size and essentially same composition.

   d. Although we usually look on the left shoulder of the titration curve to find the optimal condition for conjugation, it is possible to use right steep shoulder as well. An advantage of using the right shoulder is that streptavidin may be added in several steps with control of the conjugate size after each step.

7. DLS is an attractive technique in measuring conjugate size because this is an absolute method that does not require preliminary calibration or standards. It is fast, reliable, direct technique and the easiest method to measure particles sizes in the 20- to 1000-nm range. However, DLS is based on the principle of light scattering of moving objects that implies specific limitations on preparation of sample and its reading. There are a number of important issues that should always be taken into consideration to obtain meaningful values. First of all, it is important to possess a rudimentary knowledge of the theory to effectively use the machine. Briefly, at a moment in time, particles in solution will scatter light with a particular intensity at a set angle (at 90° in case of 90Plus Particle Sizer). If we wait for some time (Δt) and then check the scattering intensity, the intensity will change as a result in the change of particle orientation. If Δt is very small, then the intensity will not change very much, because the particles have not had enough time to move around in solution. However, as Δt increases, the chances of the intensity being the same (autocorrelating) will decrease dramatically. This dependence of intensity autocorrelation on time is directly related to the ability of particles to randomly move. If we assume the particles move according to the rules of Brownian motion, we can obtain equations that describe the speed of particle motion as a function of particle size. Hence, we can relate the decay in the autocorrelation directly to particle size. When the assumptions built into the math equations are accurate, then the DLS provides a rapid reliable means of measurement. In practical circumstances, the following points should be kept in mind when analyzing data:

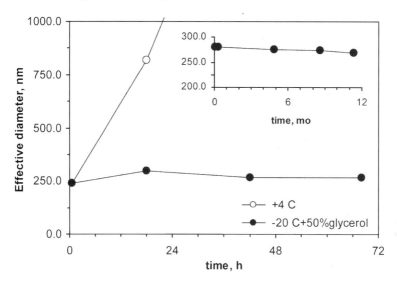

Fig. 6. Effects of viscosity on conjugate aggregation. Storage of conjugate at +4°C in PBS results in significant aggregation and imminent precipitation of immunoconjugate (open circles), whereas 50% glycerol at −20°C prevents their change in size for both short-term (closed circles) and long-term storage (insert).

a. Monitor the count rate to insure that samples are not too dilute or too concentrated for calculations (100 kcps to 1 Mcps). If the count rate is too small, then random fluctuations (e.g., dust particles) will impose very large error in the readings, and very long measurement times will be necessary to average out these occurrences. However, if concentration is too large, then particle–particle interactions become significant, and the Brownian motion is no longer valid.

b. Check at least four different fitting functions to verify particle size. To account for particle size distributions, the autocorrelator can impose different distribution functions to calculate a size and dispersity (linear, quadratic, and so on). Particle sizes calculated from each of these functions should agree seemingly well with each other. If they do not, or if dispersities are rather large (>0.2), then keep in mind that measured particle sizes are not guaranteed.

c. A simple way to evaluate homogeneity in the sample is by monitoring the shape of the decay curve at the point where the autocorrelation goes to zero. If this curve is smooth, and drops down to zero, then particles are nicely dispersed. If the curve is other than exponential and does not go to zero, particle sizes are very high and not well distributed. Typically we consider a reading is good when the autocorrelation curve is linear for 2 logs, which is rare for conjugates larger than 300–400 nm.

8. Appropriate storage of conjugates may be critical for experiments that have to be performed at a different time or location. In this case, the major obstacle is the general tendency of immunoconjugates to aggregate with time. We found that

aggregation of conjugates can be slowed down or prevented by increasing viscosity of the solution. Good results were obtained by storage of conjugates in 30% or 50% at –20°C. Under these conditions, 30% glycerol was enough to slow down aggregation for 1–2 d. However, longer incubation revealed some aggregation. Storage in 50% glycerol apparently completely prevents aggregation, as size of conjugates was stable for at least 1 yr (**Fig. 6**). Protective and enzymatic activities of catalase/anti-PECAM conjugates were practically intact after at least 1 wk. It is important to remember that glycerol affects DLS reading by changing the viscosity of solution. Thus, effects of storage in glycerol should always be compared vs. freshly prepared samples in the same concentration of glycerol.

Acknowledgments

We thank Drs. Thomas Sweitzer, Arnaud Scherpereel, and Ms. Anu P. Thomas for critically important contributions to the previous studies, which provided experimental background for development of the protocols outlined in this chapter. This work was supported by NIH SCOR in Acute Lung Injury (NHLBI HL 60290, Project 4), NHLBI RO1 HL/GM 71175-01, and the Department of Defense Grant (PR 012262) to VRM.

References

1. Muzykantov, V. R., Christofidou-Solomidou, M., Balyasnikova, I., Harshaw, D. W., Schultz, L., Fisher, A. B., et al. (1999) Streptavidin facilitates internalization and pulmonary targeting of an anti-endothelial cell antibody (platelet-endothelial cell adhesion molecule 1): a strategy for vascular immunotargeting of drugs. *Proc. Natl. Acad. Sci. USA* **96,** 2379–2384.
2. Wiewrodt, R., Thomas, A. P., Cipelletti, L., Christofidou-Solomidou, M., Weitz, D. A., Feinstein, S. I., et al. (2002) Size-dependent intracellular immunotargeting of therapeutic cargoes into endothelial cells. *Blood* **99,** 912–922.
3. Muro, S., Wiewrodt, R., Thomas, A., Koniaris, L., Albelda, S. M., Muzykantov, V. R., et al. (2003) A novel endocytic pathway induced by clustering endothelial ICAM-1 or PECAM-1. *J. Cell Sci.* **116,** 1599–1609.
4. Lasic, D. D. (1998) Novel applications of liposomes. *Trends Biotechnol.* **16,** 307–321.
5. Panyam, J. and Labhasetwar, V. (2003) Biodegradable nanoparticles for drug and gene delivery to cells and tissue. *Adv. Drug Deliv. Rev.* **55,** 329–347.
6. Muzykantov, V. R. (1997) Conjugation of catalase to a carrier antibody via a streptavidin-biotin cross-linker. *Biotechnol. Appl. Biochem.* **26,** 103–109.
7. Muzykantov, V. R. (2001) Delivery of antioxidant enzyme proteins to the lung. *Antioxid. Redox. Signal.* **3,** 39–62.
8. Atochina, E. N., Balyasnikova, I. V., Danilov, S. M., Granger, D. N., Fisher, A. B., and Muzykantov, V. R. (1998) Immunotargeting of catalase to ACE or ICAM-1 protects perfused rat lungs against oxidative stress. *Am. J. Physiol.* **275,** L806–L817.

9. Sweitzer, T. D., Thomas, A. P., Wiewrodt, R., Nakada, M. T., Branco, F., and Muzykantov, V. R. (2003) PECAM-directed immunotargeting of catalase: specific, rapid and transient protection against hydrogen peroxide. *Free Radic. Biol. Med.* **34,** 1035–1046.

10. Murciano, J. C., Muro, S., Koniaris, L., Christofidou-Solomidou, M., Harshaw, D. W., Albelda, S. M., et al. (2003) ICAM-directed vascular immunotargeting of anti-thrombotic agents to the endothelial luminal surface. *Blood,* **101,** 3977–3984.

11. Kozower, B. D., Christofidou-Solomidou, M., Sweitzer, T. D., Muro, S., Buerk, D. G., Solomides, C. C., et al. (2003) Immunotargeting of catalase to the pulmonary endothelium alleviates oxidative stress and reduces acute lung transplantation injury. *Nat. Biotechnol.* **21,** 392–398.

12. Wilchek, M. and Bayer, E. A., eds. (1990) *Avidin-Biotin Technology.* Academic Press, Inc., San Diego, CA.

13. Odian, G. (1991) *Principles of Polymerization.* John Wiley and Sons, Inc., New York, NY.

14. Osawa, M., Masuda, M., Kusano, K., and Fujiwara, K. (2002) Evidence for a role of platelet endothelial cell adhesion molecule-1 in endothelial cell mechanosignal transduction: is it a mechanoresponsive molecule? *J. Cell Biol.* **158,** 773–785.

2

Characterization of Endothelial Internalization and Targeting of Antibody–Enzyme Conjugates in Cell Cultures and in Laboratory Animals

Silvia Muro, Vladimir R. Muzykantov, and Juan-Carlos Murciano

Summary

Streptavidin–biotin conjugates of enzymes with carrier antibodies provide a versatile means for targeting selected cellular populations in cell cultures and in vivo. Both specific delivery to cells and proper subcellular addressing of enzyme cargoes are important parameters of targeting. This chapter describes methodologies for evaluating the binding and internalization of labeled conjugates directed to endothelial surface adhesion molecules in cell cultures using anti-intercellular adhesion molecule/catalase or antiplatelet endothelial cell adhesion molecule/ catalase conjugates as examples. It also describes protocols for characterization of biodistribution and pulmonary targeting of radiolabeled conjugates in rats using anti-intercellular adhesion molecule/tPA conjugates as an example. The experimental procedures, results, and notes provided may help in investigations of vascular immunotargeting of reporter, experimental, diagnostic, or therapeutic enzymes to endothelial and, perhaps, other cell types, both in vitro and in vivo.

Key Words: Endothelium; cell adhesion molecules; catalase; plasminogen activators; lung targeting.

1. Introduction

Streptavidin crosslinking of reporter and therapeutic enzymes with antibodies to endothelial cell adhesion molecules provides nanoscale conjugates useful for experimental and, perhaps, diagnostic or therapeutic vascular immunotargeting (*see* Chapter 1 and **refs.** *1–3*). Binding and appropriate subcellular addressing of antibody–enzyme conjugates to and/or into the target cells are key components for optimal design of drug-delivery systems. The size of the conjugates is an important parameter that determines the rate of intracellular uptake and, perhaps, subcellular trafficking of the conjugates *(4,5)*.

From: *Methods in Molecular Biology, vol. 283: Bioconjugation Protocols: Strategies and Methods*
Edited by: C. M. Niemeyer © Humana Press Inc., Totowa, NJ

This chapter outlines basic experimental protocols useful in the characterization of these relevant conjugates parameters. The first part (**Subheading 3.1.**) describes protocols for cell culture experiments that use fluorescent and radioisotope labeling as means to trace binding, internalization, and fate of anti-platelet endothelial cell adhesion molecule (PECAM)/catalase and anti-intercellular adhesion molecule (ICAM)/catalase conjugates. The second part (**Subheading 3.2.**) describes protocols for in vivo experiments in intact anesthetized rats using anti-ICAM/tissue-type plasminogen activator (tPA) conjugate labeled with radioisotopes. Thus, particular immunoconjugates described in this chapter are potentially useful for vascular targeting of either antioxidant (e.g., catalase to detoxify H_2O_2, **ref. 6**) or antithrombotic enzymes (e.g., tPA to dissolve fibrin, **ref. 3**). However, because of the modular nature of the conjugation and labeling procedures used, the described protocols can be used for the characterization of endothelial targeting and uptake of diverse reporter and therapeutic enzyme cargoes conjugated with a variety of carrier antibodies *(7)*. Furthermore, cell culture protocols given here for endothelial cells can be applied to other cell types of interest.

2. Materials

2.1. Equipment

1. Gamma-counter.
2. Fluorescence microscope equipped with ×40 or ×60 magnification objectives; filters compatible with fluorescein isothiocyanate (FITC; green), Texas Red (red), and UV or Alexa Fluor 350 (blue) fluorescence; digital camera; and image analysis software (ImagePro).

2.2. Reagents, Proteins, and Antibodies

1. Standard phosphate buffer, PBS, (NaH_2PO_4 20 mM, 150 mM NaCl, pH 7.4).
2. Glycine solution (50 mM glycine, 100 mM NaCl, pH 2.5) is used for elution of conjugates or antibodies bound to antigen expressed in the cells.
3. Lysis buffer (PBS containing 2% Triton X-100) is used to lyse cells and differentiate the internalized from the surface retained fractions of conjugates or antibodies.
4. PBS containing 5% bovine serum albumin (PBS-BSA) is used to increase the protein content in the analysis of free iodine label released from damaged proteins.
5. PBS containing 10% fetal bovine serum (PBS-FBS) is used to block the unspecific binding of conjugates or antibodies to the cells while providing the cell necessary nutrients.
6. Antibodies: human anti-ICAM-1 (MAb R6.5) or rat anti-ICAM-1 (MAb 1A29); human anti-PECAM-1 (MAb 62); and goat anti-mouse IgG conjugated to FITC, Texas Red, or Alexa Fluor 350.

7. Other reagents: Concentrated (100% w/v) trichloroacetic acid solution (TCA); goat serum; FITC-labeled streptavidin; tPA; catalase; paraformaldehyde; mowiol; [^{125}]iodine.

2.3. Immunoconjugates

^{125}I-labeled and nonlabeled immunoconjugates synthesized and characterized by dynamic light scattering as described in Chapter 1, where radioisotope is coupled to the cargo enzyme, not the carrier antibody, were used. In some cases, anti-ICAM/catalase conjugates based on FITC-labeled polymer were used (*4,5*).

2.4. Cells and Media

1. Human umbilical vein endothelial cells (HUVECs, Clonetics).
2. Endothelial cell growth medium (*see* Chapter 1 for details on medium composition) free of antibiotics.

3. Methods
3.1. Characterization of Immunoconjugates in Cell Culture
3.1.1. Quantitative Tracing of Radiolabeled Conjugates in HUVECs

1. Seed the cells in 24-well plates. Cultivate to confluence (approx 48 h) in the appropriate medium. Replace by fresh antibiotic-free medium 24 h before the experiment.
2. Wash cells twice by warm (37°C) culture medium. Add 0.5 μg to 1 μg of conjugate per well (i.e., specific activity 0.03 μCi/μg to 0.1 μCi/μg) in 0.5 mL of medium supplemented with 10% FBS. Incubate cells for 1–2 h at 37°C in the presence of the immunoconjugates.
3. Wash cells three times by medium to remove nonbound conjugates. Incubate cells with a glycine solution (15 min, room temperature [RT]) to elute noninternalized immunoconjugates bound to the cell surface. Using a gamma-counter, determine radioactivity in glycine-eluted fraction (*see* **Note 5**).
4. Wash cells three times by medium and incubate them for 15 min at RT with 0.5 mL of lysis buffer. Add 0.1 mL of the obtained cell lysates to 0.5 mL of PBS-BSA and sequentially add 0.2 mL of TCA and incubate 20 min at RT to precipitate proteins. Centrifuge TCA–lysate mixture (3000*g*, 10 min) and determine radioactivity in pellet and supernatant fractions.
5. Determine protein concentration in a fraction of cell lysates to normalize radioactivity values in samples per gram of total cell protein. Relative and absolute binding, internalization, and/or degradation of the immunoconjugates can be calculated as follows:

$$Total\ binding = \frac{cpm\ in\ glycine\ fraction + cpm\ in\ lysate\ pellet\ fraction + cpm\ in\ lysate\ supernatant\ fraction}{specific\ activity\ (cpm/ng\ of\ conjugate)}$$

$$\textit{Internalization percentage (if applicable)} = 100 \times \frac{\textit{cpm in lysate fraction}}{\textit{cpm in lysate fraction} + \textit{cpm in glycine-eluted fraction}}$$

$$\textit{Total internalization (if applicable)} = \frac{\textit{cpm in lysate fraction}}{\textit{specific activity (cpm/ng of conjugate)}}$$

$$\textit{Degradation percentage} = 100 \times \frac{\textit{cpm in supernatant fraction}}{\textit{cpm in glycine} + \textit{cpm in lysate supernatant} + \textit{cpm in lysate pellet fractions}}$$

$$\textit{Total degradation} = \frac{\textit{cpm supernatant fraction}}{\textit{specific activity (cpm/ng of conjugate)}}$$

3.1.2. Subcellular Detection of Immunoconjugates by Immunofluorescence

3.1.2.1. BINDING OF IMMUNOCONJUGATES TO TARGET CELLS

1. Seed the cells onto 12-mm^2 glass coverslips coated with 1% gelatin in 24-well plates. Allow cells to grow for 48 h to confluence (**Note 2**). Replace medium by fresh antibiotic-free medium 24 h before the experiment. Incubate cells for 5 min at 4°C before the experiment. Wash cells twice and replace by medium containing 10% FBS and a conjugate (1–1.5 µg of per well). Incubate cells for 30 min at 4°C to permit binding.
2. Wash cells three times with cold medium to eliminate nonbound conjugates. Fix cell by a cold solution 2% paraformaldehyde in PBS (15 min) (**Note 3**).
3. Wash cells three times with PBS and stain surface-bound conjugates by incubating fixed cells for 30 min at RT with a 4 µg/mL solution of Texas Red-labeled goat anti-mouse IgG in PBS-FBS (alternatively, use fluorescently labeled antibodies against the enzyme cargo) (**Note 1**). Wash cells three times with PBS.
4. Mount cell-containing coverslips on glass microscope slides using mowiol and incubate overnight at RT to allow the mounting media to polymerize. Observe samples by fluorescence microscopy using ×40 or ×60 objectives. Compare images of fluorescence and phase-contrast fields to confirm location of the immunoconjugate to the cell surface.

3.1.2.2. INTERNALIZATION OF IMMUNOCONJUGATES INTO TARGET CELLS

1. Seed and grow cells to confluence as described in **Subheading 3.1.2.1.**
2. Wash cells twice with 37°C prewarmed medium and add immunoconjugate and incubate with cells for 1 h at 37°C to permit binding and internalization. Fix cells and stain surface-bound conjugates as described in **Subheading 3.1.2.1.**
3. Wash cells three times with PBS and permeabilize them by 15-min incubation with a cold solution 0.2% Triton X-100 in PBS. Stain internalized conjugates by incubating permeabilized cells with FITC-labeled goat anti-mouse IgG (4 µg/mL in PBS serum).
4. Wash cells and mount coverslips on microscope slides as described in **Subheading 3.1.2.1.**

5. Take images using filters compatible with Texas Red (red) and FITC (green) in a fluorescence microscope (×40 or ×60 objective) and merge them. Surface-bound conjugates will appear yellow (double-labeled), whereas internalized conjugates will be single-labeled in green. Imaging software can be programmed to quantify relative conjugate internalization, following the formula:

$$Internalization\ percentage = 100 \times \frac{(number\ of\ green\ conjugates - number\ of\ red\ conjugates)}{number\ of\ green\ conjugates}$$

3.1.2.3. FATE OF INTRACELLULARLY DELIVERED IMMUNOCONJUGATES

1. For this type of experiments, use fluorescently labeled conjugates (i.e., based on FITC-labeled nondegradable polymer beads; see **Note 4**) prepared as previously described in detail *(4,5)*. First, incubate cells in the presence of conjugates at 4°C to permit binding to the cell surface. Then, wash nonbound immunoconjugates with cold medium, add FBS-supplemented medium, and incubate cells for the time period of interest at 37°C to permit endocytosis and intracellular trafficking of the immunoconjugates previously bound to the cell surface.
2. Wash and fix cells as in **Subheading 3.1.2.1.** followed by staining of the noninternalized conjugates for 30 min at RT with a solution 4 μg/mL goat anti-mouse IgG (i.e., labeled with Alexa Fluor 350) in PBS serum.
3. Wash the preparations three times with PBS and permeabilize cells for 15 min with a cold solution 0.2 % Triton X-100 in PBS. Incubate permeabilized cells with a solution 4 μg/mL goat anti-mouse IgG (i.e., labeled with Texas Red) in PBS serum.
4. Wash cells and mount coverslips on microscope slides as described in **Subheading 3.1.2.1.**

Inspect in a fluorescence microscope using filters compatible with FITC (green), Alexa Fluor 350 (blue), and Texas Red (red) and merge images. Immunoconjugates bound to the cell surface will appear triple-labeled as white. Nondegraded internalized conjugates will appear as double-labeled in yellow, whereas internalized counterparts with degraded protein component will be single-labeled as green.

3.2. Characterization of Immunoconjugates In Vivo

3.2.1. Biodistribution of Radiolabeled Conjugates After Intravenous Administration

1. Anesthetize rats (Sprague–Dawley) weighing 250 g using an intraperitoneal injection of 300 μL of Nembutal solution (70 mg/kg of body weight) and wait 5 min until animals are fully anesthetized (i.e., they do not react to their legs being squeezed with forceps).
2. Inject ^{125}I-labeled conjugates (approx 1–5 μg of the conjugate, 100,000–300,000 cpm per animal) via a tail vein in 0.2 mL of PBS using an insulin syringe with a

27.5-gage needle. Warming up the tail by using hot water makes the vein more visible and easy to inject.

3. One hour after injection, sacrifice anesthetized animals by dissection of the descending aorta, collect 1 mL of blood from the peritoneal cavity, and place it in a heparin-containing tube. Excise internal organs, including lung, liver, kidney, spleen, and heart; rinse in saline; blot in filter paper; weigh; and analyze for radioactivity in a gamma-counter.

4. Use radiotracing data to calculate the following parameters of conjugates behavior in vivo (for more information, *see* refs. *3*, *8*, and *9*):

   a. Percent of injected dose (%ID) characterizes total uptake in a given organ and thus it shows biodistribution and effectiveness of the immunoconjugate targeting. However, this parameter does not take into account organ sizes; thus, uptake in the liver (approx 10 g in a rat) might appear far greater than the uptake in smaller organs (e.g., lung, ~1 g).

   b. To evaluate tissue selectivity of the uptake (and compare the data obtained in different animal species, as well as organs with different sizes), calculate %ID per gram (%ID/g).

   c. The ratio between %ID/g in an organ of interest and that in blood gives the localization ratio (LR) that compensates for a difference in the blood level of circulating conjugates and allows comparison of targeting between different carriers, which may have different rates of blood clearance.

   d. By dividing the LR of a specific antibody conjugate in an organ by that of the control IgG counterpart, calculate the immunospecificity index (ISI = LR_{MAb}/LR_{IgG}), the ratio between the tissue uptake of immune and nonimmune counterparts normalized to their blood level. ISI is the most objective parameter of the targeting specificity.

3.3. Results

3.3.1. Characterization of Immunoconjugates in Cell Culture

3.3.1.1. ANALYSIS OF BINDING AND FATE OF RADIOLABELED CONJUGATES

Site-specific binding and degradation of the conjugates by cells was determined by measuring ^{125}I in fractions of glycine elution, TCA pellet, and supernatant of cell lysates obtained from HUVECs incubated with anti-PECAM/^{125}I-catalase and IgG/^{125}I-catalase conjugates as described in **Subheading 3.1.1.** The sum of the recovered ^{125}I shows total amount of catalase associated with cells and reveals the specificity of binding of anti-PECAM conjugates, using as negative control nonspecific IgG conjugates (**Fig. 1A**). A relatively minor fraction of ^{125}I was found in the supernatant after TCA precipitation of cell lysates, indicating that catalase undergoes very modest degradation within 1 h of incubation at 37°C in endothelial cells (**Fig. 1B**).

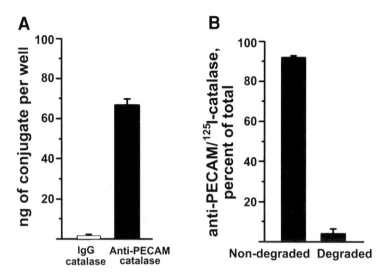

Fig. 1. Quantitative analysis of binding and degradation of radiolabeled anti-PECAM/catalase conjugates in HUVECs. HUVECs were incubated for 90 min at 37°C with anti-PECAM/^{125}I-catalase or control IgG counterpart conjugates, washed, and lysed to determine the TCA-soluble fraction of cell-bound radioactivity. The absolute amount of conjugate in the different fractions is calculated based on its specific activity as described in **Subheading 3.1.1.**

3.3.1.2. IMAGING OF BINDING, INTERNALIZATION, AND FATE
OF IMMUNOCONJUGATES BY IMMUNOFLUORESCENCE

Figure 2 shows that anti-PECAM/catalase but not IgG/catalase conjugates bind to HUVECs at 4°C, thus confirming the data obtained with ^{125}I tracing (*see* **Fig. 1**). Comparison of fluorescence and phase-contrast images indicates that anti-PECAM/catalase conjugates are located in the cell periphery, consistent with the predominant expression of PECAM-1 to the cell borders.

Moreover, in cells incubated for 1 h at 37°C with anti-PECAM/catalase conjugates, only a fraction of the conjugate was labeled before permeabilization by Texas Red-labeled secondary antibody, whereas FITC-labeled secondary antibody applied after permeabilization reveals abundant immunostaining (**Fig. 3A**). Single FITC-labeled (green) internalized conjugates are localized in the perinuclear region of the cell, whereas noninternalized double-labeled (yellow) conjugates tend to localize to the cell periphery. Semiquantitative analysis of double-labeled and single-labeled images shows that endothelial cells internalize 50% of cell-bound anti-PECAM/catalase conjugates.

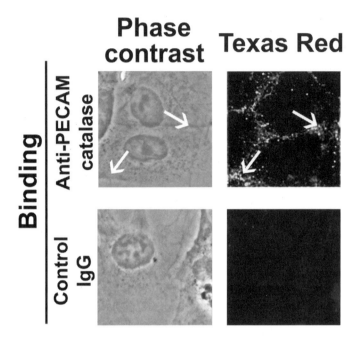

Fig. 2. Fluorescent detection of binding of anti-PECAM/catalase conjugates to HUVECs. HUVECs were incubated for 30 min at 4°C with anti-PECAM/catalase or nonspecific IgG conjugates, washed, fixed, and surface-bound anti-PECAM was stained with Texas Red goat anti-mouse IgG. The samples were analyzed by phase contrast and fluorescence microscopy. The arrows show conjugates bound to the cell surface.

To visualize and estimate degradation of internalized cargoes by fluorescence microscopy, one can retreat to use fluorescently-labeled conjugates, for example, based on FITC-labeled synthetic nanobeads used as carriers for both targeting antibodies and enzyme cargoes *(4,5)*. The advantage of this carrier is that it permits direct tracing of the conjugates in cellular compartments, including lysosomes. FITC-labeled regular immunoconjugates can also be used for this purpose, (e.g., conjugates containing FITC–streptavidin; *see* **Note 4**). A pulse-chase incubation (initial incubation 30 min at 4°C followed by removal of nonbound conjugates and incubation at 37°C), permits one to separate phases of binding, internalization, and intracellular trafficking. After internalization and fixation, surface-bound particles are counterstained using goat anti-mouse IgG conjugated to Alexa Fluor 350, followed by cell permeabilization and incubation with Texas Red-labeled goat anti-mouse IgG. This staining method

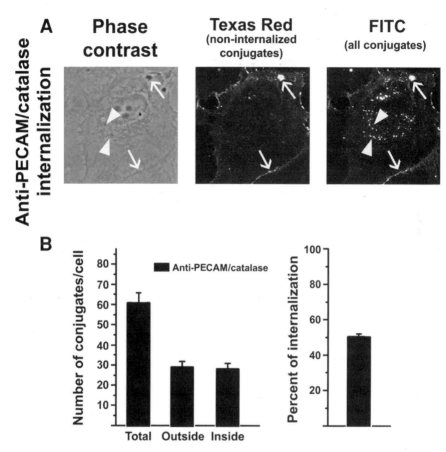

Fig. 3. Fluorescence microscopy of the uptake of anti-PECAM/catalase conjugates by HUVECs. HUVECs were incubated for 1 h at 37°C in the presence of anti-PECAM/catalase conjugates, washed, fixed, and noninternalized conjugates were stained with Texas Red-labeled goat anti-mouse IgG, followed by cell permeabilization and staining with FITC-labeled goat anti-mouse IgG. **A,** The arrows show double-labeled conjugates on the cell surface. The arrowheads show single FITC-labeled conjugates, internalized within the cell. **B,** Quantification of the experiment described above, expressed as mean and standard error (*n* = 10 fields, from two independent experiments).

(**Fig. 4**) distinguishes surface-bound (triple-stained, white), as well as internalized nondegraded (double-stained, yellow) and degraded conjugates (single-stained, green). The results of the particular experiment shown in **Fig. 4** indicate that conjugates are stable within the cell for 1–2 h and degrade 3 h after internalization.

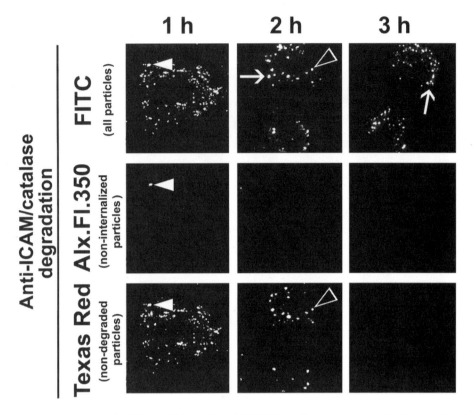

Fig. 4. Imaging of the stability of anti-ICAM/catalase nanoparticles internalized in HUVECs. HUVECs were incubated for 30 min at 4°C in the presence of FITC-labeled anti-ICAM/catalase nanoparticles to permit binding of these to the surface antigen. Then, nonbound particles were washed and the cells were incubated either for 1 h, 2 h, or 3 h at 37°C, to permit internalization and intracellular trafficking of the anti-ICAM/catalase particles. After cell fixation, noninternalized particles were stained with Alexa Fluor 350 goat anti-mouse IgG. Thereafter, the cells were permeabilized and incubated with Texas Red goat anti-mouse IgG. The samples were analyzed by fluorescence microscopy. Closed arrowheads show a triple FITC+Alexa Fluor 350+Texas Red-labeled particle, located to the cell surface. Open arrowheads show a double FITC+Texas Red-labeled particle, which indicates that the targeting antibody was not degraded after internalization within the cell. The arrow shows single FITC-labeled particles, indicating that the targeting antibody in the internalized particles has been degraded.

3.3.2. Characterization of Immunoconjugates In Vivo

3.3.2.1. BIODISTRIBUTION AND PULMONARY TARGETING OF tPA CONJUGATED WITH ANTI-ICAM

Experiments with tPA conjugated with an ICAM-1 monoclonal antibody illustrate analysis of vascular immunotargeting in vivo. **Figure 5A** shows comparison of biodistribution of radiolabeled anti-ICAM/^{125}I-tPA conjugate and its components, either ^{125}I-anti-ICAM or ^{125}I-tPA, 1 h after intravenous injection in rats. Anti-ICAM and anti-ICAM/tPA conjugate display preferential uptake in the pulmonary vasculature and significant uptake in hepatic and splenic vasculature. These highly vascularized organs (especially lungs that possess about 30% of endothelial surface in the body) represent privileged targets in agreement with the fact that ICAM is constitutively expressed on the endothelial surface *(10)*. Nonconjugated tPA shows no pulmonary targeting; in fact, its extremely rapid clearance (its half-life in rats is around 1–5 min; **ref. *11***) leads to disappearance of the tracer from blood and major organs within 1 h after injection.

3.3.2.2. COMPARISON OF BIODISTRIBUTION ATTAINED USING DIFFERENT INJECTION ROUTES

High levels of pulmonary uptake of conjugates directed against pan-endothelial determinants, such as ICAM-1, might be to the result of several reasons: (1) an extremely extended endothelial surface in the alveolar capillaries; (2) the fact that lung receives 100% of the heart blood output; or (3) the phenomenon of first-pass blood after intravenous injection. **Figure 5B** shows that injection of anti-ICAM/tPA conjugate via the left ventricle, which obviates the first pass in the lungs, produces less effective pulmonary targeting, suggesting that indeed first-pass phenomenon contributes to the pulmonary targeting. However, a high level of pulmonary uptake after left ventricle administration confirms the specificity of anti-ICAM conjugates targeting in vivo.

3.3.2.3. EVALUATION OF THE TARGETING SPECIFICITY OF IMMUNOCONJUGATES

Figure 6 illustrates the analysis of immunoconjugates biodistribution and targeting in rats 1 h after intravenous injection. A comparison of %ID/g in organs reveals that anti-ICAM/tPA conjugate but not IgG/tPA counterpart accumulates in the pulmonary vasculature. However, the blood level of anti-ICAM/tPA is lower than that of the nonimmune IgG/tPA counterpart, likely because of depletion of circulating blood pool by endothelial binding. The LR that compensates for differences in blood level reveals very high selectivity of anti-ICAM/tPA uptake in highly vascularized organs including liver (LR close to 3), spleen (LR exceeds 7), and especially lungs (LR close to 30). Calculation

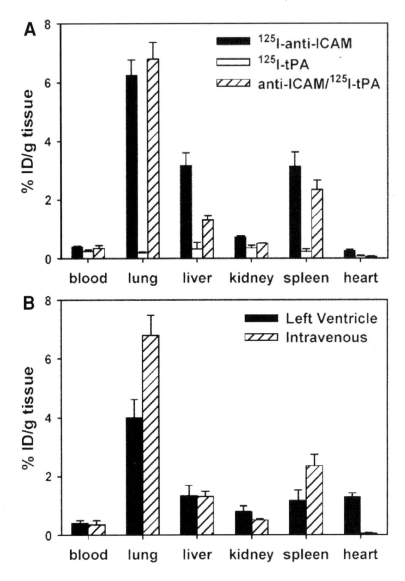

Fig. 5. Biodistribution of immunoconjugates and free components in vivo. Tracer amounts of radiolabeled proteins (approx 1 µg of radioactive material per sample) were injected intravascularly in anesthetized rats. After 1 h, animals were sacrificed and blood and organs extracted and analyzed for radioactivity. **A,** ^{125}I-anti-ICAM (black bars) or anti-ICAM/125I-tPA (hatched bars), but not free ^{125}I-tPA (white bars) accumulate in the lung, liver, and spleen after intravenous injection. **B,** Comparison of biodistribution of anti-ICAM/^{125}I-tPA after injections via the tail vein (hatched bars) or the left ventricle (black bars). Data are presented as mean ± SD, *n* = 4–9 animals per determination.

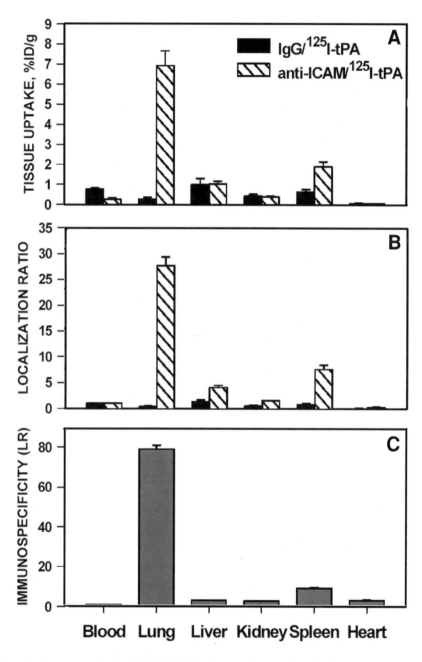

Fig. 6. Analysis of anti-ICAM/tPA biodistribution in vivo. Radioactivity in organs was analyzed 1 h after intravenous injection anti-ICAM/^{125}I-tPA (hatched bars) or control nonspecific IgG/^{125}I-tPA (black bars). The data (mean ± SD, n = 4–9) is presented as: (**A**) % ID/g of tissue; (**B**) LR; and (**C**) ISI. Adapted from **ref. 3**.

of ISI reveals that anti-ICAM/tPA accumulates in the lungs almost 100 times higher than IgG/tPA counterpart, thus confirming high specificity of targeting.

4. Notes

1. Uptake and trafficking of immunoconjugates within the target cells can be studied by tracing the antibody carrier, the enzymatic cargo, or both moieties. The protocols described in **Subheadings 3.1.2.1.**, **3.1.2.2.**, and **3.1.2.3.** trace antibody moieties using secondary antibodies against murine IgG. The same protocols can be used to trace enzyme cargo, for example, using an antibody to catalase. Moreover, conjugates directly labeled with a fluorescent probe, such as the ones based on fluorescent-labeled nanobeads or streptavidin crosslinker, are optimal because they can be visualized without additional staining. There are some specific factors that may require adjustment and optimization of the described protocols to be applied to particular conjugates and target cells of interest. Some general considerations are given below.

2. Many cell types do not adhere well to glass surfaces. Coating coverslips with a proadhesive protein (i.e., fibronectin, vitronectin, collagen) before cell seeding helps to solve this problem. A 1-h incubation with 1% gelatin solution in PBS followed by a 1-h incubation to dry coverslips up is a generic choice. The density of seeding of each cell type must be adjusted to reach confluence within the first 48 h after seeding to avoid repeated division cycles that can lead to detachment. For example, optimal density for HUVEC is 7×10^4 cell per 24 wells when seeded 48 h before the experiment. Moreover, cells tend to detach from any substrate at $4°C$; thus, cold incubation should be minimal to permit binding of the conjugates. To avoid excessive detachment, pour washing medium gently and slowly on the well wall rather than directly on the cells. Glycine elution of membrane-bound conjugates may also provoke cell detachment and incubation time must be minimal (do not exceed 15 min). Inspect cell morphology and monolayer integrity by phase contrast microscopy and terminate "high-risk" exposures at the first signs of cell retraction, rounding or detachment.

3. Fixation of cells with 2% paraformaldehyde solution (10–15 min) is generally used when preparing samples for immunofluorescence, but the concentration must be optimized and can be lowed to 0.5 % to 1% if necessary to avoid disruption of the plasma membrane and partial cell permeabilization. In addition, the concentrations and incubation times of labeled antibodies given above are arbitrary and should be adjusted for particular preparations. To block nonspecific binding of labeled antibodies, preincubate fixed cells with a solution 10–20% serum of a corresponding animal species before immunostaining. To reduce nonspecific binding of the immunoconjugates (e.g., to control cells that do not express a target antigen), use incubation media containing 2–4% BSA.

4. Adjust settings for acquisition and processing of fluorescence images to optimize visualization. For instance, in the case that fluorescent signal was low, rational increase of the exposure time or brightness postacquisition can be performed, although preserving the specificity of the signal and the legitimacy of the image.

This approach helps to colocalize fluorescent signals obtained from different objects when labeled with fluorescent probes at different intensity, such as staining of a highly fluorescent FITC-labeled conjugate using secondary antibody that is relatively poorly labeled with Texas Red. Merging the images taken under similar acquisition parameters will show FITC signal masking Texas Red on the same object, not permitting visualization of a real double-labeled object and, therefore, leading to misinterpretation of the result. In addition, the choice of the fluorescent probes to reveal colocalizing objects should be made such that colors resulting from merged images permit an easy interpretation of the results. For instance, colocalization of green and red results in yellow and the three colors can be readily interpreted. However, colocalization of green and blue results in a light, bluish shade, not clearly distinguishable from the two parental colors.

5. Finally, the data on internalization and degradation of the radiolabeled conjugates should be analyzed and interpreted cautiously. For instance, multimeric conjugates can bind to cells with such high avidity that resulting large antibody/antigen clusters are difficult to disrupt by glycine elution, providing false-positive internalization result. Visualization of the uptake using double-fluorescence based techniques permits to circumvent this artifact.

Acknowledgments

The authors thank Drs. Michael Koval and Steven Albelda for contributions to the previous studies, which provided experimental background for the development of the protocols outlined in this chapter. This work was supported by NIH SCOR in Acute Lung Injury (NHLBI HL 60290, Project 4), NHLBI RO1 HL/GM 71175–01, and the Department of Defense Grant (PR 012262) to VRM.

References

1. Muzykantov, V. R., Christofidou-Solomidou, M., Balyasnikova, I., Harshaw, D. W., Schultz, L., Fisher, A. B., et al. (1999) Streptavidin facilitates internalization and pulmonary targeting of an anti-endothelial cell antibody (platelet-endothelial cell adhesion molecule 1): a strategy for vascular immunotargeting of drugs. *Proc. Natl. Acad. Sci. USA* **96,** 2379–2384.
2. Scherpereel, A., Wiewrodt, R., Christofidou-Solomidou, M., Gervais, R., Murciano, J. C., Albelda, S. M., et al. (2001) Cell-selective intracellular delivery of a foreign enzyme to endothelium in vivo using vascular immunotargeting. *FASEB J.* **15,** 416–426.
3. Murciano, J. C., Muro, S., Koniaris, L., Christofidou-Solomidou, M., Harshaw, D. W., Albelda, S. M., et al. (2003) ICAM-directed vascular immunotargeting of antithrombotic agents to the endothelial luminal surface. *Blood* **101,** 3977–3984.
4. Wiewrodt, R., Thomas, A. P., Cipelletti, L., Christofidou-Solomidou, M., Weitz, D. A., Feinstein, S. I., et al. (2002) Size-dependent intracellular immunotargeting of therapeutic cargoes into endothelial cells. *Blood* **99,** 912–922.

5. Muro, S., Wiewrodt, R., Thomas, A., Koniaris, L., Albelda, S. M., Muzykantov, V. R., et al. (2003) A novel endocytic pathway induced by clustering endothelial ICAM-1 or PECAM-1. *J. Cell Sci.* **116,** 1599–1609.

6. Kozower, B. D., Christofidou-Solomidou, M., Sweitzer, T. D., Muro, S., Buerk, D. G., Solomides, C. C., et al. (2003) Immunotargeting of catalase to the pulmonary endothelium alleviates oxidative stress and reduces acute lung transplantation injury. *Nat. Biotechnol.* **21,** 392–398.

7. Muzykantov, V. R., Atochina, E. N., Ischiropoulos, H., Danilov, S. M., and Fisher, A. B. (1996) Immunotargeting of antioxidant enzyme to the pulmonary endothelium. *Proc. Natl. Acad. Sci. USA* **93,** 5213–5218.

8. Danilov, S. M., Gavrilyuk, V. D., Franke, F. E., Pauls, K., Harshaw, D. W., et al. (2001) Lung uptake of antibodies to endothelial antigens: key determinants of vascular immunotargeting. *Am. J. Physiol. Lung Cell Mol. Physiol.* **280,** L1335–L1347.

9. Murciano, J. C., Harshaw, D., Neschis, D. G., Koniaris, L., Bdeir, K., Medinilla, S., et al. (2002) Platelets inhibit the lysis of pulmonary microemboli. *Am. J. Physiol. Lung Cell Mol. Physiol.* **282,** L529–L539.

10. Panes, J., Perry, M. A., Anderson, D. C., Manning, A., Leone, B., Cepinskas, G., et al. (1995) Regional differences in constitutive and induced ICAM-1 expression in vivo. *Am. J. Physiol.* **269,** H1955–H1964.

11. Kuiper, J., Otter, M., Rijken, D. C., and van Berkel, T. J. (1988) Characterization of the interaction in vivo of tissue-type plasminogen activator with liver cells. *J. Biol. Chem.* **263,** 18,220–18,224.

3

Smart Polymer–Streptavidin Conjugates

Patrick S. Stayton, Zhongli Ding, and Allan S. Hoffman

Summary

The conjugation of stimuli-responsive, or "smart," polymers to streptavidin is described. The polymer is synthesized with a thiol-reactive end-group, which is used to end-graft the polymer to cysteine or lysine side-chains that are genetically engineered into controlled positions on the streptavidin surface. The conjugation positions are chosen on the basis of their location relative to the binding site, together with the criteria that they be solvent accessible and thus reactive. The polymer composition can be controlled to impart responsiveness to temperature, pH, and/or specific wavelengths of light. These signals are sent to the polymer, which serves as an antennae and actuator to gate biotin or biotinylated protein association with the streptavidin binding sites. The molecular switching and gating activity is directed by the reversible polymer transition between a hydrophilic, expanded coil and a more hydrophobic, collapsed state that is smaller in volume. The differences in the polymer steric properties serve to block or allow ligand access to the binding site. The control of polymer molecular weight is a particularly important design parameter for these molecular gates.

Key Words: Smart polymer; streptavidin; molecular gate; group transfer polymerization; bioconjugate.

1. Introduction

A new approach to constructing molecular gates is described that uses stimuli-responsive polymers to regulate protein-binding events *(1,2)*. The polymers respond to signals such as small changes in temperature, pH, and/or light by transitioning between an expanded hydrophilic state and a collapsed hydrophobic state. The properties of the gates are a combination of the properties of the stimuli-responsive polymers and the conjugation position where they are grafted onto the protein. The monomer composition determines the stimuli-sensitivity of the polymer, with the molecular weight also playing an important role in the gating activity. The composition can be tailored to control the specific transition range, for example, to manipulate at which pH the polymer

From: *Methods in Molecular Biology, vol. 283: Bioconjugation Protocols: Strategies and Methods*
Edited by: C. M. Niemeyer © Humana Press Inc., Totowa, NJ

collapse and expansion occurs or at which temperature the transition occurs. The optimal polymer molecular weight is determined by the spatial relationship between the conjugation site and the binding site, as well as by the size of the ligand to be gated. When these design criteria are matched, the gate can serve as a size-selective barrier to ligand binding.

Streptavidin has been a good model system to develop the molecular gate concept. The convenient genetic engineering system, high-resolution structural information, and absence of native cysteine residues allow the design and construction of appropriate polymer conjugation positions. Chemical synthesis techniques provide routes to minimize the polydispersity of the stimuli-responsive polymers molecular weights. Here, we detail an illustrative example where a temperature-responsive polymer was used to construct size-selective molecular gates for streptavidin and biotinylated target proteins.

2. Materials

1. Tris-(2-carboxyethyl) phosphine (HCl) (Pierce, Rockford, IL).
2. BioMag® Biotin-beads (PerSeptive Diagnostics, Cambridge, MA).
3. Tosyl-activated magnetic microbeads (Dynabeads® M-280, Dynal, Inc. Lake Success, NY).
4. *Kpn*I and *Xba* restriction enzymes (New England Biolabs, Beverly, MA).
5. T4 DNA Ligase (New England Biolabs).
6. Qiaex II kit (Qiagen, Santa Clarita, CA).
7. pET21a plasmid (Novagen, Madison, WI).
8. *N,N*-Diethylacrylamide (DEAAm; PolySciences Inc.,Warrington, PA).
9. CaH$_2$ (Sigma, St. Louis, MO).
10. 1-Methoxyl-1-(trimethylsiloxyl)-2-methyl-1-propene (Lancaster, Windham, NH).
11. 2-(Trimethylsiloxy)ethyl methacrylate (Aldrich, Milwaukee, WI).
12. Tetrahydrofuran (THF; Fisher Scientific, Fair Lawn, NJ).
13. Dyedeoxy sequencing kit (PE Applied Biosystems, Foster City, CA).
14. Iminobiotin-agarose resin suspension (Sigma).
15. Centriprep-10 with a MW cutoff of 10,000 Dalton (Amicon).
16. Mutagenic oligonucleotides (Integrated DNA Technologies Inc., Coralville, IA).
17. Sodium benzophenone ketyl (Aldrich).
18. Ttetrabutylammonium acetate (Aldrich).
19. Core streptavidin gene in pUC18 plasmid.

3. Methods

The key design parameter for the polymer in the temperature-sensitive molecular gate application is molecular weight. Chain transfer free radical polymerization is typically used to synthesize the stimuli-responsive polymers, such as poly(*N*-isopropylacrylamide). These synthetic techniques typically yield polymer chains with a broad polydispersity of molecular weights. The streptavidin conjugates would thus also reflect this polydispersity in the size of the polymer chains, which complicates the steric gating performance. To obtain narrow molecular weight distributions, a group transfer living polymerization strategy was thus used that is applicable to fully N-substituted acrylamides (*3–7*) such as N,N-diethylacrylamide. Poly(N,N-diethylacrylamide) (PDEAAm) has a temperature responsiveness that is similar to poly(N-isopropylacrylamide. This synthetic procedure is given below, along with protein conjugation and purification protocols, and an immobilization method on magnetic beads.

3.1. Polymer Synthesis

1. The monomer DEAAm was purified by stirring over CaH_2 for 24 h at room temperature followed by two distillations under reduced pressure (bp. 95°C/12 mmHg). The purified monomer was kept in a septum-sealed flask filled with argon at 4°C and used within 1 wk. (*see* **Note 1**).
2. The initiator, 1-methoxyl-1-(trimethylsiloxyl)-2-methyl-1-propene, was distilled twice under reduced pressure (bp. 84°C/40 mmHg) and stored in an argon atmosphere at 4°C.
3. The capping reagent, 2-(trimethylsiloxy)ethyl methacrylate was purified by two distillations under reduced pressure (bp. 57°C/1 mmHg).
4. The purity of monomer, initiator, and capping reagent were characterized by gas chromatography-mass spectrometry, consisting of a gas chromatography (Hewlett Packard 5890A) and mass selective detector (Hewlett Packard 5971A).
5. THF was purified by refluxing in the presence of sodium benzophenone ketyl, followed by distillation.
6. Dry THF (30 mL) was directly transferred from the distillation reservoir to a septum-sealed reactor filled with Argon. The desired amount of initiator, 1-methoxyl-1-(trimethylsiloxyl)-2-methyl-1-propene, was added to the reactor using a syringe. The reactor was chilled in an icewater bath. At an initiator/tetrabutylammonium acetate catalyst ratio of 10, the polymerization was very successful, yielding a uniform polymer with an 85% yield. Variations on ratios of initiator-to-catalyst ratios are used to vary the polymer molecular weight (**Table 1**).
7. Ten milliliters of DEAAm were added dropwise to the reactor using a syringe followed by addition of the catalyst–THF solution. The reaction was conducted in the icewater bath for 4 h and monitored by GPC analysis of the reaction solution.
8. The capping reagent, 2-(trimethylsiloxy) ethyl methacrylate, was added to the reactor to introduce a hydroxyl end group to the polymer. The solution was then stirred at room temperature for 20 h.
9. To stop the polymerization and deprotect the hydroxyl group, 1 mL of methanol:hydrochloric acid (1:1) mixture was added to the solution. The solution was stirred at room temperature for 4 h.

Table 1
Effect of the Ratio of Monomer/Initiator
on the Molecular Weight and the Polydispersity of the Polymer

Molar ratio of monomer/initiator	Theoretical molecular weight	Measured Mn	Polydispersity
100/5	2,540	3300	1.15
100/2.56	5,000	6800	1.15
100/1.27	10,000	12,800	1.18
100/0.637	20,000	3250	2.00
100/0.42	30,000	1310	3.75

Initiator/catalyst ratio, 100/10; initiator, 1-methoxyl-1-(trimethylsiloxyl)-2-methyl-1-propene; catalyst, tetrabutylammonium acetate.

10. The polymer preparation was isolated by precipitation in *n*-hexane twice and dried in a vacuum oven at room temperature.

11. The hydroxyl groups were reacted with divinylsulfone in alkaline conditions. The divinylsulfone was added at 10-fold molar excess to the hydroxyl groups in the polymer in 30 mL of anhydrous dichloromethane, with a total polymer weight of 5 g. The solution was purged with nitrogen at room temperature for 30 min before adding 0.03 g of potassium tert-butyloxide. The reaction was conducted at room temperature for 16 h under a nitrogen atmosphere. The derivatized PDEAAm was isolated by precipitation in diethyl ether (*see* **Note 2**).

3.2. Streptavidin Site-Directed Mutagenesis

1. A double lysine E51K/N118K streptavidin mutant was constructed by a combination of site-directed cassette mutagenesis and polymerase chain reaction (PCR) mutagenesis. The synthetic gene for core streptavidin inserted in pUC18 was used as the template *(8)*. A segment was removed from the gene by restriction enzyme digestion using *Kpn*I and *Xba*I. A pair of complementary oligonucleotides (sequences are shown in **Fig. 1**), which encode a lysine at position 51, were annealed by heating to 95°C and cooling to room temperature. The annealed cassette was then ligated with T4 DNA ligase into the *Kpn*I and *Xba*I doubly cut plasmid. The mutated streptavidin E51K gene sequence was confirmed by automated dyedeoxy sequencing and used as a template for the subsequent PCR mutagenesis.

2. In a two-step PCR mutagenesis protocol, the pUC18/streptavidin E51K construct was used as the template. A synthetic oligonucleotide (sequence shown in **Fig. 1**) encoding the N118K mutation was used as a primer with a universal pUC18 primer to generate a PCR product containing the mutation. This product and a second universal pUC18 primer were used as primers in the second PCR step to produce the E51K/N118K streptavidin gene. The double mutant sequence was confirmed by automated dyedeoxy sequencing.

A E51K Cassette

5' CTACGAATCCGCTGTTGGTAACGCTA**AAT** 3'

3' CATGGATGCTTAGGCGACAACCATTGCGA**TTT**AGATC 5'

B N118K Primer

5'GAAGCT**AAA**GCGTGGAAA3'

Fig. 1. Synthetic oligonucleotides for (**A**) for the site-directed cassette mutagenesis to create N51K and (**B**) for the PCR mutagenesis to create E118K. The bold-face type indicates the coding sequence encoding the amino acid change.

3. The E51K/N118K streptavidin gene was then inserted to pET21a expression vector, for protein production in *Escherichia coli* strain BL21. The recombinant protein was purified by iminobiotin affinity chromatography and shown to be the correct molecular weight by electron-spray mass spectrometry.

3.3. Polymer–Protein Conjugation and Purification

1. PDEAAm with a vinyl sulfone end group was conjugated to E51K/N118K streptavidin in 100 mM sodium tetraborate buffer, pH 9.5, containing 50 mM of sodium chloride. The polymer:protein ratio was typically 50:1. The conjugation was conducted at 4°C for 16–20 h with gentle stirring.
2. The conjugate was separated from unconjugated streptavidin by thermally induced precipitation. The conjugation solution was centrifuged at 15,000 rpm (31,000g) for 15 min at 37°C. The precipitated conjugate was then redissolved in fresh sodium phosphate buffer (100 mM, pH 7.0), and the thermally induced precipitation was repeated a total of three times (*see* **Note 3**).
3. The conjugate was purified from free polymer by affinity chromatography. A column was packed with 10 mL of iminobiotin-agarose resin suspension that was equilibrated with loading buffer (50 mM Na$_2$CO$_3$, 500 mM NaCl, pH 11). The conjugate purified in **step 2** was dissolved in loading buffer and slowly added to the iminobiotin column. The column was washed with 10 column volumes of loading buffer and then the bound conjugate was eluted with elution buffer (100 mM acetic acid). The eluate was fractionated and protein concentration was determined spectrophotometrically. The recovered conjugate was concentrated and the buffer exchanged using a Centriprep-10 with a molecular weight cutoff of 10,000 Daltons at 4°C.

3.4. Conjugate Immobilization

1. The immobilization of the purified conjugates to magnetic beads was accomplished by either streptavidin association to biotinylated beads or by direct cova-

lent conjugation to amine-reactive beads. For biotin-based immobilization, 400 µg of conjugate was typically added to 1 mg of biotinylated magnetic beads (BioMag Biotin-beads) in 1 mL of sodium phosphate buffer, pH 7.4. The mixture was incubated at 4°C for 4 h. The immobilized protein was quantified by analyzing the depletion of the protein from the supernatant. The beads were then washed with sodium phosphate buffer, pH 7.4, followed by sodium phosphate buffer plus 0.2 wt% bovine serum albumin (to remove excess and loosely bound streptavidin and to prevent nonspecific protein adsorption in later experiments). The washed beads were finally suspended in 100 mM sodium phosphate buffer, pH 7.4, containing 50 mM NaCl, 5 mM ethylenediamine tetraacetic acid, and 0.2 wt% bovine serum albumin.

2. For covalent immobilization, tosyl-activated magnetic beads (TMB) were used. The conjugate was dissolved in immobilization buffer (100 mM sodium tetraborate buffer, pH 9.5) to a concentration of 100 µg/mL and then mixed with TMB at a protein/TMB ratio of 100 µg/1 mg. The mixture was rotated end-to-end at 37°C for 16 h. The beads were then washed with 100 mM sodium phosphate buffer, pH 7.4. The efficiency of immobilization was estimated from the protein depletion from the supernatant, quantified spectroscopically (λ = 280 nm).

4. Notes

1. Glassware, syringes, and syringe needles were baked in an oven at 125°C for at least 20 h and cooled to room temperature in an argon atmosphere immediately before use. All the liquid reagents were transferred using syringes under the protection of argon. The catalysts were measured in an argon-filled glove bag and dissolved in THF for later use.

2. The molecular weight and polydispersity of the polymers were determined by GPC in THF against polystyrene standards. The polymer composition was determined by ^{1}H-NMR analysis (Bruker 200 MHz). Vapor pressure osmometry was also used to determine the number average molecular weight. Polymers were dissolved in methanol at concentrations of 10, 20, 30, and 40 g/kg. The osmotic pressures of the solutions were determined in reference to methanol, and the number average molecular weights of the polymers were obtained from a calibration curve established from standards. Proton NMR (200 MHz, Bruker) was used to characterize the polymers with hydroxyl or vinyl sulfone groups. Ellman's reagent was used to analyze the content of vinyl sulfone in the polymers *(9)*. The polymer was reacted with excess cystamine followed by titration of the remaining cystamine with Ellman's reagent.

3. Because of the high local concentration of protein and polymer and high temperature during thermally induced precipitation, nonspecific reactions between primary amines and the remaining vinyl sulfone groups might occur. To avoid the nonspecific reactions during thermally induced precipitation, the pH of the conjugate solutions was adjusted to 7.0 by adding 1.0 N HCl before heating.

Acknowledgments

This work was supported by the National Institutes of Health through Grant No. GM 53771.

References

1. Ding, Z., Fong, R. B., Long, C. J., Hoffman, A. S., and Stayton, P. S. (2001) Size-dependent control of the binding of biotinylated proteins to streptavidin using a polymer shield. *Nature* **411,** 59–62.
2. Shimoboji, T., Larenas, E., Fowler, T., Kulkarni, S., Hoffman, A. S., and Stayton, P. S. (2002) Photo-responsive polymer-enzyme switches. *Proc. Natl. Acad. Sci. USA* **99,** 16,592–16,596.
3. Sogah, D. Y., Hertler, W. R., Webster, O. W., and Cohen, G. M. (1987) Group transfer polymerization. Polymerization of acrylic monomers. *Macromolecules* **20,** 1473–1488
4. Odian, G. (1991) *Principles of Polymerization,* 3d ed. John Wiley & Sons, Inc., New York.
5. Okano, T., Jacobs, B. H., and Kim, S. K. (1990) Thermally on-off switching polymers for drug permeation and release. *J. Control Release* **11,** 255–265.
6. Freitag, R., Baltes, T., Eggert, M., Schuger, K., and Bahr, U. (1994) Synthesis and characterization of a highly uniform, thermosensitive affinity macroligand precursor suitable for carbodiimide coupling. *Bioseparation* **4,** 353–367.
7. Bromberg, L. and Levin, G. (1998) Poly(amino acid)-b-poly(*N, N*-diethyl-acrylamide)-b-poly(amino acid) conjugates of well-defined structure. *Bioconj. Chem.* **9,** 40–49.
8. Chilkoti, A., Tan, P. H., and Stayton, P. S. (1995) Site-directed mutagenesis studies of the high affinity streptavidin-biotin complex: contributions of tryptophan residues 79, 108, and 120. *Proc. Natl. Acad. Sci. USA* **92,** 1754–1758.
9. Riddles, P. W., Blakeley, R. L. and Zerner, B. (1983) Reassessment of Ellman's reagent. *Methods Enzymol.* **91,** 49–60.

4

Conjugates of Peptides and Proteins to Polyethylene Glycols

Margherita Morpurgo and Francesco M. Veronese

Summary

This chapter provides a critical overview of the technology presently available in the field of protein PEGylation. The chemistry of the polymer and of its reactive derivatives is discussed and presented together with several protocols used to obtain PEG–protein conjugates. The coupling protocols are critically discussed on the basis of the properties of the protein to be modified and those desired for the final product. Methods for product purification and characterization are also provided. The overall information provided will guide the reader toward all of the critical steps involved in the preparation of PEG–protein adducts.

Key Words: Poly(ethylene glycol); protein; peptide; bioconjugation; chemical coupling; methods.

1. Introduction

Since the pioneering experiments by Abuchowski et al. *(1,2)*, the past 30 years of research on covalent attachment of poly(ethyleneglycol) (PEG) to proteins have clearly demonstrated the usefulness of this strategy to improve the therapeutic value of peptide and protein drugs. Because of this, several PEG reagents have been developed that allow a better design of bioconjugates. Moreover, various techniques have been optimized for conjugate purification and characterization, which are needed for pharmaceutical products approval. As a consequence of this growing interest, many activated polymers have become available on the market, thus giving access to PEGylation to those laboratories that do not have the chemical facilities required for the reactive PEG preparation.

However, designing PEG conjugates is not a straightforward procedure and many factors need to be considered. In fact, the activity, stability, pharmacokinetics, biodistribution, and immunogencity of bioconjugates may depend on

From: *Methods in Molecular Biology, vol. 283: Bioconjugation Protocols: Strategies and Methods*
Edited by: C. M. Niemeyer © Humana Press Inc., Totowa, NJ

several parameters, such as polymer size and shape, the site of conjugation, the number of PEG molecules attached, the chemistry of linkage, and the amino acids involved *(3,4)*. The reagents now available allow selective coupling to most common functional groups present on proteins, namely primary amines, thiol residues, carboxylic functions, or sugar residues. However, selective modification at a precise location in a protein (for example, to preserve its activity) is a difficult task and it is not possible to provide a unique strategy for this purpose.

Although the aim of this chapter is to give practical protocols for protein PEGylation, it is important to point out that every protein is characterized by its own reactivity and susceptibility to conjugation, as already shown by classic protein chemistry studies. Therefore, we can only suggest general guidelines as derived from the rationalization of results obtained in our laboratory or published in the literature *(5–20)*.

2. Materials

2.1. Equipment

1. Spectrophotometric (UV-VIS) and fluorometric equipment.
2. Gel filtration and ion-exchange chromatography systems.
3. Ultrafiltration equipment.
4. Sodium dodecyl sulfate-polyacrylamide gel electrophoresis (SDS-PAGE) electrophoresis.
5. Nuclear magnetic resonance (NMR).

2.2. Reagents

1. PEG reagent(s) (*see* **Note 1**).
2. 10X Gly-Gly (20 mM): 5.28 mg glycil-glycine in 2 mL of 0.2 M borate buffer, pH 8.
3. Borate pH 8: 0.2 M borate buffer, pH 8.
4. Borate, pH 9: 0.1 M borate buffer, pH 9.3.
5. Trinitrobenzenesulfonic acid (TNBS) solution 1: 0.03 M 2,4,6, trinitrobenzene sulfonic acid in dd-H_2O (10 mg/mL).
6. TNBS solution 2: 1% trinitrobenzenesulfonic in dd-H_2O (1 mg/mL).
7. Phosphate-ethylenediamine tetraacetic acid (EDTA), pH 7: 0.1 M phosphate, EDTA 1 mM, pH 7.
8. Ellman's reagent: 10mM 5,5'-dithio-bis-(2-nitrobenzoic acid) in phosphate–EDTA pH 7.
9. Phosphate EDTA, pH 6: 0.1 M phosphate, EDTA 1 mM, pH 6.
10. Cysteine solution (4 mM): dissolve 4.84 mg of cysteine in 1 mL of phosphate-EDTA, pH 6; then, dilute this solution 1:10.
11. Hydroxylamine solution: 2 M NH_2OH in water, pH 7.4.
12. Bicarbonate, pH 8.5: 4% $NaHCO_3$, pH 8,5.

13. Phosphate-EDTA, pH 7.4: phosphate 0.1 M, EDTA 5 mM, pH 7.4.
14. SDS solution: 10% SDS in H_2O.
15. BaCl$_2$ solution for iodine test: 5% BaCl$_2$ in 1 M HCl
16. Iodine solution: 2 g KI + 1.27 g I$_2$ in 100 mL H_2O.
17. Ferrothiocianate solution (0.1 N): 16.2 g anhydrous FeCl3 + 30.4 g NH$_4$SCN in 1 L H_2O.
18. High-performance liquid chromatography (HPLC) eluting buffer A: 2 mM acetic acid, pH 4.5.
19. HPLC eluting buffer B: 20 mM ammonium acetate, 50 mM NaCl, pH 6.4.

3. Methods

3.1. Characterization of PEG Reagents

Before conducting any coupling reaction, it is generally advised that one verify the amount of active moiety in the activated PEG batch. This is important especially when using a reactive PEG from a vial that was opened previously (*see* **Note 2**). For this control, several methods are available depending upon the PEG reactive moiety, among which are NMR and colorimetric tests.

3.1.1. NMR of PEG Derivatives

Any reactive PEG can be analyzed by both ^{1}H- and ^{13}C-NMR analysis. Both techniques are straightforward and accurate. However, NMR normally requires a high amount of reagent, which is not easily recovered after the analysis. The PEG sample is dissolved in the desired deuterated solvent at 3–5% (w/v) concentration (depending on the molecular weight of the polymer backbone). Particular care needs to be taken in the instrument shim. The best solvents are deuterated dimethylsulfoxide, that is, DMSO-d$_6$, and deuterated chloroform (CDCl$_3$), but others can also be used. As a common procedure, the intensity of the signals of the reactive group are compared with those of the backbone chain (CH2)$_2$ or, better, the terminal methoxy residue *(7,16,17)*.

3.1.2. Colorimetric Assays

Colorimetric assays are used to determine the degree of activation of amine- or thiol-reactive PEGs and PEG-thiols. These assays are as accurate as NMR and often require the consumption of less material. More precisely, the reactivity of PEG-*N*-hydroxysuccinimidyl (PEG-*NHS*) esters can be assayed by the "glycil-glycine" test (*see* **Subheading 3.1.2.1.**) whereas the amount of free thiols can be determined by direct Ellman's assay *(21,22*; **Subheading 3.1.2.2.**). An indirect Ellman's assay can be used to determine the amount of thiol reactive groups (maleimide, vinylsulfone, pyridylthione).

3.1.2.1. Glycil–Glycine Test to Evaluate the Activation of Acylating Groups (NHS and Benzotriazole Reagents)

This test is performed at room temperature.

1. Prepare two vials, mark them 1 and 2, and add, to each of them, 50 µL of 10X gly-gly and 450 µL of borate pH, 8.0.
2. To sample 1, while stirring, add 1 equivalent of PEG-*NHS* or other acylating group (1 µmol; e.g., if the MW of mPEG-*NHS* is 5000, add 5 mg and double amount for the 10,000 Da of PEG) in powder form.
3. To sample 2, add nonreactive PEG-OH of any molecular weight in powder form (the same amount in weight as PEG-*NHS* of sample 1; e.g., if mPEG-*NHS* in sample 1 is 5 KDa, then add 5 mg of PEG-OH of any molecular weight).
4. Wait 1 h.
5. Perform the TNBS test according to the method of Snyder and Sobocinsky (*see* **Subheading 3.1.2.1.1., ref. 23**).

3.1.2.1.1. TNBS Test According to Snyder and Sobocinsky for NH₂ Quantification

Perform the test at room temperature, in duplicate:

1. In separate test tubes, mix 100 µL of sample 1, or sample 2, or borate, pH 8.0 (for the blank); 900 µL of borate, pH 9.3; and 40 µL of TNBS solution 1.
2. Wait 30 min and then, using a spectrophotometer, read the absorbances at 420 nm.
3. Compute: 100 – [(Abs from sample B) – (Abs from blank)]/[(Abs from sample A) – (Abs from blank)] × 100, where Abs is absorbance. This value corresponds to the degree of activation of PEG-NHS in the batch.

3.1.2.2. Direct Ellman's Assay for SH Quantification

This assay *(21,22)* is used to determine free SH groups in solution and it may be followed to monitor any SH bearing molecule, this being a protein, a polymer, or a low molecular weight compound.

1. Dissolve PEG-SH (or any SH-carrying compound) in oxygen-free phosphate-EDTA, pH 7.0, at a final concentration of 2 m*M* (if PEG-SH has a MW of 5000, then dissolve it at 10 mg/mL). This solution should be used immediately after preparation.

Perform the test at room temperature, in duplicate:

2. In separate test tubes, mix 30 µL of PEG-SH solution (or the solution containing SH) or 30 µL of phosphate EDTA, pH 7.0 (as blank), and 970 µL of phosphate-EDTA, pH 7.0.
3. Add 50 µL of Ellman's reagent.
4. Incubate at room temperature for 15 min, then read Abs at 412 nm.

5. Compute the following: Abs sample – Abs blank. Thiol concentration is calculated knowing that the molar absorbivity (ε) of the 5-thio-nitro benzoic acid that forms quantitatively from is SH residue is 14,150 *(22)*.

3.1.2.3. INDIRECT ELLMAN'S ASSAY (TO EVALUATE THIOL-REACTIVE PEGS)

1. Dissolve the reactive PEG in phosphate-EDTA, pH 6.0, at a final concentration of 2 m*M* (e.g., if the PEG molecular weight is 5000 Daltons, then dissolve it at 10 mg/mL).

Perform the test at room temperature, in duplicate:

2. In separate test tubes, mix 500 µL of reactive PEG solution or phosphate-EDTA, pH 6.0 (as a blank), and 500 µL of cysteine solution.
3. Wait 30 min.
4. Perform, on both samples, the direct Ellman's assay as described previously (**Subheading 3.1.2.2., step 2**).

The amount of thio-reactive mPEG is calculated from the difference in thiol concentration in the two test tubes.

3.2. PEG–Protein Conjugation Protocols

3.2.1. General information

The concentration of protein in the reaction mixture is important in determining the final yield of coupling. The best coupling yields are normally obtained at protein concentrations of 1–5 mg/mL. At higher protein concentrations, higher yields may be expected, whereas the yield drops significantly at protein concentration below 1 mg/mL. Therefore, it is important to know the protein concentration before the coupling procedure. This can be determined by amino acid analysis, UV, or classic colorimetric tests *(24)*. It may be that a protein/peptide is only weakly soluble in water or at the reaction pH because of a high hydrophobicity or tendency to aggregation. In this case, the solubility may be increased by the addition of urea or guanidinium HCl (1–2 *M*), SDS, or organic solvents. Coupling yields also depend on the PEG molecular weight (*see* **Note 3**).

Most PEG reagents developed for protein coupling react rapidly with their targets so that the risk of protein degradation is low even if the reaction is performed at room temperature. In most cases, the reaction is completed within 1–2 h at room temperature. Coupling can also be performed at 4°C. In such a case, one has to keep in mind that any reaction is significantly slower at this temperature. As a rule of thumb, the reaction rate doubles when the temperature is increased by 10°C.

It is also important to remind that high-molecular-weight PEGs (>10,000 Daltons) lead to conjugates that are more difficult to purify and characterize than those with low molecular weight ones. In fact, for high-molecular-weight PEG conjugates, the polymer properties become dominant over the protein ones.

Considering that it is not possible to provide a unique protocol that will satisfy all of the readers' needs, we are providing here only general guidelines. For further practical details, we redirect the interested reader to the information collected in **Table 1**, where we report the reaction conditions used by several authors and the coupling yield they obtained.

3.2.2. Coupling to Primary Amines

Primary amines are good nucleophiles and are the residues most commonly used as a target in PEG coupling technology. In peptide and protein, they are present at the level of the lysine side chains (ε-NH$_2$), a common hydrophilic amino acid that usually represents 10% of the protein total composition and is commonly exposed to the solvent. Primary amines are also present at the α-terminus of protein chains (α-NH$_2$) when this residue is not masked by post-translational modification.

Although the pK_a of ε-NH$_2$ is around 10.5, the one of the α-NH$_2$ depends on the type of α-terminal amino acid and varies from 8.95 for lysine to10.78 for cysteine *(25)*. If in a protein the pK_a values of ε- and α-NH$_2$ are significantly different, it is possible to target the coupling to one of the two types of amine by changing the reaction pH. As a general rule, at pH values above 8.0, the ε-NH$_3$ groups react first, whereas at approx pH 5–6, α-NH$_2$ is the most reactive.

Several reagents are available, and coupling conditions can be designed so that primary amines become coupled to PEG via an acyl (using any PEG-*NHS*, benzotriazole, or *p*-nitrophenylcarbonate) or an alkyl (PEG-aldehyde, tresyl, or epoxide) bond. In the latter case, the positive charge of the amine is preserved. Among the acylating PEGs, the rate of amino reaction depends upon the composition of the active moiety (*see* **Table 2**). For example, PEG succinimidyl carboxymethyl (PEG-O-CH$_2$-COO-*NHS*) is the most reactive (and most sensitive to moisture), whereas PEG succinimidyl butyrrate (PEG-O-(CH$_2$)$_3$-COO-*NHS*) is less reactive. PEG–aldehyde is even less reactive (Morpurgo, M., and Harris, J. M., personal communication; **ref. 26**). Special NH$_2$ reactive PEG reagents have been developed for specific purposes. For example, various PEGs having a special amino acid as spacer between the polymer backbone and the reactive function have been developed to allow accurate characterization of the products in terms of number and location of the polymer chains in the conjugate *(27)*. More precisely, the use of *n*-leucine or β-alanine as spacers (both are non-natural amino acids and they are not present in pro-

Table 1
PEGylation Reaction Conditions and Yields as Derived From the Literature

Type of reactive PEG	Target residue on protein	Reaction conditions	[Protein] in the coupling (mg/mL)	Protein Mw (kDa)	PEG Mw (kDa)	PEG/protein in the reaction (mole:mole)	PEG/protein in the product	Yield (%) PEG added (/PEG bound)	Ref.
NHS	α/ε-NH$_2$	Aqueous, pH 8.5	1.5	13.7	5	27.5	5–6	20%	*16*
NHS	α/ε-NH$_2$	Aqueous, pH 8.5 (borate)	10	25	5	10–30	5–7	~30%	*53*
NHS	α/ε-NH$_2$	Aqueous, pH 8.5	4	23	5	40	8	20%	*16*
NHS (n = 0)	α/ε-NH$_2$	50 m*M* Phosphate, pH 7.4, 4°C	1.5–2	39	5	66.5	4.5	6.7%	*71*
NHS	α/ε-NH$_2$	Aqueous, pH 8.5	6	141	5	276	49	17.7%	*16*
NHS	α/ε-NH$_2$	Aqueous, pH 8.5	2.5	250	5	570	48	8.4%	*16*
NHS	α/ε-NH$_2$	Aqueous, pH 8.5	1.5	137	10	55	6	11%	*16*
NHS	α/ε-NH$_2$	Aqueous, pH 8.5	4	23	10	40	9	22.5%	*16*
NHS	α/ε-NH$_2$	Aqueous, pH 8.5	6	141	10	304	37	12%	*16*
NHS	α/ε-NH$_2$	Aqueous, pH 8.5	2.5	250	10	1120	42–43	3.8%	*16*
NHS (n = 0)	α/ε-NH$_2$	Aqueous, pH 7.0	2	3.2	12	2	1	50%	*72*
NHS	αNH$_2$	Aqueous, pH 6.5	5	15	6	12	~1	8%	*37*
Aldehyde	αNH$_2$	Aqueous, pH 5.0 20 m*M* NaCNBH$_3$, 16 h, 4°C	5	15	6	7.5	~1	13%	*37*
Aldehyde	αNH$_2$	Aqueous, pH 5.0 20 m*M* NaCNBH$_3$, 16 h, 4°C	>1	16.5	20	5	~1	20%	*54*

(continued)

Table 1 (*Continued*)
PEGylation Reaction Conditions and Yields as Derived From the Literature

Type of reactive PEG	Target residue on protein	Reaction conditions	[Protein] in the coupling (mg/mL)	Protein Mw (kDa)	PEG Mw (kDa)	PEG/protein in the reaction (mole:mole)	PEG/protein in the product	Yield (%) PEG added (/PEG bound)	Ref.
Aldehyde	α or ε NH$_2$	Ethanol/phosphate, pH 6.0 (50/50), then 4 NaCNBH$_3$ at pH 8.0, 20 h	2	0.9	5	5	1	20%	73
Aldehyde	α and ε NH$_2$	0.1 M HEPES, pH 7.5; 20 mM NaCNBH$_3$, 22°C; 3.5 h; then 50 mM ethanolamine	5	150	5	150	7.7	5%	15
Aldehyde	α and ε NH$_2$	0.1 M HEPES, pH 7.5; 20 mM NaCNBH$_3$, 22°C; 23 h; then 50 mM ethanolamine	5	150	5	150	15	10%	15
Aldehyde	α and ε NH$_2$	0.1 M HEPES, pH 7.5; 20 mM NaCNBH$_3$, 0°C; 3.5 h; then 50 mM ethanolamine	5	150	5	150	4	2.7%	15
Aldehyde	α and ε NH$_2$	0.1 M HEPES, pH 7.5; 20 mM NaCNBH$_3$, 0°C; 24 h; then 50 mM ethanolamine	5	150	5	150	6.5	4.3%	15

Hydrazide	COOH	0.01 M Pyridine/water, then EDAC	1.33	15	2	105	6–7	6.2%	*46*
Hydrazide	COOH	0.01 M Pyridine/water, then EDAC	1.33	15	5	210	5–7	2.85%	*45*
Hydrazide	Carbohydrates	Sugars were oxidized (periodate) Then, incubation at pH 6.2 (1 h), followed by an excess NaCNBH$_3$	3	150	5	50	2–6	4–12%	*46*
Vinyl-sulfone	SH	0.1 M Borate 1 EDTA, pH 8.0, 90', RT	2.7	13.4 (8-SH)	3.4	80	8	(*)	*17*
Vinyl-sulfone	SH	60 mM Phosphate, pH 7.5, 8 h, RT, then quench with 1 mM DTT	6	18 (1–4 SH)	20	2–8	1–4	(*)	*43*
Maleimide	SH	30 mM MES, pH 6.5, 90' RT	6	18 (1–4 SH)	5	1–4	1–4	(*)	*43*
Maleimide	SH	PBS, pH 7.4, 4°C overnight	32	64.5 (4-SH)	5, 10, 20	10	0–3	(*)	*20*

(*) The yield (%) is not significant due to the fact that complete SH modification is generally achieved.

Table 2
Reactivity (Aminolysis and Hydrolysis in 0.1 *M* Phosphate Buffer, pH 8.0)
of the *N*-Hydroxysuccinimidyl Esters of Several PEG-COOH Molecules *(26–27)*

Type of PEG-X-NHS		Hydrolysis t_2 (25°C) (min)	Hydrolysis t_2 (10°C) (min)	Aminolysis t_2 (10°C) (min)
PEG-O-	$n = 0$	20.3	92	15.2
(CH2)n-	$n = 1$	0.75	2	0.4
COO-NHS	$n = 2$	17.1	67	9.5
	$n = 3$	23.3	100	10.3
PEG-O-CO-n-leu-COO-NHS		5.4	16	4

Each reagent differs by the length or type of the spacer between the polymer chain and the activated function. Hydrolysis half lives were calculated assuming a pseudo first-order mechanism with respect to the *NHS* residue; Aminolysis half life was calculated on a low-molecular-weight substrate (*N*-acetyl-L-lysine methyl ester 0.3 m*M*) and using 10- or 20-fold excess reactive PEG to mimic a first-order mechanism.

teins) allows the precise quantification of the number of PEG molecules/protein simply by conducting an amino acid analysis of the purified product. Besides, by using mPEG-methionine-*n*-leucine (or β-alanine), the mPEG moiety may be removed by CNBr treatment, thus leaving the *n*-leu (or β-alanine) residue on the protein for the exact localization of the PEGylation site (*see* **Subheading 3.4.2.2.**). Moreover, a branched PEG molecule (PEG₂) containing a lysine residue bridging two linear mPEG chains is used to increase the polymeric hindrance at the protein coupling site *(16)*. Also in this case, lysine content as determined by amino acid analysis accounts for the number of PEG molecules bound to the modified protein. In any of these cases, the carboxylic function of the amino acid is activated as *NHS* ester.

The modification of primary amines can be followed by colorimetric assays, among which the most common one is based on the reaction with TNBS (*see* **Subheading 3.2.2.3.**). This test may be conducted directly in the bulk reaction mixture without any previous purification from excess reagent or leaving group. It is noteworthy that this method of evaluation, even if very useful and followed by many researchers, is not very accurate. Any result obtained by this method on the crude reaction mixture should be confirmed on the purified product, possibly by an alternative method.

3.2.2.1. REACTION WITH *NHS* ESTERS OR CARBONATES

Reaction of primary amines with *NHS* reagents leads to the formation of acyl bonds (amide for esters, urethane for carbonates). The water hydroxyl anion (OH⁻) of the aqueous buffer used as solvent in the coupling competes with the primary amines for the reaction. Therefore, an excess of activated PEG is usually needed. The ratio PEG-to-amino groups may range from 2–3 to 100 depending upon the PEG and protein amine reactivity.

Coupling leads to the loss of the amine positive charge and, as a general rule, the effect of charge loss on biological activity is more significant in peptides than in proteins. Sometimes, unexpected coupling can also occur at tyrosine *(28)* or histidine residues *(29–33)*. In one case, modification of serine with the formation of an ester was also reported *(29)*. Any undesired adduct to tyrosines can be cleaved by hydroxylamine treatment *(34)*. Modification of these residues is not detectable by the TNBS assay. Tyrosine modification may be revealed by change in the protein UV absorption spectrum because acylation of tyrosine hydroxyl functions induces significant changes in its absorption between 250 and 300 nm *(34)*. In any case, the involvement of such residues in conjugation can be identified indirectly by mass spectrometry, amino acid analysis, or peptide characterization after enzymatic digestion, as described later in this chapter.

1. Prepare the protein solution (1–5 mg/mL) in aqueous buffer having a pH around 7–0 (phosphate; HEPES [4-(2-hydroxyethyl)-1-piperazine-1-ethansulfonic acid]; borate). Do not use *N*-(Tris-hydroxymethyl)-aminomethane or any primary amine containing buffer because they will compete for the coupling. Keep aside a small portion of protein solution to perform further tests.
2. While gently stirring, add the PEG reagent in a powder form (or dissolved in the minimal amount of dimethylsulfoxide). To choose the amount of PEG for the reaction, refer to **Table 1**.
3. Allow the reaction to stir gently for at least 1 h (at room temperature) or longer time if at 4°C.
4. If desired, quench the reaction with hydroxylamine by adding 250 µL of hydroxylamine solution to each milliliter of reaction mixture.
5. Perform TNBS test as described below (**Subheading 3.2.2.3.**), but only in the absence of NH₂OH.
6. Purify the product and characterize as described below.

3.2.2.2. Modification With PEG–aldehyde

Reaction of PEG–aldehyde with primary amines yields stable secondary amines by a two step process: in the first step, Shiff bases are formed, which are then reduced in the second step with NaCNBH₃ to yield stable secondary amines. Shiff bases are in equilibrium with the free species (aldehyde and primary amines), and optimal pH for their formation is approx 5.0–6.0. Reaction at the α-NH₂ terminus will be kinetically favored at this pH, and this strategy is often chosen for selective modification at this loaction. In the past, PEG-acetaldehyde *(15,35)* was the most used aldehyde form. However, recently the propionaldehyde derivative was shown to give better results thanks to its better stability *(36–39)*.

1. Prepare a protein solution at 1–5 mg/mL in a buffer having a pH around 5.0–6.0 (phosphate, HEPES).
2. Add PEG–aldehyde. In order to choose the amount of PEG for the reaction, refer to **Table 1**.
3. Add NaCNBH3 to a final concentration of 20 mM. Lower amounts of reducing agent have also demonstrated to be effective (*see* **Table 1**).
4. Let the reaction be at 4°C for 20 h under gentle stirring.
5. Purify the product and characterize as described in the next section.

3.2.2.3. EVALUATION OF THE DEGREE OF AMINE MODIFICATION
BY AMINE-DETECTING COLORIMETRIC TEST (TNBS TEST ACCORDING TO HABEEB)

This method *(40)* is based on the evaluation of unreacted primary amines in the reaction mixture. A sample of unmodified protein solution (at the same concentration as the one in the reaction mixture) is needed as a reference.

1. In duplicate, mix 250 µL of protein solution to be assayed (ideally approx 40 µg of protein, both unmodified [N] and PEGylated [P] protein at the same concentration) in bicarbonate, pH 8.5 (or 250 µL of buffer as the blank [B]), with 250 µL of TNBS solution 2.
2. Incubate for 2 h at 40°C.
3. Add 250 µL of SDS solution.
4. Wait for 15 min.
5. Add 250 µL of 1 N HCl.
6. Using a spectrophotometer, read sample Abs at 335 nm.
7. Subtract Abs of blank (B) from Abs of protein samples N and P). The degree (%) of amine substitution is calculated as follows: { 1 – [Abs (P) – Abs (B)]/[Abs (N) – Abs (B)]} × 100.

3.2.3. Coupling to Thiol Groups

Thiols are present in proteins at the level of cysteine side chains. This amino acid is not common in proteins and, often, it cannot be modified for its role in catalysis or recognition processes. Therefore, coupling to thiols was proposed for the modification of recombinant proteins where cysteine residues are introduced in tailored positions to direct the PEG chains to specific locations *(41,42)*.

The thiol function is a good nucleophile, and several chemical approaches are available for its selective modification without affecting other protein functional groups.

Among the PEG reagents, the most commonly used are PEG–maleimide *(41)*, PEG–vinylsulfone (PEG-VS; **ref. *17***), and PEG–pyridyldisulfide *(13)*.

Maleimide and pyridyl disulfide react faster than vinylsulfone, there are advantages to using the latter reagent *(43)*. When performing PEGylation of thiols, total modification is generally desired. The reaction can be monitored by titrating the thiols in solution using the Ellman's assay (*see* **Subheading 3.1.2.2.**).

It should be kept in mind that reaction of primary amines can also occur with PEG–maleimide or PEG–VS even if at a slower rate than thiols. Reaction takes place via amine addition to the reactive double bond and it may become significant at pH above 8.5. It is also important to remind that when PEG–pyridyldisulfide is used the protein will be linked to the polymer through a disulfide bond (S–S), which is unstable to reducing agents. In the case of both PEG–VS and PEG–maleimide, a thioether bond is formed, which is more stable. In the first case, care must be taken to avoid the presence of any reducing agent in all of the steps of conjugate preparation and purification.

3.2.3.1. COUPLING WITH PEG–MALEIMIDE

1. Prepare a protein solution (1–5 mg/mL) in oxygen-free buffer having a pH of approx 5.0–6.0 (either phosphate or HEPES) and containing 1–5 m*M* EDTA as a scavenger for thiol oxidation. To follow the reaction by thiol titration using the Ellman's assay, keep aside a small portion of protein solution as unmodified protein reference solution (N).
2. Add PEG–maleimide. Generally, one equivalent of reagent for each thiol present in solution is sufficient (*see* **Table 1**). Excess reagent can also be used because no side reactions usually occur at this pH.
3. Allow the reaction to stir at room temperature for 1–2 h.
4. The reaction can be followed by Ellman's assay (**Subheadings 3.2.3.3.** and **3.1.2.2.**).

3.2.3.2. COUPLING WITH PEG–VS

1. Prepare a protein solution 1–5 mg/mL in oxygen-free buffer having a pH approx 7.0–8.0 (either borate or tris) containing 1–5 m*M* EDTA as a scavenger. Do not use buffers having pH > 8.0 because an amine reaction can occur (even if slow). If you intend to follow the reaction by thiol titration using the Ellman's assay, keep aside a small portion of protein solution as unmodified protein reference solution (N).
2. Add PEG–VS. The VS group is less reactive than maleimide. To accelerate the reaction, more than one equivalent (2–10) of reagent for each thiol may be used (*see* **Table 1**).
3. Allow the reaction to stir at room temperature (or 4°C) for 1.5–8 h, depending on the amount of PEG reagent used. A longer time may be required if the reaction is performed either at 4°C or with a lower ratio of PEG–VS/thiol.
4. It is advisable to follow the reaction by Ellman's assay (**Subheading 3.2.3.3.**; *see also* **Subheadings 3.1.2.2.** and **3.1.2.3.**)

5. When the reaction is complete, purify the product and characterize as described below.

3.2.3.3. ELMANN'S ASSAY

Titration of protein thiols may be conducted in a similar fashion as previously described (*see* **Subheading 3.1.2.2.**).

1. At scheduled times, in separate tubes, add the following:
   a. An aliquot (x µL) of the reaction mixture corresponding to 0.2 µmol of initial SH;
   b. x µL of original protein solution (without PEG) that was kept aside;
   c. x µL of reaction buffer (blank) and dilute with $(800 - x)$ µL of phosphate–EDTA buffer, pH 7.4.
2. To each tube, add 50 µL of Ellman's reagent solution;
3. Incubate at room temperature for 15 min, then read Abs at 412 nm
4. Compute the following: $100 - [(\text{Abs sample b} - \text{Abs blank})/(\text{Abs sample a} - \text{Abs blank})] \times 100$. This value will give the degree of thiol modification.

3.2.4. Coupling to Carboxylic Groups Using PEG–Azide

Direct amide formation between protein carboxylic groups and amino–PEGs cannot be performed because protein crosslinking would occur. However, an indirect procedure was devised by Zalipsky *(7,44)* where protein carboxyl groups are coupled to PEG–hydrazide at low pH values *(3,4)*. At this pH, protein amino groups are protonated (therefore not reactive) whereas PEG–hydrazide NH_2 residues, having lower pK_a, are available for coupling. Only a few examples are reported in the literature and the reader is addressed to the original articles *(7,44,45)*.

3.2.5. Coupling to Sugar Residues

As in the case reported above, few data are available in the literature on the PEGylation at the protein sugars *(46)*. In the most common approach, the protein sugars are oxidized with periodate, producing reactive aldehyde residues that may react with PEG–amine in a similar way as previously described for PEG–aldehyde coupling to primary amines (**Subheading 3.2.2.2.** and **Table 1**). However, this approach is not satisfactory because undesired crosslinking can easily occur.

3.2.6. Enzyme Mediate Glutamine-Directed Modification

An interesting approach for the selective PEGylation at the level of glutamine residues was recently described by Sato et al. *(47,48)*. This strategy relies on the use of PEG–NH_2 together with the enzyme transglutaminase. The enzyme promotes an exchange between the glutamine NH_2 with the polymer terminal amine. For practical details, we refer the reader to the related publications *(47,48)*.

3.2.7. Active Site Protection

To maintain the biomolecule activity after PEG coupling, modification at a specific location is preferable than random PEGylation. General strategies can be suggested, such as modifying the reaction solvent or pH to expose specific protein residues or affect their reactivity and, finally, favor the coupling to specific locations. In fact, the reaction pH influences the reactivity of amino acid residues (*see* histidine modification in α-interferon; **ref. *33***) whereas changes in the reaction solvent may affect the structure and accessibility of specific residues, as in the case of insulin *(49)* or growth hormone releasing factor (GRF) *(50)*. Alternatively, it is possible to conduct the PEGylation in the presence of substrates or reversible inhibitors that may protect the active site during the coupling *(51,52)*.

3.3. Purification of Conjugates

A typical PEGylation reaction often produces a mixture of heterogeneous PEGylated compounds, which differ by the number and location of the PEG molecules. Moreover, a crude reaction mixture also contains several low and high-molecular-weight byproducts. For preliminary studies on PEGylated compounds, a purification from low-molecular-weight byproducts (such as the leaving groups) and unreacted PEG, plus a rough estimation of the degree of modification, are often sufficient. For more advanced studies (such as in the development of bioconjugates as drugs), a higher degree of purification and characterization are required. In this case, both the isolation and the identification of each species produced in the reaction are therefore needed.

Purification of PEG conjugates from low-molecular-weight compounds and unreacted PEG can usually be achieved by ultrafiltration or dialysis. Ultrafiltration may be the fastest method, but it is always important to verify that no conjugate escapes through the ultrafiltration membranes: this can be easily verified by analyzing the filtrate for protein content. It is also important to note that the large hydrodynamic radius of PEG may be responsible for its low filtration through the ultrafiltration membranes.

Removal of unreacted PEG and separation of different PEGylated products can be achieved by chromatography on either Fast Protein Liquid Chromotograpy (FPLC) or HPLC systems. Either gel filtration (Superose 6 or 12; TSK 250; SP-Sepharose, **ref. *53***, or other columns) or ion exchange chromatography (TSK gel SP-5PW cation exchange; **ref. *54***) can be used. Proteins can be visualized by UV at 220 nm or 280 nm, whereas PEG molecules can be identified in collected fractions by iodine assay (**Subheading 3.4.1.2.**).

3.3.1. Fractionation of PEGylated Products
by Ion-Exchange Chromatography

PEG coupling is generally accompanied by changes in the protein isoelectric point, and ion-exchange chromatography, thanks to the high-loading capacity of resin, is probably the most successful method for the fractionation of PEGylated products in preparative amounts. Not only conjugates with different degree of modification but positional isomers also can often be separated. Strong cation exchange resins (sulfonic) are usually used with either standard FPLC (monoQ columns, Pharmacia) or HPLC (Toso Haas TSK-gel-SP-SPW) systems. Gradients of either pH or salt (NaCl) or both may be used as eluting systems. For a 25- × 15-cm Toso column, a standard elution protocol may be as follows:

1. Equilibrate the column with HPLC eluting buffer A.
2. Inject the PEGylation reaction mixture and elute for 20 min with buffer A at a flow rate of 5 mL/min.
3. Apply a linear gradient up to 100% of HPLC eluting buffer B for 100 min and collect fractions. An excess of PEG is generally not retained in such conditions whereas the different PEGylated protein species will be sequentially eluted according to their degree of modification, with the unmodified protein eluting as the last one.
4. Pool the protein containing fractions and concentrate by lyophylization for further analysis. Note that high amount of salts may be present in the eluted samples. Salts must be removed before complete dryness. This can be accomplished by ultrafiltration or gel filtration.

3.4. Characterization of Products

3.4.1. Amount of PEG–Protein

Several methods *(55,56)* are available to determine the amount of PEG attached to the protein, among which are SDS electrophoresis *(57)*, capillary electrophoresis *(58,59)*, matrix-assisted laser desorption/ionization time-of-flight (MALDI-TOF) mass spectrometry *(60–62)*, PEG–protein titration by ^{1}H-NMR, iodine *(63)* or ammonium ferrothiocyanate *(64)*, amino acid analysis *(65,66)*, colorimetric evaluation of unreacted primary amines by TNBS *(40)* or fluorescamine *(67,68)*. Note that electrospray ionization mass spectrometry is not easily applicable for the characterization of PEG–protein adducts. In fact, the polydispersivity of PEG makes the data interpretation rather complicated.

PEG–protein titration, amino acid analysis, and NH_2 titration by TNBS or fluorescamine allow quantification of the overall number of PEG chains attached to the protein in the sample. SDS electrophoresis and MALDI mass spectroscopy visualize each PEGylated species in the sample even if they still fail to differentiate among the different positional isomers. This information would be useful for the precise characterization of the product when multiple

PEGylation sites are available. Furthermore, SDS electrophoresis is not applicable for high-molecular-weight conjugates because of the large contribution of the highly hydrated PEG molecule to the conjugate hydrodynamic radius.

The information obtained with ^{1}H-NMR, iodine (or ammonium ferrothiocyanate) assay, or amino acid analysis (only useful if PEG-*n*-leucine or β-alanine have been used) is complementary to the one obtained by the colorimetric assays because it is independent of the type of amino acid involved in the binding. Therefore, those PEG chains that are linked to "unexpected" amino acids (e.g., tyrosine or histidine) can also be revealed. Both methods require prior purification of the conjugates from unreacted PEG and nonmodified protein using gel filtration or ion-exchange chromatography.

SDS electrophoresis generally overestimates the molecular weight of PEGylated species because of the large hydration volume of the polymer chain. PEGylated standards should be used for an accurate estimation of the molecular weight. Gradient electrophoresis (4–20%) allows better separation between different PEGylated species.

3.4.1.1. ^{1}H-NMR OF PEG–PROTEIN CONJUGATES

The overall amount of PEG in a sample can be determined by NMR. Quantitative ^{1}H-NMR is conducted by comparing the intensity of signals from the PEG backbone (or from the terminal methoxy residue) to the one of an internal standard (e.g., acetone, acetonitrile) added at a fixed concentration. PEG samples of know concentration (0.25–1 mg/mL) need to be prepared to build a calibration curve *(12,15,46,69)*.

1. Prepare an internal standard stock solution (15 µL of CH_3CN in 700 µL of D_2O).
2. Prepare PEG standard solutions at known concentrations in D_2O (0.25, 0.75, 1, 1.25, and 1.5 mg/mL).
3. Mix 700 µL of each PEG solution with 10 µL of internal standard stock solution.
4. Perform NMR analysis and measure the intensity of signals of PEG backbone (3.64 ppm) and CH_3CN (1.98 ppm); build a calibration curve by plotting the ratio between such signal intensities.
5. Freeze dry the amount of PEGylated protein to be used in the NMR assay and redissolve it in D_2O. Calculate protein concentration by UV absorbance.
6. Add 10 µL of internal standard stock solution to each 700 µL of protein/D_2O solution and run NMR analysis. Again, calculate the ratio between 3.64 and 1.68 ppm signal intensities.
7. Compute the PEG concentration form the calibration curve obtained as above.

3.4.1.2. IODINE ASSAY

This assay *(63)*, which is based on a nonspecific interaction of iodine with the PEG backbone, provides both qualitative and quantitative information. When conducted properly, it is very reliable and sensitive, allowing the quantification of as low as 1 μg/mL of PEG, independently of its molecular weight. Quantitative information can be obtained provided that a calibration curve is properly built in parallel with sample measurements. It is important to note that the absorbance values in the assay change significantly with time, even seconds. When quantitative data are searched, the assay should be conducted with the aid of a timer in order to allow the exact time of interaction between each PEG sample and the reagent mix before the spectrophotometric evaluation.

1. Mix $(500 - x)$ μL of MilliQ grade water, with 250 μL of $BaCl_2$ solution and 250 μL of iodine solution.
2. To $(1000 - x)$ μL of the above mix, in triplicate, add x μL of the PEG solution to be tested.
3. In parallel, build a calibration curve by using between 2.5 and 40 μg of PEG in the x μL.
4. Wait for 15 min at room temperature.
5. Read the Abs at 535 nm against a blank solution prepared with the same buffer or solution containing the PEG samples.

3.4.1.3. AMMONIUM FERROTHIOCIANATE ASSAY

This assay is based on the partitioning of a chromophore from an aqueous ammonium ferrothiocianate solution into a chloroform phase in the presence of PEG. Extraction of the chromophore into the organic phase occurs only in the presence of PEG, and the efficiency of extraction depends on polymer concentration. Therefore, a linear correlation between the organic phase color intensity (Abs at 510 nm) and the PEG concentration in the assay is obtained. Each unknown samples should be analyzed in parallel to solutions at known PEG (having the same molecular weight) concentration.

This assay was never tested in our laboratory. Therefore, we only provide the protocol as described in the original publication *(64)*.

1. Prepare a series of microcentrifuge polypropylene tubes and, to each of them, add equal volumes (0.5 mL) of ferrothiocianate solution and chloroform.
2. To each tube, add 50 μL of each PEG solution to be analyzed. Mix vigorously for 30 min, and then centrifuge the tubes for 2 min at 3000*g*.
3. Separate the organic and aqueous phases and read spectrophotometrically the Abs (510 nm) of the organic one.
4. Calculate the PEG concentration in the unknown sample on the basis of a calibration curve that is built in parallel.

The assay gives a linear response up to maximum 200 μg of PEG in the assay.

3.4.1.4. DIRECT EVALUATION OF PEG–PROTEIN RATIO BY AMINO ACID ANALYSIS

This method can be followed if PEG-Nle, PEG-β-Ala, or the PEG_2-lysine (PEG_2), activated as *NHS* esters, were used for the coupling. An aliquot of the purified sample is hydrolyzed by acidic treatment in a closed vial (e.g., 22 h, 110°C, 6 N HCl; or Pico Tag equipment [Waters]), and amino acid analysis is performed. The degree of modification is determined by the ratio between Nle (or β-Ala) content (or the increase of lysine amount) with respect to stable amino acids in the protein. The molar increase of the spacer amino acid (Nle, β-Ala, or Lys) corresponds to the molar amount of PEG in the sample.

3.4.2. Localization of PEG Attachment

Two methods are described. The first is an indirect one (**Subheading 3.4.2.1.**) that is based on the fact that proteolytic digestion cannot take place at the level of those amino acids where PEG chains are bound. Therefore, the site of conjugation can be identified by comparing the HPLC or MALDI-TOF fingerprinting of the native protein and the PEG–conjugate digestion mixtures *(70)*.

The second method (**Subheading 3.4.2.2.**) can only be conducted if special PEG reagents have been used for the coupling, namely those containing an unusual amino acid (Nleu or β-Ala), linked with a methionine as spacers between the polymer chain and the reactive residue (*NHS*; **ref. 27**). The main advantage of the first method is its rapidity of execution. However, risks of false positives exist because of the incomplete protein digestion inferred by the polymer surrounding the protein surface. The second method is more precise because it is based on a standard procedure of sequence analysis, historically developed to reveal posttranslational modifications of proteins. This method will also identify, when present, couplings to "unexpected" residues different to lysines.

3.4.2.1. "INDIRECT" LOCALIZATION OF PEG

1. Perform a proteolytic digestion on both nonmodified and PEG-conjugated protein. Use a few milligrams of each sample. Typically, the proteolytic enzyme (normally, TPCK trypsin) is incubated for 2 h at 37°C in the appropriate buffer [for trypsin, use 50 mM N-(Tris-hydroxymethyl)-aminomethane, 20 mM CaCl$_2$, pH 7.6], at a proteolytic enzyme:protein w:w ratio of about 1:100. Other proteolytic enzymes with different digestion specificity can be used according to the sequence requirements of the protein under investigation.
2. Analyze and fractionate both native and PEG-conjugated digestion mixtures by reverse-phase or ion exchange-HPLC, as commonly conducted in protein sequence analysis. Collect the peaks identified by UV.
3. Compare the elution patterns of the modified and native digests. The identity of those peptides that are missing in the PEGylated protein digest can be identified by analyzing the corresponding peak in the non-PEGylated digest. For this purpose, amino acid analysis, sequence, or mass spectrometry can be used. In the case of simple peptides of up to 30–40 amino acids, identification can be performed by

direct sequence analysis (Edman degradation) of the PEG–peptide. In this case, the PEGylated residues will be revealed as "missing" amino acids.

3.4.2.2. "Direct" Localization of PEG

1. Perform a CNBr treatment on a few milligrams of nonmodified and PEGylated samples that were previously freeze dried *(27)*.
2. Dilute each mixture with water and dry it by evaporation. Repeat this procedure two- to threefold.
3. Perform chromatographic analysis and fractionation as described in **Subheading 3.4.2.1.**
4. Compare the elution patterns of the modified and native digests. Analyze by MALDI or ion-spray mass spectrometry the new peaks appearing in the modified protein digest as compared with the nonmodified sample. Alternatively, perform amino acid analysis to reveal the presence of the reporter amino acid.

4. Notes

1. PEG reagents used in protein modification must be monofunctional, namely only one extreme of the polymer is reactive whereas the other one is capped with a stable methoxy residue (mPEG). The term mPEG defines the monofunctional methoxy end-capped PEG. Nevertheless, for simplicity, we are using the term PEG here to define any poly(ethyleneglycol) molecule, independently of its terminal functionality or shape.

   The polymer is formed by the anion polymerization (started by CH_3O^-) of ethylene oxide in dry conditions. When traces of water are present in the polymerization reaction, some of the polymer chains grow at both ends leading to a bifunctional product (HO-PEG-OH) as a contaminant having twice the molecular weight of the desired mPEG. Bifunctional polymers may cause undesired crosslinkings when used in conjugation and commercial mPEG-OH products must not contain more than 5–10% of such a contaminant. The presence of the bifunctional form can be detected by size-exclusion HPLC using a refraction index detector.

   Several PEGs having reasonable purity may be purchased from common chemical suppliers, Sigma-Aldrich in particular (St Louis, MO).Ultrapure PEGs for standard purposes may be obtained from Polymer Laboratories (Church Stretton, Shropshire, UK), whereas LCC Engineering and Trading GmbH (Egerkingen, SW) supplies activated monodisperse PEGs of up to 850 Daltons in molecular weight, and a few polydisperse PEGs of higher molecular weights. The richest catalogs of functionalized products are from PEG-Shop (SunBio Inc., Anyang City, South Korea) and Nektar, originally Shearwater Polymers Inc. (Huntsville, AL). The last company is the most well-known and traditional supplier from which custom-made products are also available. The great majority of the hydroxyl and activated PEGs used so far in research or for the production of already-approved PEGylated drugs come from this source. Abundant literature and several recent reviews are available on this subject that can help a reader interested in preparing his or her own PEG reagents *(5–20)*.

2. The polymeric backbone of PEG is relatively stable at room temperature. However, upon long storage oxidation can occur with formation of peroxide groups ending by breaking of the polymer chain. Moreover, because most of the reactive PEGs are also sensitive to moisture, we generally recommended storing any PEG in a dry, oxygen-free environment at low temperature.

   Furthermore, those reagents that are sensitive to moisture or oxygen can rapidly loose their reactivity upon storage, especially if the compounds have been kept in the wrong environment. For example, *N*-hydroxysuccinimide esters (NHS) are extremely sensitive to hydrolysis even by air moisture whereas PEG–SH can easily dimerize to PEG-S-S-PEG.

3. The amount of reagent to be added to the protein in the conjugation reaction mixture depends on the reactivity of the PEG derivative and the functional group on the protein. As a general guideline, independently of the type of reactive function in the PEG, the larger and more sterically hindered is the polymer chain, the less reactive is its functional end-group.

References

1. Abuchowski, A., McCoy, J. R., Palczuk, N. C., van Es, T., and Davis, F. F. (1977) Effect of covalent attachment of polyethylene glycol on immunogenicity and circulating life of bovine liver catalase. *J. Biol. Chem.* **252,** 3582–3586.
2. Abuchowski, A., van Es, T., Palczuk, N. C., and Davis, F. F. (1977) Alteration of immunological properties of bovine serum albumin by covalent attachment of polyethylene glycol. *J. Biol. Chem.* **252,** 3578–3581.
3. Francis, G. E., Fisher, D., Delgado, C., Malik, F., Gardiner, A., and Neale, D. (1998) PEGylation of cytokines and other therapeutic proteins and peptides: the importance of biological optimisation of coupling techniques. *Int. J. Hematol.* **68,** 1–18.
4. Lee, L. S., Conover, C., Shi, C., Whitlow, M., and Filpula, D. (1999) Prolonged circulating lives of single-chain Fv proteins conjugated with polyethylene glycol: a comparison of conjugation chemistries and compounds. *Bioconjug. Chem.* **10,** 973–981.
5. Bückmann A. F., Morr, M., and Johansson, G. (1981) Functionalization of poly(ethylene glycol) and monomethoxy-poly(ethylene glycol). *Makromol. Chem.* **182,** 1379–1384
6. Harris, J. M., Struck, E. C., Case, M. G., Paley , M. S., Yalpani, M., van Alstine, J. M., et al. (1984) Synthesis and characterization of PEG derivatives. *J. Polymer Sci: Polymer Chem. Ed.* **22,** 341–352.
7. Zalipsky, S. (1995) Functionalized poly(ethylene glycol) for preparation of biologically relevant conjugates. *Bioconjug. Chem.* 6, 150–165.
8. Veronese, F. M. and Morpurgo, M. (1999) Bioconjugation in pharmaceutical chemistry. *Farmaco* **54,** 497–516.

9. Veronese, F. M. (2001) Peptide and protein PEGylation: a review of problems and solutions. *Biomaterials* **22,** 405–417.
10. Roberts, M. J., Bentley, M. D., and Harris, J. M. (2002) Chemistry for peptide and protein PEGylation. *Adv. Drug Deliv. Rev.* **54,** 459–476.
11. Veronese, F. M., Largajolli, R., Boccu, E., Benassi, C. A., and Schiavon, O. (1985) Surface modification of proteins. Activation of monomethoxy-polyethylene glycols by phenylchloroformates and modification of ribonuclease and superoxide dismutase. *Appl. Biochem. Biotechnol.* **11,** 141–152.
12. Jackson, C. J., Charlton, J. L., Kuzminski, K., Lang, G. M., and Sehon, A. H. (1987) Synthesis, isolation, and characterization of conjugates of ovalbumin with monomethoxypolyethylene glycol using cyanuric chloride as the coupling agent. *Anal. Biochem.* **165,** 114–127.
13. Woghiren, C., Sharma, B., and Stein, S. (1993) Protected thiol-polyethylene glycol: a new activated polymer for reversible protein modification. *Bioconjug. Chem.* **4,** 314–318.
14. Miron, T. and Wilchek, M. (1991) A simplified method for the preparation of succinimidyl carbonate polyethylene glycol for coupling to proteins. *Bioconjug. Chem.* **4,** 568–569.
15. Chamow, S. M., Kogant, T. P., Venuti, M., Gadek, T., Harris, R. J., Peers, D. H., et al. (1994) Modification of CD4 immunoadhesin with monomethoxy-poly(ethyleneglycol) aldehyde via reductive alkylation. *Bioconjug. Chem.* **5,** 133–140.
16. Monfardini, C., Schiavon, O., Caliceti, P., Morpurgo, M., Harris, J. M., and Veronese, F. M. (1995) A branched monomethoxypoly(ethylene glycol) for protein modification. *Bioconjug Chem.* **6,** 62–69.
17. Morpurgo, M., Veronese, F. M., Kachensky, D., and Harris, J. M. (1996) Preparation and Characterization of Poly(ethylene glycol) Vinyl Sulfone. *Bioconjug. Chem.* **7,** 363–368.
18. Dolence, E. K., Hu, C., Tsang R., Sanders C. G., and Osaki, S. (1997) Electrophilic polyethylene oxides for the modification of polysaccharides, polypeptides (proteins) and surfaces, US Patent no. 5,650,234.
19. Guiotto, A., Pozzobon, M., Sanavio, C., Schiavon, O., Orsolini, P., and Veronese, F. M. (2002) An improved procedure for the synthesis of branched polyethylene glycols (PEGs) with the reporter dipeptide Met-beta Ala for protein conjugation. *Bioorg. Med. Chem. Lett.* **12,** 177–180.
20. Manjula, B. N., Tsai, A., Upadhya, R., Perumalsamy, K., Smith, P. K., Malavalli, A., et al. (2003) Site-specific PEGylation of hemoglobin at cys-93(beta): correlation between the colligative properties of the PEGylated protein and the length of the conjugated PEG chain. *Bioconjug Chem.* **14,** 464–472.
21. Riddles, P. W., Blakeley, R. L., and Zerner, B. (1979) Ellman's reagent: 5, 5'-dithiobis(2-nitrobenzoic acid)—a reexamination. *Anal. Biochem.* **94,** 75–81.
22. Riddles, P. W., Blakeley, R. L., and Zerner, B. (1983) Reassessment of Ellman's reagent. *Methods Enzymol.* **91,** 49–60.

23. Snyder, S. L. and Sobocinski, P. Z. (1975) An improved 2,4,6-trinitro-benzenesulfonic acid method for the determination of amines. *Anal Biochem.* **64,** 284–288.

24. Stoscheck, C. M. (1990) Quantitation of Protein. *Methods Enzymol.* **182,** 50–69.

25. Lehninger, A. L. (1970) *Biochemistry: the Molecular Basis of Cell Structure and Function.* Worth Publishers Inc., New York, NY.

26. Harris, J. M., Guo, Z., Fang, L., and Morpurgo, M. (1995) PEG-protein tethering for pharmaceutical applications. Proceeding s of the 7th International Symposium on Recent advances in drug delivery systems. Salt Lake City, Utah, February 27th-March 2nd, 1995.

27. Veronese, F. M. Sacca, B., de Laureto, P. P., Sergi, M., Caliceti, P., Schiavon, O., et al. (2001) New PEGs for peptide and protein modification, suitable for identification of the PEGylation site. *Bioconjug. Chem.* **12,** 62–70.

28. Orsatti, L. and Veronese, F. M. (1999) An unusual coupling of poly(ethylene glycol) to tyrosine residues in epidermal growth factor. *J. Bioactive Compatible Polymers* **14,** 429–436.

29. El-Tayar, N., Zhao, X., and Bentley, M (2002). PEG-LHRH analog conjugates. US Patent Application no. 20020183257.

30. Holmquist, B., Blumberg, S., and Vallee, B. L. (1976) Superactivation of neutral proteases: acylation with N-hydroxysuccinimide esters. *Biochemistry* **15,** 4675–4680.

31. Wang, Y. S., Youngster, S., Bausch, J., Zhang, R., McNemar, C., and Wyss, D. F. (2000) Identification of the major positional isomer of pegylated interferon alpha-2b. *Biochemistry* **39,** 10,634–10,640.

32. Wylie, D. C., Voloch, M., Lee, S., Liu, Y. H., Cannon-Carlson, S., Cutler, C., et al. (2001) Carboxyalkylated histidine is a pH-dependent product of pegylation with SC-PEG. *Pharm. Res.* **18,** 1354–1360.

33. Sivakolundu, S. G. and Mabrouk, P. A. (2003) Proton NMR study of chemically modified horse heart ferricytochrome c confirms the presence of histidine and lysine-ligated conformers in 30% acetonitrile solution. *J. Inorg. Biochem.* **94,** 381–385.

34. Riordan, J. F. and Vallee, B. L. (1972) O-acetyl tyrosine. *Methods Enzymol.* **25,** 500–506.

35. Bentley, M. D., Roberts, M. J., and Harris, J. M. (1998) Reductive amination using poly(ethylene glycol) acetaldehyde hydrate generated in situ: applications to chitosan and lysozyme. *J. Pharm. Sci.* **87,** 1446–1449.

36. Sherman, M. R., Williams, L. D., Saifer M. G.P., French, J. A., Kwak, L. W., and Oppenheim, J. J. (1997) *Poly(ethylene glycol) Chemistry and Biological Applications* (Harris, J. M., and Zalipsky, S., eds.), ACS, Washington, DC.

37. Kinstler, O. B., Brems, D. N., Lauren, S. L., Paige, A. G., Hamburger, J. B., and Treuheit, M. J. (1996) Characterization and stability of N-terminally PEGylated rhG-CSF. *Pharm. Res.* **13,** 996–1002.

38. Kinstler, O., Molineux, G., Treuheit, M., Ladd, D., and Gegg, C. (2002) Mono-N-terminal poly(ethylene glycol)-protein conjugates *Adv. Drug Deliv. Rev.* **54,** 477–485.

39. Lee, H., Jang, I. H., Ryu, S. H., and Park, T. G. (2003) N-terminal site-specific mono-PEGylation of epidermal growth factor. *Pharm. Res.* **20,** 818–825.

40. Habeeb, A. F. S. A. (1966) Determination of free amino groups in proteins by trinitrobenzenesulphonic acid. *Anal. Biochem.* **14,** 328–336.

41. Goodson, R. J. and Katre, N. V. (1990) Site-directed pegylation of recombinant interleukin-2 at its glycosylation site. *Biotechnology* **8,** 343–346.

42. Benhar, I., Wang, Q. C., FitzGerald, D., and Pastan, I. (1994) Pseudomonas exotoxin A mutants. Replacement of surface-exposed residues in domain III with cysteine residues that can be modified with polyethylene glycol in a site-specific manner. *J. Biol. Chem.* **269,** 13,398–13,404.

43. Pepinsky, R. B., Shapiro, R. I., Wang, S., Chakraborty, A., Gill, A., Lepage, D. J., et al. (2002) Long-acting forms of Sonic hedgehog with improved pharmacokinetic and pharmacodynamic properties are efficacious in a nerve injury model. *J. Pharm. Sci.* **91,** 371–387.

44. Zalipsky, S. and Menon-Rudolph, S. (1997) Hydrazide derivatives of polyethylene glycols) and their bioconjugates, in *Poly(ethyleneglycol) Chemistry and Biological Applications* (Harris, J. M., and Zalipsky, S., eds.), ACS, Washington, DC, pp. 318–341.

45. Sakane, T. and Pardridge, W. M. (1997) Carboxyl-directed pegylation of brain-derived neurotrophic factor markedly reduces systemic clearance with minimal loss of biologic activity. *Pharm. Res.* **14,** 1085–1091.

46. Larson, R. S., Menard, V., Jacobs, H., and Kim, S. W. (2001) Physicochemical characterization of poly(ethylene glycol)-modified anti-GAD antibodies. *Bioconjug. Chem.* **12,** 861–869.

47. Sato, H., Yamamoto, K., Hayashi, E., and Takahara, Y. (2000) Transglutaminase-mediated dual and site-specific incorporation of poly(ethylene glycol) derivatives into a chimeric interleukin-2. *Bioconjug. Chem.* **11,** 502–509.

48. Sato, H. (2002) Enzymatic procedure for site-specific pegylation of proteins. *Adv. Drug Deliv. Rev.* 54, 487–504.

49. Uchio, T., Baudys, M., Liu, F. Song, S. C., and Kim, S. W. (1999) Site-specific insulin conjugates with enhanced stability and extended action profile. *Adv. Drug Deliv. Rev.* **35,** 289–306.

50. Digilio, G., Barbero, L., Bracco, C., Corpillo, D., Esposito, P., Piquet, G., et al. (2003) NMR structure of two novel polyethylene glycol conjugates of the human growth hormone-releasing factor, hGRF(1–29)-NH$_2$. *J. Am. Chem. Soc.* **125,** 3458–3470.

51. Caliceti, P., Schiavon, O., Sartore, L., Monfardini, C., and Veronese, F. M. (1993) Active site protection of proteolytic enzymes by poly(ethylene glycol) surface modification *J. Bioactive Compatible Polymers* **8,** 41–50.

52. Caliceti, P., Morpurgo, M., Schiavon, O., Monfardini, C., and Verronese, F. M. (1994) Preservation of thrombolytic activity of urokinase modified with monomethoxypoly(ethylene glycol) *J. Bioactive Compatible Polymers* **9,** 252–266.

53. Clark, R., Olson, K., Fuh, G., Marian, M., Mortensen, D., Teshima, G., et al. (1996) Long-acting growth hormones produced by conjugation with polyethylene glycol. *J. Biol. Chem.* **271,** 21,969–21,977.
54. Guerra, P. I., Acklin, C., Kosky, A. A., Davis, J. M., Treuheit, M. J., and Brems, D. N. (1998) PEGylation prevents the N-terminal degradation of megakaryocyte growth and development factor. *Pharm. Res.* **15,** 1822–1827.
55. Snider, J., Neville, C., Yuan, L. C., and Bullock, J. (1992) Characterization of the heterogeneity of polyethylene glycol-modified superoxide dismutase by chromatographic and electrophoretic techniques. *J. Chromatogr.* **599,** 141–155.
56. Bullock, J., Chowdhury, S., Severdia, A., Sweeney, J., Johnston, D., and Pachla, L. (1997) Comparison of results of various methods used to determine the extent of modification of methoxy polyethylene glycol 5000-modified bovine cupri-zinc superoxide dismutase. *Anal. Biochem.* **254,** 254–262.
57. Zimmerman, S. B. and Murphy, L. D. (1996) Electrophoresis of polyethylene glycols and related materials as sodium dodecyl sulfate complexes. *Anal. Biochem.* **234,** 190–193.
58. Na, D. H., Park, M. O., Choi, S. Y., Kim, Y. S., Lee, S. S., Yoo, S. D., et al. (2001) Identification of the modifying sites of mono-PEGylated salmon calcitonins by capillary electrophoresis and MALDI-TOF mass spectrometry. *J. Chromatogr. B Biomed. Sci. Appl.* **754,** 259–263.
59. Li, W., Wang, Y., Zhu, X., Li, M., and Su, Z. (2002) Preparation and characterization of PEGylated adducts of recombinant human tumor necrosis factor-alpha from Escherichia coli. *J. Biotechnol.* **92,** 251–258.
60. Vestling, M. M., Murphy, C. M., Keller, D. A., Fenselau, C., Dedinas, J., Ladd, D. L., and Olsen, M. A. (1993) A strategy for characterization of polyethylene glycol-derivatized proteins. A mass spectrometric analysis of the attachment sites in polyethylene glycol-derivatized superoxide dismutase. *Drug Metab. Dispos.* 21, 911–917.
61. Watson, E., Shah, B., DePrince, R., Hendren, R. W., and Nelson, R. (1994) Matrix-assisted laser desorption mass spectrometric analysis of a pegylated recombinant protein. *Biotechniques* 16, 278–281.
62. Chowdhury, S. K., Doleman, M., and Johnston, D. (1995) Fingerprinting proteins coupled with polymers by mass spectrometry: Investigation of polyethylene glycol-conjugated superoxide dismutase. *J. Am. Soc. Mass. Spectrom.* **6,** 478–487.
63 Sims, G. E., and Snape, T. J. (1980) A method for the estimation of polyethylene glycol in plasma protein fractions. *Anal. Biochem.* **107,** 60–63.
64. Nag, A., Mitra, G., and Ghosh, P. C. (1996) A colorimetric assay for estimation of polyethylene glycol and polyethylene glycolated protein using ammonium ferrothiocyanate. *Anal. Biochem.* **237,** 224–231.
65. Sartore, L., Caliceti, P., Schiavon, O., and Veronese, F. M. (1991-a). Enzyme modification by MPEG with an amino acid or peptide as spacer arms. *Appl. Biochem. Biotechnol.* 27, 45–54.
66. Sartore, L., Caliceti, P., Schiavon, O., Monfardini, C., and Veronese, F. M. (1991) Accurate evaluation method of the polymer content in monomethoxy-(polyethylene glycol) modified proteins based on amino acid analysis. *Appl. Biochem. Biotechnol* **31,** 213–222.

67. Udenfriend, S., Stein, S., Bohlen, P., Dairman, W., Leimgruber, W., and Weigele, M. (1972) Fluorescamine: a reagent for assay of amino acids, peptides, proteins, and primary amines in the picomole range. *Science* **178**, 871–872.
68. Stocks, S. J., Jones, A. J., Ramey, C. W., and Brooks, D. E. (1986) A fluorometric assay of the degree of modification of protein primary amines with polyethylene glycol. *Anal. Biochem.* **154**, 232–234.
69. Mabrouk, P. A. (1994) Effect of pegylation on the structure and function of horse cytochrome c. *Bioconjug. Chem.* **5**, 236–241.
70. Lee, K. C., Moon, S. C., Park, M. O., Lee, J. T., Na, D. H., Yoo, S. D., et al. (1999) Isolation, characterization, and stability of positional isomers of mono-PEGylated salmon calcitonins. *Pharm. Res.* **16**, 813–818.
71. Fang, J., Sawa, T., Akaike, T., and Maeda, H. (2002) Tumor-targeted delivery of polyethylene glycol-conjugated D-amino acid oxidase for antitumor therapy via enzymatic generation of hydrogen peroxide. *Cancer Res.* **62**, 3138–3143.
72. Lee, K. C., Tak, K. K., Park, M. O., Lee, J. T., Woo, B. H., Yoo, S. D., et al. (1999b) Preparation and characterization of polyethylene-glycol-modified salmon calcitonins. *Pharm. Dev. Technol.* **4**, 269–275.
73. Morpurgo, M., Monfardini, C., Hofland, L. J., Sergi, M., Orsolini, P., Dumont, J. M., et al. (2002) Selective Alkylation and Acylation of alpha and epsilon Amino Groups with PEG in a Somatostatin Analogue: Tailored Chemistry for Optimized Bioconjugates. *Bioconjug. Chem.* **13**, 1238–1243.

5

Chemical Production of Bispecific Antibodies

Robert F. Graziano and Paul Guptill

Summary

This chapter discusses two related methods for creating Fab' × Fab' chemically linked BsAb. Both methods require the generation of purified $F(ab')_2$ fragments of each antibody and use reagents that react with the free thiols generated upon reduction of interheavy chain disulfide bonds of the $F(ab')_2$ fragments. Upon reduction, the resulting Fabs are then recombined to form a Fab' × Fab' BsAb. 5,5'-dithiobis (2-nitrobenzoic acid) (DTNB) acts to regenerate disulfide bonds between the two Fabs, whereas o-phenylenedimaleimide (o-PDM) acts to form a thioether bond between the two Fabs. After coupling, the bispecific antibody is purified from the uncoupled Fabs by size-exclusion chromatography. The advantages and disadvantages of each conjugation method are discussed.

Key Words: Bispecific antibodies; Fab fragment; antibody heavy chain; chemical conjugation; DTNB; o-PDM; Ellman's reagent; antibody hinge regions.

1. Introduction

Over the past two decades, bispecific antibodies (BsAb)—molecules combining two or more antibodies with different antigenic specificities—have been developed as tools for basic research as well as for clinical studies (for reviews, *see* **refs. *1–8***). A number of methods for producing BsAb have been developed. BsAb can be produced biologically by fusing two hybridoma lines, yielding quadromas that are capable of secreting BsAb. However, because of the various potential combinations of heavy and light chain pairing, only a small percentage of the molecules being secreted will have the appropriate bispecificity *(9)*. BsAb can also be generated genetically, and a variety of genetic techniques have been used to create bispecific molecules *(10)*. A third way to create BsAb is by chemical means. Nisonoff and Rivers pioneered the production of chemically linked BsAb over 40 yr ago *(11)*. Since then, many methods using a variety of homobifunctional and heterobifunctional chemical

From: *Methods in Molecular Biology, vol. 283: Bioconjugation Protocols: Strategies and Methods*
Edited by: C. M. Niemeyer © Humana Press Inc., Totowa, NJ

reagents have been developed, leading to the refinement of chemically linked BsAb production *(12)*. In this chapter, we discuss two related methods for creating Fab' × Fab' chemically linked BsAb, methods that were originally described by Brennan et al. *(13)* and by Glennie et al. *(14)*. Both methods use reagents that react with the free thiols generated upon reduction of inter heavy chain disulfide bonds. The two Fabs are then recombined to form a Fab' × Fab' BsAb. 5,5'-dithiobis (2-nitrobenzoic acid) (DTNB) acts to regenerate disulfide bonds between the two Fabs, whereas *o*-phenylenedimaleimide (*o*-PDM) acts to form a thioether bond between the two Fabs. The advantages and disadvantages of each of these methods will be discussed.

2. Materials

1. 0.1 N NaOH.
2. 2 M Tris, pH 9.0.
3. 20% Ethanol.
4. SACE: 50 mM Na acetate, 0.5 mM ethylenediamine tetraacetic acid, pH 5.3.
5. Phosphate-buffered saline (PBS), pH 7.4.
6. Dimethyl formamide.
7. 500 mM Mercaptoethanolamine (MEA).
8. 500 mM Iodoacetamide (IAA).
9. PE: 0.1 M sodium phosphate, 25 mM ethylenediamine tetraacetic acid, pH 7.4.
10. *o*-PDM.
11. DTNB (also known as Ellman's reagent).
12. TSK 3000 analytical size-exclusion high-performance liquid chromatography (HPLC) columns (Tosohaus).
13. HPLC equipment.
14. Chromatography equipment.
15. Sephadex G-25 gel (Pharmacia).
16. Superdex 200 gel (Pharmacia).
17. Sodium dodecyl sulfate-polyacrylamide gel electrophoresis (SDS-PAGE) equipment.
18. Purified F(ab')$_2$ fragments of antibodies to be coupled.
19. Spectrophotometer.

3. Methods

The methods described below outline the generation of Fab' × Fab' BsAb using two similar methods. Both start from purified F(ab')$_2$ fragments of each antibody to be coupled. Generation of F(ab')$_2$ fragments is normally performed by pepsin digestion of the whole antibody; this method has been well-described and is beyond the scope of this chapter. The DTNB method of coupling will be described first followed by the *o*-PDM method. **Figure 1** shows schematically the reactions involved in the DTNB coupling procedure and **Fig. 2** the reactions involved in the *o*-PDM coupling procedure.

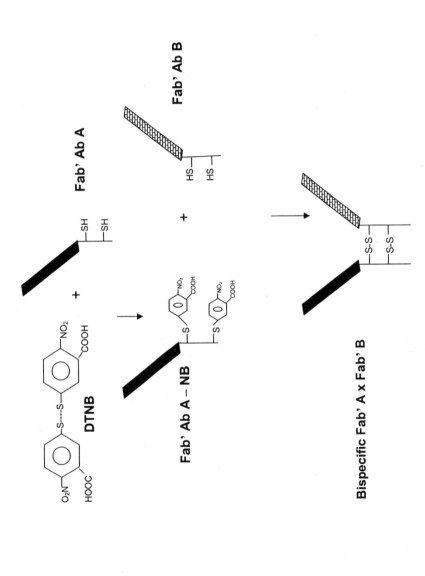

Fig. 1. Creating a BsAb using DTNB. The Fab' fragment of antibody "A" is reacted with DTNB resulting in a Fab'A – NB derivative. The Fab A – NB is reacted with a free Fab' "B" fragment resulting in the formation of a disulfide linked Fab'A × Fab'B BsAb.

A

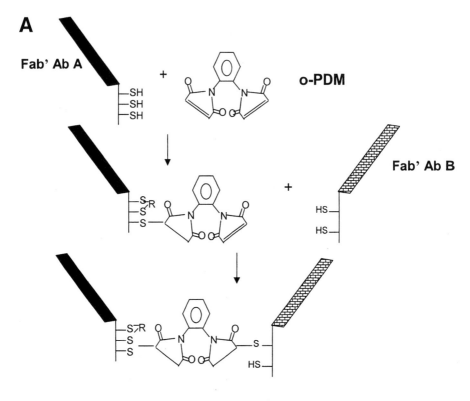

Fab' A x Fab' B Bispecific Antibody

Fig. 2. Creating a BsAb using *o*-PDM. **(A)** The Fab' fragment of antibody "A" is reacted with *o*-PDM. This results in the vicinal dithiols complexed with *o*-PDM (R), and one of the SH groups bound to *o*-PDM with a free maleimide group remaining. The Fab A–*o*-PDM is reacted with a free Fab' "B" fragment, resulting in the formation

3.1. DTNB Coupling

The method begins with the mild reduction of the inter-heavy chain disulfide bonds of the F(ab')$_2$ fragment of antibody "A" (2–30 mg/mL of PBS). Before beginning the reduction, remove a sample of the F(ab')$_2$ fragment and analyze it by size-exclusion HPLC to obtain baseline retention time (**Fig. 3A**).

3.1.1. Reduction and DTNB Derivation of Antibody "A"

1. To the F(ab')$_2$ fragment of Ab "A," add enough volume of the 500 m*M* MEA to obtain a final MEA concentration of 30 m*M* (*see* **Note 1**). Incubate the solution at 30°C for 30–60 min.

B

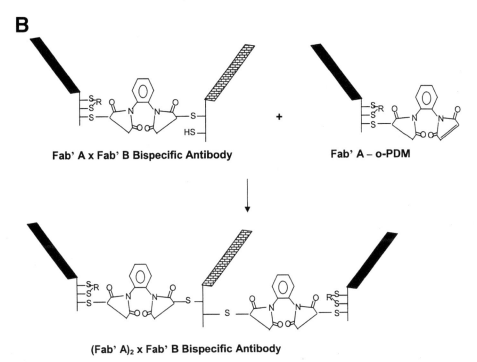

Fab' A x Fab' B Bispecific Antibody   Fab' A – o-PDM

(Fab' A)₂ x Fab' B Bispecific Antibody

Fig. 2. (*continued*) of a thioether linked BsAb. (**B**) Creating a bispecific antibody using *o*-PDM that is bivalent for one specificity and monovalent for the second. Excess Fab' A–*o*-PDM can react with the remaining free sulfhydrals on Fab' B, resulting in the formation of F(ab')₃ BsAb.

2. To monitor the progress of the reduction, remove an aliquot of the mixture, mix it with an equal volume of 500 mM IAA solution, and inject onto the TSK 3000 analytical column. The IAA will serve to alkylate the free sulfhydral groups and prevent reoxidation of the Fab'.

3. Once it has been determined that >95% of the F(ab')₂ has been reduced to Fab', as determined by a shift in HPLC retention time (**Fig. 4A**), the Fab' is buffer exchanged by running it through a G-25 Sephadex column that has been equilibrated in PE buffer. The mixture is loaded onto the column and the protein peak is collected. Care should be taken to make sure that no free MEA remains in the Fab' (*see* **Note 2**).

4. Determine the volume of the Fab' Ab "A" and add to 1/6 of the volume of a 35 mM solution of DTNB: final DTNB concentration of 5 mM. Incubate for 30–60 min at room temperature. Remove free DTNB from the derivatized protein by G-25 Sephadex chromatography. The DTNB-derivatized Fab' is stable and may be stored in this form.

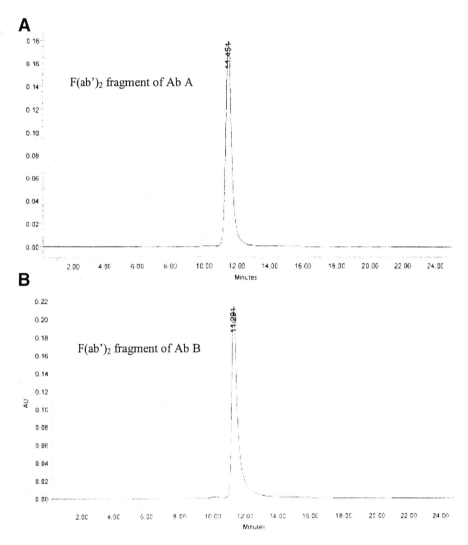

Fig. 3. Size-exclusion HPLC profiles of the F(ab')$_2$ fragments of Abs to be coupled.

3.1.2. Reduction of F(ab')$_2$ Ab "B"

1. To the F(ab')$_2$ fragment of Ab "B," add enough volume of the 500 mM MEA to obtain a final MEA concentration of 30 mM (*see* **Note 1**). Incubate the solution at 30°C for 30–60 min.
2. To monitor the progress of the reduction, remove an aliquot of the mixture, mix it with an equal volume of 500 mM IAA solution, and inject onto the TSK 3000 analytical column. The IAA will serve to alkylate the free sulfhydral groups and prevent reoxidation of the Fab'.

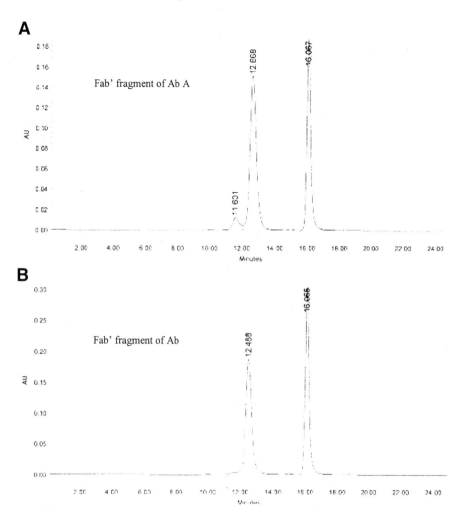

Fig. 4. Size-exclusion HPLC profiles of the Fab' fragments of Abs to be coupled.

3. Once it has been determined that >95% of the F(ab')$_2$ has been reduced to Fab', as determined by a shift in HPLC retention time (**Fig. 4B**), the Fab' is buffer exchanged by running it through a G-25 Sephadex column that has been equilibrated in PE buffer. The mixture is loaded onto the column and the protein peak is collected. Care should be taken to make sure that no free MEA remains in the Fab' (*see* **Note 2**).

3.1.3. Conjugation Reaction

1. Mix equal molar ratios of Fab'A–DTNB with Fab' B.
. 2. Incubate for 1 h at 37°C. Monitor the conjugation reaction by analytic HPLC.

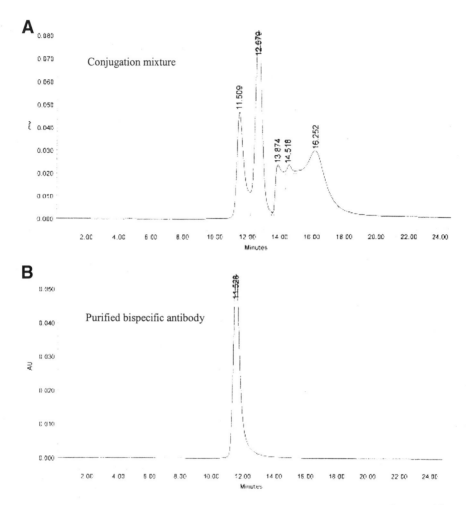

Fig. 5. **(A)** Size-exclusion HPLC profile of conjugation mixture before purification. **(B)** Size-exclusion HPLC profile of Superdex purified bispecific antibody generated using DTNB.

Figure 5A shows the mixture of bispecific $F(ab')_2$ molecules (retention time = 11.509 min) and the uncoupled Fab' molecules (retention time = 12.679) in this reaction.

3. Place the reaction mixture at 4°C and incubate further overnight.
4. Purify the bispecific fraction from the uncoupled Fabs by running it over a Superdex 200 column that has been equilibrated in PBS (*see* **Note 3**). **Figure 5B** shows an HPLC profile of the Superdex purified bispecific antibody. **Figure 6** (A, nonreduced; B, reduced) shows a Coomassie blue-stained SDS-PAGE of a purified DTNB-generated BsAb. The purified BsAb migrates as a single band at

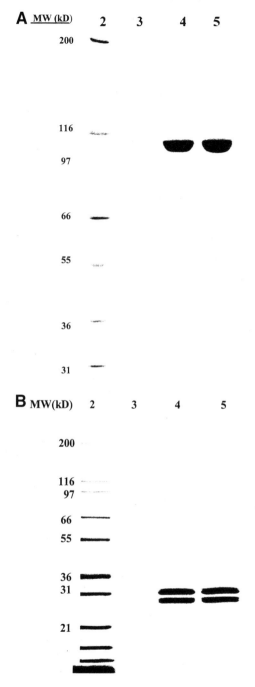

Fig. 6. Analysis of DTNB constructed BsAb by SDS-PAGE. Gels were run under nonreducing (**A**) or reducing conditions (**B**). Lane 2 contains molecular weight markers, lane 3 is empty, and lanes 4 and 5 were loaded with 2 μg of DTNB conjugated BsAb.

an apparent molecular weight (MW) of 100–110 kDa under nonreducing conditions (**Fig. 6A**). Under reducing conditions the BsAb migrates as two bands, one representing the Fd fragment at apparent MW of 31–33 kDa and the second representing the light chain at an apparent MW of 25 kDa (**Fig. 6B**).

3.2. o-PDM Coupling

o-PDM, like DTNB, reacts with free sulfhydral groups that are generated upon reduction of the interheavy chain disulfide bonds of F(ab')$_2$ Ab fragments. One significant disadvantage of using *o*-PDM is its requirement of having an odd number of interheavy chain bonds in the Ab to be maleimidated (*see* **Fig. 2**; further advantages and disadvantages of the DTNB vs the *o*-PDM method are discussed in the **Note 4**). The condition for reduction of the F(ab')$_2$ fragments of the antibodies and buffer exchange to be coupled are essentially the same as for the DTNB reaction with the exception that the G-25 columns are equilibrated with SACE buffer instead of PE buffer. In addition, the *o*-PDM treatment of the appropriate Ab, the G-25 buffer exchange, and the conjugation reaction must be done on ice (between 0 and 4°C).

3.2.1. Reduction and o-PDM Derivitization of Antibody "A"

1. To the F(ab')$_2$ fragment of Ab "A," add enough volume from the 500 m*M* of MEA to obtain a final MEA concentration of 30 m*M* (*see* **Note 1**). Incubate the solution at 30°C for 30–60 min.
2. To monitor the progress of the reduction remove an aliquot of the mixture, mix it with an equal volume of 500 m*M* IAA solution and inject onto the TSK 3000 analytical column. The IAA will serve to alkylate the free sulfhydral groups and prevent reoxidation of the Fab'.
3. Once it has been determined that >95% of the F(ab')$_2$ has been reduced to Fab', as determined by a shift in HPLC retention time (**Fig. 4A**), the Fab' is buffer exchanged by running it through a G-25 Sephadex column that has been equilibrated in SACE buffer. The mixture is loaded onto the column and the protein peak is collected. Care should be taken to make sure that no free MEA remains in the Fab' (*see* **Note 2**). Store the Fab' on ice.

3.2.2. Reduction of F(ab')$_2$ Ab "B"

1. To the F(ab')$_2$ fragment of Ab "B," add enough volume of the 500 m*M* MEA to obtain a final MEA concentration of 30 m*M* (*see* **Note 1**). Incubate the solution at 30°C for 30–60 min.
2. To monitor the progress of the reduction, remove an aliquot of the mixture, mix it with an equal volume of 500 m*M* IAA solution, and inject onto the TSK 3000 analytical column. The IAA will serve to alkylate the free sulfhydral groups and prevent reoxidation of the Fab'.
3. Once it has been determined that >95% of the F(ab')$_2$ has been reduced to Fab', as determined by a shift in HPLC retention time (**Fig. 4B**), the Fab' is buffer

exchanged by running it through a G-25 Sephadex column that has been equilibrated in SACE buffer. The mixture is loaded onto the column and the protein
peak is collected. Care should be taken to make sure that no free MEA remains in
the Fab' (*see* **Note 2**). Measure the volume of the buffer exchanged Fab' Ab B and
store on ice.

3.2.3. Generating an o-PDM Derivative of Fab' Ab "B"

1. Make a 12 mM (3.22 mg/mL) solution of *o*-PDM by dissolving the *o*-PDM into
   dimethyl formamide that has been chilled on ice. The total volume of the *o*-PDM
   solution required is equal to half of the volume of Fab' Ab B.
2. Mix the chilled *o*-PDM solution into the chilled Fab' Ab B fragment with
   gentle swirling. The final *o*-PDM concentration will be 4 mM. Incubate for
   30 min on ice.
3. Load the mixture onto a chilled, water-jacketed G-25 column that has been equilibrated with SACE. The column should be chilled by having ice water pumped
   through the jacket.
4. Collect the protein peak taking care that no free *o*-PDM remains with the Fab'–*o*-
   PDM Ab B. Determine the protein concentration of the derivatized Fab' by taking the OD$_{280}$.

3.2.4. Conjugation Reaction

1. Mix Fab'Ab B–*o*-PDM and Fab' Ab A at an equal molar ratio.
2. Incubate for at least 12 h on ice. Monitor the progress of the conjugation reaction
   by size-exclusion HPLC.
3. Reduce any F(ab')$_2$ homodimers that may have formed during the conjugation
   reaction by warming the mixture to 30°C, adding a volume of 500 mM MEA
   such that the final MEA concentration is 1 mM, and incubating it for 30 min.
4. Alkylate the resulting Fab-SH by adding IAA to a final concentration of 25 mM
   and incubating it at room temperature for 30–60 min.
5. Purify the bispecific fraction from the uncoupled Fabs by running it over a
   Superdex 200 column that has been equilibrated in PBS (*see* **Note 6**). **Figure 7**
   shows an HPLC profile of the Superdex purified bispecific antibody. Two major
   peaks are shown in the figure at retention times of 10.250 and 10.867 min,
   respectively. The identity of these peaks is discussed in **Note 7**. **Figure 8** (A,
   nonreduced; B, reduced) shows a Coomassie blue-stained SDS-PAGE of a purified *o*-PDM-generated BsAb. The primary species migrates at an apparent MW
   of 100–110 kDa under nonreducing conditions and represents the F(ab')$_2$ BsAb.
   However, several bands appear on the nonreduced gel. The nature of these
   species and the bands shown on under reducing conditions are discussed in
   Note 5.

4. Notes

1. The MEA concentration required to reduce F(ab')$_2$ to Fab' may vary from Ab to
   Ab. Small-scale trial reduction should be performed to determine the optimal

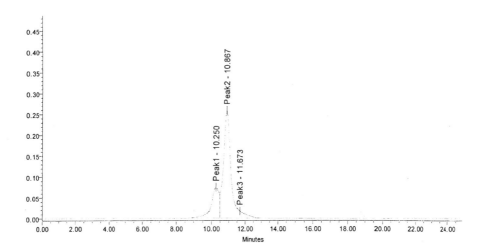

Fig. 7. Size-exclusion HPLC profile of Superdex purified bispecific antibody generated using *o*-PDM.

reducing conditions. A range of MEA concentrations from 1 m*M* to 50 m*M* should be assessed. After reduction, the fragments can be analyzed by SDS-PAGE. Conditions should be chosen such that efficient reduction of the inter heavy chain disulfides in achieved without extensive reduction of heavy-light chain disulfide bonds.

2. To ensure that clean separation of Fab' from free MEA is achieved after the G-25 steps, aliquots corresponding to various points in the G-25 profile can be tested for free SH groups using the Ellman's test *(15)*. Ideally, fractions corresponding to Fab' will be positive in the Ellman's test, fractions corresponding to those following the protein elution but before the elution of free MEA will be negative, and fractions containing free MEA will be strongly positive. These results confirm efficient separation of the Fab' from free MEA. In general, the volume of the G-25 and Superdex columns should be 10× the volume of the sample to be loaded.

3. G-25 and Superdex columns can be sanitized by running at least three column volumes of 0.1 *N* NaOH through them before use. The columns should be adequately equilibrated in the appropriate buffer before loading the sample. Sanitizing should remove undesirable contaminating endotoxin. Of course, care should be taken to make buffers using endotoxin-free reagents.

4. Although the DTNB and the *o*-PDM methods of generating BsAb are similar, important differences exist which are critical when choosing one method over the other. Both methods create BsAb that are coupled at a defined site, the hinge region sulfhydral, which should not affect the affinity of the respective Fabs. BsAb created using DTNB can be purified to homogeneity functionally as well as biochemically (**Figs. 5** and **6**). BsAb created using the *o*-PDM method may be more stable because of the formation of a thioether bond *(12)*, and yields are

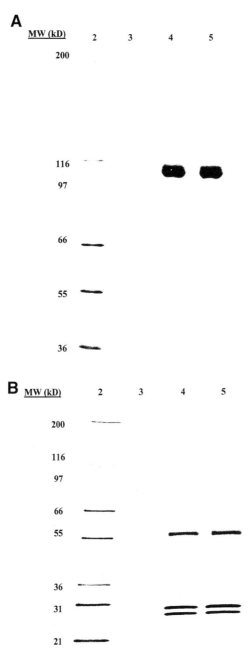

Fig. 8. Analysis of *o*-PDM constructed BsAb by SDS-PAGE. Gels were run under nonreducing (**A**) or reducing conditions (**B**). Lane 2 contains molecular weight markers, lane 3 is empty, and lanes 4 and 5 were loaded with 2 µg of *o*-PDM conjugated BsAb.

generally better than those generated by DTNB. However, it is more difficult to purify *o*-PDM-generated BsAb to biochemical homogeneity (**Figs. 7** and **8**). Another distinct disadvantage of the *o*-PDM method is the necessity to have an odd number of inter heavy chain disulfide bonds in the antibody molecule to be maleimidated. This prevents its application in the construction of human–human BsAb unless the hinge region of the human antibody is altered.

5. Upon SDS-PAGE analysis of *o*-PDM generated BsAb, several species are often observed. The primary species migrates at an apparent MW of 100–110 kDa under nonreducing conditions and represents the $F(ab')_2$ BsAb. The band that migrates at an apparent MW of 130–150 kDa represents the $F(ab')_3$ BsAb. Other species may represent $F(ab')_3$ or $F(ab')_2$ species that may have lost a noncovalently linked light chain *(14)*. Under reducing conditions, four bands are observed with this BsAb. The band that migrates at an apparent MW of approx 90 kDa most likely represents three heavy chains from the $F(ab')_3$ BsAb species, which are linked covalently by thioether bonds. The second band that migrates at an apparent MW of approx 65 kDa most likely represents two heavy chains from the $F(ab')_2$ BsAb species, which are linked covalently by thioether bonds. The bands running at 25 kDa and at 28 kDa represent the light chains from each of the Abs that were coupled, which, in this instance, run at slightly different apparent MW.

6. Purification of bispecific antibody by size-exclusion chromatography using Superdex 200 gel normally gives adequate separation of the bispecific Ab from uncoupled free Fabs and small molecules (*o*-PDM, IAA). However, other methods that have been developed to purify bispecific of interest such as affinity or ion exchange chromatography may be employed. Fractions can be analyzed by SDS-PAGE or by HPLC before appropriate pooling.

7. After purification of *o*-PDM-linked BsAb, two peaks are seen on size-exclusion HPLC (**Fig. 7**). The first peak likely consists of two Fab' fragments of the *o*-PDM-treated antibody (Ab "B" in this example) linked to one Fab' fragment of the untreated Fab' (Ab "A" in this example). The second peak (retention time = 10.867 min) likely consists of one Fab' of Ab "B" linked to one Fab' fragment of Ab "A" (*see* **Fig. 2**; **refs.** *16* and *17*). The ratio of the $F(ab)_3$ BsAb to the $F(ab')_2$ BsAb can be altered by adjusting the ratio of *o*-PDM Ab added to the free Fab' upon conjugation or by carefully pooling the fractions from the Superdex column.

Acknowledgments

We are grateful to Dr. Aditya Mandel for providing data shown in **Figs. 6–8**, and to Kim Wunder for her expert assistance in preparation of the manuscript.

References

1. Fanger, M. W. (1995) *Bispecific Antibodies*, R.G. Landes Co., Austin, TX.
2. Fanger, M. W., Morganelli, P. M., and Guyre, P. M. (1992) Bispecific antibodies. *Crit. Rev. Immunol.* **12,** 101–124.
3. Goldenberg, D. M. (2003) Advancing role of radiolabeled antibodies in the therapy of cancer. *Cancer Immunol. Immunother.* **52,** 281–296.

4. Peipp, M. and Valerius, T. (2002) Bispecific antibodies targeting cancer cells. *Biochem. Soc. Trans.* **30**, 507–511.

5. Segal, D. M., Qian, J. H., Mezzanzanica, D., Garrido, M. A., Titus, J. A., Andrew, S. M., et al. (1992) Targeting of anti-tumor responses with bispecific antibodies. *Immunobiology* **185**, 390–402.

6. Segal, D. M., Weiner, G. J., and Weiner, L. M. (1999) Bispecific antibodies in cancer therapy. *Curr. Opin. Immunol.* **11**, 558–562.

7. van Spriel, A. B., van Ojik, H. H., and van De Winkel, J. G. (2000) Immunotherapeutic perspective for bispecific antibodies. *Immunol. Today* **21**, 391–397.

8. Weiner, L. M. (2000) Bispecific antibodies in cancer therapy. *Cancer J.* **6(Suppl 3)**, S265–S271.

9. Suresh, M. R., Cuello, A. C., and Milstein, C. (1986) Bispecific monoclonal antibodies from hybrid hybridomas. *Methods Enzymol.* **121**, 210–228.

10. Chamow, S. M. and Ashkenazi, A. (1999) *Antibody Fusion Proteins*, Wiley-Liss, New York, NY.

11. Nisonoff, A. and Rivers, M. M. (1961) Recombination of a mixture of univalent antibody fragments of different specificity. *Arch. Biochem. Biophys.* 460–462.

12. Graziano, R., Somasundaram, C., and Goldstein, J. (1995) The production of bispecific antibodies, in *Bispecific Antibodies* (Fanger, M., ed.), R.G. Landes Company, Austin, TX, pp. 1–26.

13. Brennan, M., Davison, P. F., and Paulus, H. (1985) Preparation of bispecific antibodies by chemical recombination of monoclonal immunoglobulin G1 fragments. *Science* **229**, 81–83.

14. Glennie, M., McBride, H., Worth, A., and Stevenson, G. (1987) Preparation and performance of bispecific F(ab' gamma)2 antibody containing thioether-linked Fab' gamma fragments. *J. Immunol.* **139**, 2367–2375.

15. Ellman, G. L. (1959) Tissue sulfhydral groups. *Arch. Biochem. Biophys.* 70–78.

16. Tutt, A., Greenman, J., Stevenson, G. T., and Glennie, M. J. (1991) Bispecific F(ab'gamma)3 antibody derivatives for redirecting unprimed cytotoxic T cells. *Eur. J. Immunol.* **21**, 1351–1358.

17. Tutt, A., Stevenson, G., and Glennie, M. (1991) Trispecific F(ab')3 derivatives that use cooperative signaling via the TCR/CD3 complex and CD2 to activate and redirect resting cytotoxic T cells. *J. Immunol.* **147**, 60–69.

6

Preparation of Immunoconjugates Using Antibody Oligosaccharide Moieties

Carl-Wilhelm Vogel

Summary

Heterobifunctional crosslinking reagents are small molecular weight chemicals containing two different reactive groups that have become important tools in generating conjugates of two different biomolecules, such as two proteins. The resulting bioconjugates are hybrid molecules or proteins, a new category of biomolecules that exhibit the combined functions of the two parent biomolecules. An important category of hybrid proteins are conjugates of antibodies with other effector molecules, such as drugs or toxins. These antibody conjugates or immunoconjugates have a variety of the applications in medicine, with particular emphasis on the treatment of cancer. The most commonly used heterobifunctional crosslinking reagents for the synthesis of antibody conjugates contain an N-hydroxysuccinimide ester moiety, which allows derivatization of amino groups in proteins. The chemical modification of a functionally important amino group in the antigen-binding region of an antibody causes impairment or loss of the antigen binding function, resulting in a defective antibody conjugate that lacks one of its component functions. Furthermore, even if the chemical derivatization does not affect the antigen binding function, the subsequent coupling of an effector protein at or near the antigen-binding region can also cause the loss of the antigen binding function for steric reasons. In this chapter, heterobifunctional crosslinking reagents are described that allow the generation of antibody conjugates where the effector proteins are coupled to the antibody carbohydrate moieties. Because antibody carbohydrate moieties are distal from the antigen-binding region, the use of carbohydrate-directed heterobifunctional crosslinking reagents, such as S-(2-thiopyridyl)-L-cysteine hydrazide (TPCH), prevents inactivation of the antigen-binding function. The synthesis of two carbohydrate-directed heterobifunctional crosslinking reagents is described. Coupling protocols for the preparation of antibody conjugates with effector proteins of different sizes using carbohydrate-directed heterobifunctional crosslinking reagents are also provided.

Key Words: Antibody conjugates; carbohydrate-directed derivatization; crosslinking; crosslinking reagents; heterobifunctional crosslinking reagents; hybrid proteins; immunoconjugates; immunotoxins; oligosaccharide moieties; protein derivatization; protein–protein conjugation; site-directed conjugation; regio-specific conjugation.

From: *Methods in Molecular Biology, vol. 283: Bioconjugation Protocols: Strategies and Methods*
Edited by: C. M. Niemeyer © Humana Press Inc., Totowa, NJ

1. Introduction

Derivatization, coupling, and immobilization of biomolecules—and biological macromolecules in particular—have been the subject of intense research for at least two decades with the intent of developing new applications for biological molecules in biotechnology and medicine. Conjugates consisting of two or more biological macromolecules can be created by recombinant means if the biomolecules involved are proteins, or they can be generated by chemical means. Bioconjugates represent a novel and interesting category of chemicals because they represent hybrids of biological molecules that do not exist in nature but are synthesized by combining two or more naturally occurring biological macromolecules into a new chemical compound. The need for derivatization, coupling, and/or immobilization of biological macromolecules has made it necessary to develop a host of novel chemical procedures that take into consideration the fact that many biological macromolecules are water soluble and insoluble in organic solvents, offer only a limited variety of potentially reactive chemical groups, and will maintain their biological function, in most cases, only within a very limited range of pH and temperature.

An important category of semisynthetic hybrid proteins are conjugates of antibodies with a host of other molecules, such as drugs, toxins, chelating reagents, enzymes, and biological response modifiers *(1–3)*. These antibody conjugates, often referred to as immunoconjugates, have a variety of applications in medicine, with particular emphasis on the diagnosis or treatment of cancer. Several antibody conjugates are used successfully in cancer therapy *(4,5)*.

The field of immunoconjugate research has received a significant boost with the availability of heterobifunctional crosslinking reagents. These are small-molecular-weight chemicals that contain two different reactive groups, each of which is able to react with a chemically different functional group in a biological macromolecule *(6,7)*. Approximately 100 heterobifunctional crosslinking reagents have been prepared and, for the most part, are commercially available. The vast majority of heterobifunctional reagents contain a chemical moiety that reacts with amino groups in proteins and a second chemical moiety that reacts with free sulfhydryl groups. The amino-reactive group is almost exclusively an *N*-hydroxysuccinimide ester, whereas three sulfhydryl-reactive groups are commonly used: the pyridyldithio group, the maleimide group, and an aliphatic halide (iodide). This is not the place to provide a detailed review of the various properties of the different heterobifunctional crosslinking reagents, the chemical nature of the resulting intermolecular crosslinks, and their chemical and biochemical properties. Suffice it to say that heterobifunctional crosslinking reagents containing the pyridyldithio group generate an intermolecular crosslink with a disulfide bond, whereas reagents containing maleimide or halide groups result in the formation of a thioether bond. Other differences relating to the chemical nature of the intermolecular crosslink include length, charge, solubility, aromaticity, and stability (to reduction, enzymatic cleavage, and pH) *(6,7)*.

One advantage inherent to all heterobifunctional crosslinking reagents is the fact that they result in the formation of heteroconjugates, which means that the resulting conjugates contain at least one molecule each of the two biomolecules to be coupled. The design of the heterobifunctional reagents prevents the formation of homoconjugates, that is, the formation of conjugates consisting of only one of the two protein species intended to be coupled. However, heterobifunctional crosslinking reagents do not generate hybrid proteins consisting of only one protein molecule each of the two molecular species to be coupled. This is a consequence of the fact that proteins usually contain multiple amino groups (amino-terminal amino groups and ε-amino groups of lysine residues). For example, an immunoglobulin G antibody molecule has at least 70 amino groups. Accordingly, derivatization of a protein with an amino group-directed heterobifunctional crosslinking reagent results in the modification of several or even many amino groups which, in turn, allows for the subsequent coupling of multiple protein molecules of the second coupling partner with free sulfhydryl groups. If the free sulfhydryl group-containing protein contains only one (usually naturally occurring) free sulfhydryl group, its coupling to the amino group-derivatized protein results in a mixture of hybrid proteins of the molecular composition 1:1, 1:2, 1:3, 1:4, 1:5, and so on. If a protein does not contain one (or more) natural free sulfhydryl groups, these can be introduced by a crosslinking reagent (e.g., *N*-succinimidyl-3-(2-pyridyldithio)-propionate [SPDP*]) (**Fig. 1**). SPDP derivatizes amino groups, resulting in the introduction of pyridyldithio groups. Subsequent reduction of the pyridyldithio groups results in free sulfhydryl groups *(8)*. When a protein with introduced free sulfhydryl groups is coupled to another protein derivatized with sulfhydryl-reactive groups, the resulting conjugates represent mixtures of hybrid proteins of the molecular composition 1:1, 1:2, 2:1, 2:2, 1:3, 3:1, 2:3, 3:2, 3:3, and so on.

As much as heterobifunctional crosslinking reagents with one of the reactive groups being an amino-reactive group allow for easy and multiple derivatization of proteins, they exhibit one inherent drawback. Because proteins, depending on

---

*Abbreviations for this chaper are as follows: BMPH, *N*-(β-maleimidopropionic acid) hydrazide; CVF, cobra venom factor; DTT, dithiothreitol; EMCH, *N*-(ε-maleimidocaproic acid) hydrazide; GMBS, *N*-(γ-maleimidobutyryl-*N*-hydroxysuccinimide ester; KMUH, *N*-(κ-maleimidoundecanoic acid) hydrazide; M₂C₂H, 4-(*N*-maleimidomethyl)-cyclohexane-1-carboxyl hydrazide; MPBH, 4-(4-*N*-maleimidophenyl)-butyric acid hydrazide; PBS, phosphate-buffered saline; PDPH, 3-(2-pyridyldithio)-propionyl hydrazide; SDS, sodium dodecyl sulfate; SMCC, succinimidyl-4-(*N*-maleimidomethyl)-cyclohexane-1-carboxylate; SMPB, succinimidyl-4-(*p*-maleimidophenyl)-butyrate; SPDP, *N*-succinimidyl-3-(2-pyridyldithio)-propionate; TPCH, *S*-(2-thiopyridyl)-L-cysteine hydrazide; TPMPH, *S*-(2-thiopyridyl)-3-mercaptropropionic acid hydrazide.

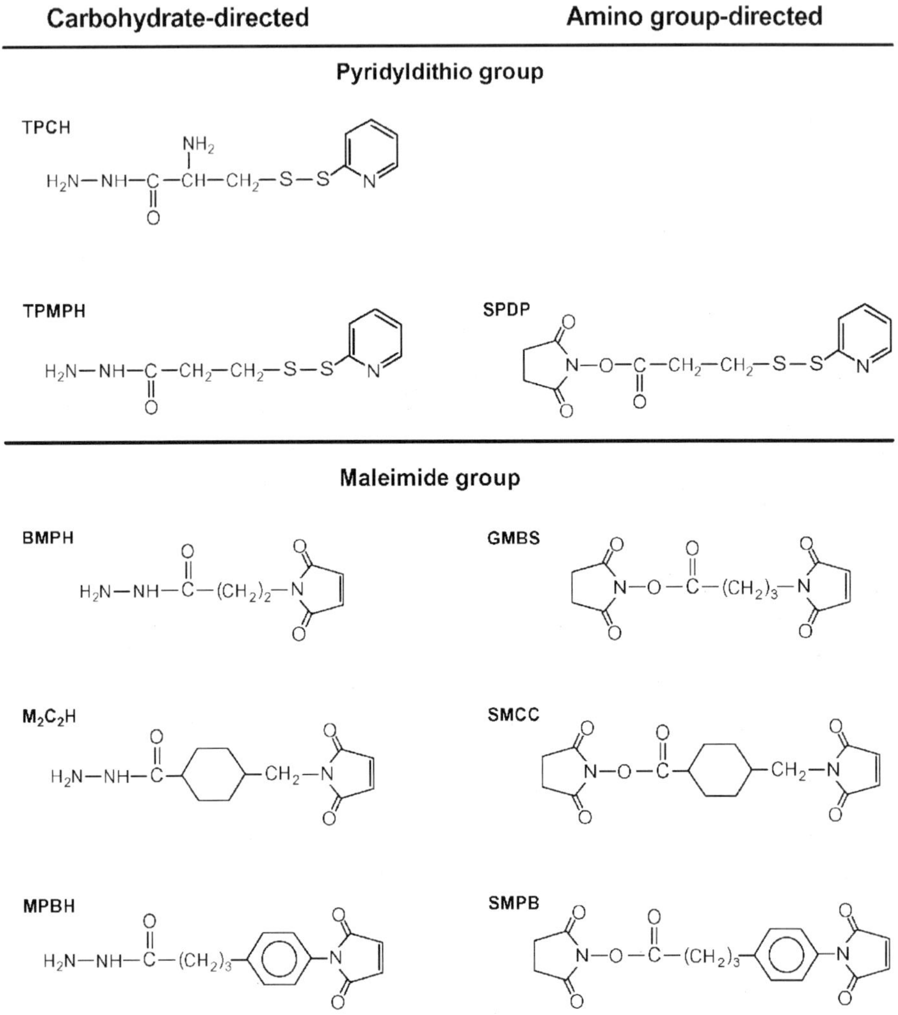

Carbohydrate-directed   **Amino group-directed**

Pyridyldithio group

TPCH

TPMPH

SPDP

Maleimide group

BMPH

GMBS

M₂C₂H

SMCC

MPBH

SMPB

Fig. 1. Chemical structures of corresponding carbohydrate-directed (left) and amino group-directed (right) heterobifunctional crosslinking reagents. Upper panel, crosslinking reagents introducing a pyridyldithio group. Lower panel, crosslinking reagents introducing a maleimide group.

their size, usually have several-to-many amino groups available for derivatization by the crosslinker, the derivatization of one or several amino groups in a given protein can lead to functional inactivation of the protein. Whereas a protein derivatized at a functionally important amino group can still be incorporated into a hybrid protein, the derivatized protein has lost its function, resulting in the creation of a hybrid protein that lacks one of its component functions.

In addition to functional inactivation of a protein by direct chemical modification of functionally important amino groups, additional functional inactivation of a coupling partner can be caused by steric hindrance after incorporation of a protein into a hybrid protein. For example, in the case of antibodies, both chemical derivatization with the crosslinker at the antigen-binding site and conjugation of the coupling partner at or near the antigen-binding site will impair the antigen-binding function of the particular Fab component of the resulting antibody conjugate. One successful approach to avoid both chemical and steric inactivation of the antigen-binding function of an antibody is to couple the other protein to the oligosaccharide moieties of antibodies, which are located distal to the antigen binding sites.

In this chapter, the generation of antibody conjugates with other proteins is described using heterobifunctional crosslinking reagents where one reactive group is a hydrazide that binds to aldehyde groups generated in the oligosaccharide moieties of antibodies by periodate oxidation of cis-diol groups (e.g., S-(2-thiopyridyl)-L-cysteine hydrazide [TPCH]; **Fig. 1**). This crosslinking approach prevents the functional inactivation of antibody-binding sites as will be shown further below *(9,10)*.

2. Materials

2.1. Chemicals

1. All chemicals were obtained from Aldrich (Milwaukee, WI).
2. With the exception of TPCH, all heterobifunctional crosslinking reagents mentioned in the manuscript are commercially available from Pierce (Rockford, IL). The company sells S-(2-thiopyridyl)-3-mercaptopropionic acid hydrazide (TPMPH) using the acronym PDPH, which is derived from the alternate chemical name 3-(2-pyridyldithio)-propionyl hydrazide for TPMPH.
3. N,N'-bis-(tert-butyloxycarbonyl)-L-cystine dimethyl ester was obtained by protecting the amino groups of L-cystine methyl ester with di-tert-butyl pyrocarbonate *(11)*.
4. S-(2-thiopyridyl)-3-mercaptopropionic acid was prepared from 3-mercaptopropionic acid by thiol/disulfide exchange with 2,2'-dipyridyl disulfide *(8)*.
5. Silica gel (mesh 32–60 and 70–230) and Fractogel 55S and HW65F were from EMD Chemicals (Gibbstown, NJ).
6. Prepacked Sephadex G-25 columns (PD-10 columns) were from Amersham Biosciences (Piscataway, NJ).

2.2. Proteins

1. Human monoclonal IgM antibody 16–88, derived from a patient immunized with autologous human colon carcinoma cells, was used in this study *(12,13)*. The antibody is obtained from hollow fiber culture and purified by gel filtration and ion-exchange chromatography *(14)*.
2. Cobra venom factor (CVF) was purified from lyophilized cobra venom (Serpentarium Laboratories, Punta Gorda, FL) as described previously *(15)*.

3. Ricin A-chain was obtained from Inland Laboratories (Austin, TX).
4. Barley toxin, purified as described *(16)*, was a gift from Organon Teknika Corporation/Biotechnology Research Institute (Rockville, MD).

3. Methods
3.1. Synthesis of TPCH

The synthesis of TPCH is a five-step process *(9,17)*.

3.1.1. Preparation of N,N'-Di-(Tert-Butyloxycarbonyl)-L-Cystine Dihydrazide

1. A solution of 2.47 g (5.28 mmol) of *N*,*N*'-di-(tert-butyloxycarbonyl)-L-cystine dimethyl ester in 50 mL of methanol is treated dropwise with 10 mL of anhydrous hydrazine at room temperature.
2. The solution is maintained at room temperature for 2 h, over which time a fine white material precipitates.
3. The solution is cooled to 0°C for 30 min, and the product is collected by filtration and washed with ice-cold methanol to provide white crystals.

The yield is 2.14 g (86.8%). ^{1}H-NMR (CD$_3$COCD$_3$): 9 (br s, 1H, exchangeable with D$_2$O), 5.62 (br d, 1H, exchangeable with D$_2$O), 4.84 (m, 1H), 3.5 (br, 1H, exchangeable with D$_2$O), 2.92 (br, 2H), 1.45 (s, 9H) ppm.

3.1.2. Preparation of Tetra-(Tert-Butyloxycarbonyl)-L-Cystine Dihydrazide

1. A suspension of 10.40 g (22.22 mmol) of *N*,*N*'-di-(tert-butyloxycarbonyl)-L-cystine dihydrazide in 180 mL of ethanol is treated with 20 mL of diisopropylethylamine and warmed to reflux. The suspension dissolves upon warming.
2. 9.70 g (44.44 mmol) of di-(tert-butyl)-dicarbonate is added portionwise.
3. The clear, colorless solution is refluxed for 30 min and then allowed to cool to room temperature. After 20 min, the product begins to crystallize from solution.
4. The mixture is stored at room temperature for 1 h and then cooled to 0°C for 1 h. The white crystalline product is collected by filtration and washed with ice-cold ethanol.

The yield is 10.80 g (72.8%). ^{1}H-NMR is very complex because of the apparent restricted rotation about the three amide-type bonds. At least three rotamers can be identified in the spectrum.

3.1.3. Preparation of Di-(Tert-Butyloxycarbonyl)-L-Cysteine Hydrazide

1. Zinc dust (3 g) is added in portions over 2 h to a suspension of 10.80 g (16.17 mmol) tetra-(tert-butyloxycarbonyl)-L-cystine dihydrazide in 40 mL of acetic acid containing 6 mL of water. Gradually, the suspension dissolves.

2. After 2 h the solution is concentrated under reduced pressure, and the residue is partitioned between methylene chloride and saturated aqueous sodium bicarbonate.
3. The methylene chloride is dried over sodium sulfate and concentrated to a viscous glass.

The yield is 10 g (92.6%). ^{1}H-NMR is very complex because of the apparent restricted rotation about the three amide-type bonds. At least three rotamers can be identified in the spectrum.

3.1.4. Preparation of Di-(Tert-Butyloxycarbonyl)-S-(2-Thiopyridyl)-L-Cysteine Hydrazide

1. 6.57 g (29.85 mmol) of 2,2'-dipyridyl disulfide is added portionwise to a solution of 5 g (14.93 mmol) di-(tert-butyloxycarbonyl)-L-cysteine hydrazide in 75 mL of methanol at room temperature.
2. This solution is maintained at room temperature for 24 h and then concentrated in vacuo to a yellow syrup.
3. The crude product is taken up in 400 mL of methanol, and 20 g of silica gel (32–60 mesh) is added.
4. The crude product is adsorbed onto the silica gel by evaporation of the solvent, and the impregnated gel is placed atop a 95- × 55-mm column of silica gel (32–60 mesh).
5. The product is isolated by eluting with ethyl acetate:hexane (35:65, v/v). Fractions (100 mL each) containing product are pooled and concentrated to provide a colorless glass.

The yield is 3.5 g (52.8%). ^{1}H-NMR (CDCl$_3$) is complex because of the presence of at least two rotamers in solution: 9.54 (br, 0.25H, exchangeable with D$_2$O), 8.60 (br, 0.75H, exchangeable with D$_2$O), 6.553 (br, 1H, exchangeable with D$_2$O), 5.786 (m, 0.5H), 4.925 (m, 0.5H), 4.526 (br s, 1H), 3.384 (m, 1H), 2.910 (m, 1H), 1.456 (br s, 9H), 1.408 (br, 9H) ppm.

3.1.5. Preparation of TPCH

1. A solution of 1.15 g (2.58 mmol) of di-(tert-butyloxycarbonyl)-S-(2-thiopyridyl)-L-cysteine hydrazide in 15 mL of ethyl acetate is cooled to 0°C.
2. 25 mL of a saturated solution of anhydrous hydrogen chloride in ethyl acetate is added slowly. After 30 min, a white crystalline material begins to separate.
3. The mixture is stirred at room temperature for 4 h.
4. The mixture is filtered under argon, washed with ethyl acetate, dried under argon, and then under vacuum to provide hygroscopic white crystals.

The yield is 830 mg (91%). ^{1}H-NMR (D$_2$O): 8.63 (m, 1H), 8.31 (m, 1H), 8.14 (m, 1H), 7.74 (m, 1H), 4.47 (m, 1H), 3.45 (m, 2H) ppm. ^{13}C-NMR (DMSO-d$_6$) 166, 157, 149, 140, 122, 121, 50, 21 ppm. The melting point of TPCH is 155–162°C.

3.2. Synthesis of TPMPH

The synthesis of TPMPH is a two-step process *(9)*.

3.2.1. Preparation of N-Tert-Butyloxycarbonyl-S-(2-Thiopyridyl)-3-Mercaptopropionic Acid Hydrazide

1. S-(2-thiopyridyl)-3-mercaptopropionic acid (4 mmol; 0.86 g) is mixed with 4 mmol (0.83 g) of 1,3-dicyclohexylcarbodiimide in 10 mL of anhydrous dichloromethane at an ice bath temperature.
2. After the addition of 4 mmol (0.53 g) of tert-butyl carbazate, the reaction is warmed to room temperature and 50 mL of ice water is added.
3. The product is extracted with chloroform (20 mL, 3×).
4. The chloroform solution is chromatographed on silica gel (70–230 mesh) using chloroform:methanol (9:1, v/v), resulting in 0.55 g (42% yield) of the product.

3.2.2. Preparation of TPMPH

1. N-tert-butyloxycarbonyl-S-(2-thiopyridyl)-3-mercaptopropionic acid hydrazide (0.55 g, 1.67 mmol) is treated with ethyl acetate (25 mL), previously saturated with HCl, at 0°C for 30 min.
2. TPMPH is precipitated with ether (10 mL) and recrystallized (ethanol/ether).

The yield is 0.42 g (96%). ^{1}H-NMR spectroscopy: 8.40 (m, 1H, pyridyl), 7.83 (m, 1H, pyridyl), 7.22 (m, 1H, pyridyl), 3.05 (t, 2H, $J_{H,H}$ = 7 Hz), 2.57 (t, 2H, $J_{H,H}$ = 7 Hz) ppm. The melting point of TPMPH is 168–170°C.

3.3. Antibody Derivatization With Heterobifunctional Crosslinkers

3.3.1. Antibody Derivatization With Carbohydrate-Directed Crosslinkers TPCH or TPMPH

1. Oxidation of the IgM antibody 16–88 (2 mg/mL) is performed with 1 mM Na metaperiodate at 0°C for 15 min in 0.1 M Na acetate, pH 5.5, in the presence of 15 mM TPCH.
2. The reaction mixture is subjected to size-exclusion chromatography on Sephadex G-25 (PD-10 column) equilibrated with phosphate-buffered saline (PBS) (10 mM Na phosphate, 100 mM NaCl), pH 8.0. The eluted antibody (1.3 mg/mL), derivatized in the carbohydrate moieties with pyridyldithio groups, is ready for coupling to effector molecules *(9)*.

The extent of antibody derivatization with TPCH has virtually no measurable effect on the antigen binding activity of the antibody (*see* **Note 1**). However, TPCH derivatization affects the complement-activating activity of the antibody (*see* **Note 2**). It is important to perform the periodate oxidation in the presence of the TPCH crosslinker (*see* **Note 3**). The protein concentration should be above 1 mg/mL (*see* **Note 4**).

To achieve a higher degree of derivatization with TPCH, the eluted antibody (1.3 mg/mL) is further incubated in the presence of 10 mM TPCH in PBS, pH 8.0, for up to 150 min at 25°C, and then resubjected to gel filtration on Sephadex G-25 equilibrated with PBS, pH 7.2 (*see* **Note 3**; **ref. 9**).

The method for antibody derivatization with TPMPH is identical to the method for derivatization with TPCH *(9)*. Other carbohydrate-directed heterobifunctional crosslinking reagents with a maleimide function as sulfhydryl-reactive group have been described (*see* **Note 5**).

3.3.2. Antibody Derivatization With Amino Group-Directed Crosslinker SPDP

1. The IgM antibody 16-88 (2 mg, 2.2 nmol) is incubated at 25°C for 30 min with 10 nmol of SPDP in a total volume of 1 mL of PBS, pH 7.2.
2. The pyridyldithio-derivatized antibody is purified by gel filtration on a G-25 Sephadex column in PBS, pH 7.2 *(10)*.

The extent of antibody derivatization with SPDP affects the antigen-binding activity of the antibody (*see* **Note 1**). The SPDP concentration needs to be adjusted depending on the molecular weight of the protein to be derivatized (*see* **Note 4**).

3.4. Preparation of Antibody Conjugates

3.4.1. Preparation of Antibody Conjugates With CVF

Pyridyldithio groups are introduced into CVF by incubating the protein (2 mg/mL) with 100 µM SPDP for 30 min at 25°C in PBS, pH 7.5. The pyridyldithio-derivatized CVF is purified by gel filtration on a G-25 Sephadex column in PBS, pH 7.2 *(18,19)*. SPDP-derivatized CVF (2.8 pyridyldithio groups/CVF molecule) is incubated in the presence of 50 mM dithiothreitol (DTT) for 20 min at 25°C to reduce the pyridyldithio groups, subjected to gel filtration chromatography on Sephadex G-25 equilibrated with deaerated PBS, pH 7.2, and then used immediately for conjugation to either TPCH-modified or SPDP-modified antibody.

The reaction mixture containing crosslinker-derivatized antibody (1.3 mg) and sulfhydryl-derivatized CVF (1 mg) in a total volume of 1 mL of PBS, pH 7.2, is flushed with nitrogen and incubated for 15 h at 25°C and then for 24 h at 4°C. After purification of the antibody conjugates by size-exclusion chromatography using a Fractogel HW65F column (1.5 × 115 cm) equilibrated in PBS, pH 7.2, the fractions are pooled, concentrated by ultrafiltration, and stored at 4°C *(10)*.

In contrast to antibody conjugates prepared with SPDP, the antigen-binding activity of conjugates prepared with TPCH is virtually unaffected (*see* **Note 1**). The protein concentration during the coupling reaction should be above 1 mg/mL (*see* **Note 4**). Once free sulfhydryl groups are introduced, the proteins should be immediately subjected to coupling. The coupling reaction vial should be flushed with nitrogen. Once coupled, the antibody conjugates exhibit good stability (*see* **Note 6**). The coupling of effector proteins is somewhat less efficient using TPCH compared to SPDP (*see* **Note 7**). Furthermore, the activity of large effector proteins such as CVF can be compromised when coupled to TPCH-derivatized antibody compared to SPDP-derivatized antibody (*see* **Note 8**).

3.4.2. Preparation of Antibody Conjugates With Ricin A-Chain

Ricin A-chain at a final concentration of 3.3 mg/mL in 0.1 M Na acetate, pH 4.5, is freshly reduced in the presence of 50 mM DTT for 60 min at 25°C. After removal of DTT by size-exclusion chromatography on Sephadex G-25 equilibrated with deaerated PBS, pH 7.2, the eluted ricin A-chain is immediately used for conjugation to TPCH-derivatized IgM antibody 16–88 (approx 7 pyridyldithio groups/antibody molecule). Ricin A-chain (1.33 mg, approx 44 nmol) is incubated with 4 mg (approx 4.4 nmol) of TPCH-derivatized antibody in 4.7 mL of PBS, pH 7.2. After flushing with nitrogen, the reaction mixture is incubated for 15 h at 25°C and then subjected to size-exclusion chromatography on a Fractogel 55S column (1.2 × 100 cm) equilibrated with 40 mM Na phosphate, 150 mM NaCl, pH 7.2. The fractions of the first peak, containing the IgM conjugates, are pooled, concentrated by ultrafiltration, and stored at 4°C *(9)*. The coupling efficiency of ricin A-chain is relatively low compared to barley toxin, a protein of similar size (*see* **Note 7**).

3.4.3. Preparation of Antibody Conjugates With Barley Toxin

Barley toxin (5 mg, 165 nmol) is derivatized with 840 nmol SPDP in a total volume of 1.9 mL of PBS, pH 7.2. After 30 min at 25°C the pyridyldithio-derivatized barley toxin is purified by gel filtration on Sephadex G-25 in 0.1 M Na acetate, pH 4.5, and concentrated using a centrifugal microconcentrator (M_r 30,000 cutoff). Subsequently, the pyridyldithio groups of the SPDP-derivatized barley toxin (4.8 pyridyldithio groups/toxin molecule) are reduced for 20 min at 25°C in the presence of 50 mM DTT. After removal of DTT by gel filtration on Sephadex G-25 equilibrated with deaerated PBS, pH 7.2, the eluted sulfhydryl-derivatized barley toxin is immediately used for conjugation to TPCH-derivatized IgM antibody 16–88 (7 pyridyldithio groups/antibody molecule). Sulfhydryl-derivatized barley toxin (1.3 mg) is incubated with 4 mg of TPCH-derivatized antibody in 3.7 mL of PBS, pH 7.2. The incubation and subsequent purification of the antibody conjugates is performed as described above for ricin A-chain *(9)*.

3.5. Determination of Incorporated Pyridyldithio Groups

The number of pyridyldithio groups incorporated into a protein by derivatization with TPCH, TPMPM, or SPDP is determined by spectrophotometric measurement of released pyridine-2-thione at 343 nm after reduction of the derivatized protein with 5 mM DTT (final concentration) *(8)*. The extinction coefficient of pyridine-2-thione at 343 nm is 8.08×10^3 M/cm. The protein content is determined by its absorbance at 280 nm, corrected for the contribution of the introduced pyridyldithio groups using the empirical formula:

$$A_{280} \text{ (protein)} = A_{280} \text{ (observed)} - (B \times 5.1 \times 10^3)$$

where B is the molar concentration of the pyridine-2-thione in the solution.

3.6. Determination of the Stoichiometry of Effector Proteins to Antibody

Sodium dodecyl sulfate (SDS) gradient polyacrylamide gel electrophoresis is a good method to get a rough estimate of the molecular composition of the hybrid protein mixture after conjugation *(18–20)*. To determine the average ratio of effector molecule per antibody molecule, ^{125}I-labeled effector protein (at approx 1.5×10^6 cpm/mg) needs to be used for conjugation. Stoichiometries can then be determined from the difference in specific radioactivity before and after conjugation. The toxin to antibody stoichiometry for smaller effector molecules such as ricin A-chain or barley toxin can also be obtained by scanning gel densitometry after SDS polyacrylamide gel electrophoresis under reducing conditions *(9,10)*.

3.7. Other Methods

1. The antigen binding activity of IgM antibody 16–88 is determined in a competitive binding assay with ^{125}I-labeled antibody using microtiter plates with immobilized tumor antigen *(10)*.
2. The complement activating activity of IgM antibody 16–88 is determined in a modified complement fixation assay using immobilized tumor antigen extract and human serum as a complement source. Sensitized sheep erythrocytes are used to determine the remaining serum complement activity *(10)*.
3. The CVF hemolytic activity is determined in a bystander lysis assay using guinea pig erythrocytes *(15)*.
4. Barley toxin activity is determined in a cell-free reticulocyte assay based on the inhibitory activity of barley toxin on protein translation *(9)*.
5. Proteins are radiolabeled with Na^{125}I using immobilized chloramine-T (IodoBeads; **ref. 21**).
6. SDS gradient (5–15% w/v) polyacrylamide gel electrophoresis is performed under reducing conditions followed by Coomassie staining *(19)*.
7. Protein concentrations are determined by the Lowry method (*see* **Note 9**; **ref. 22**).

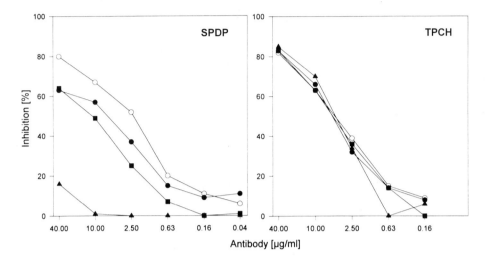

Fig. 2. Effect of SPDP derivatization (left panel) and TPCH derivatization (right panel) on the antigen-binding activity of IgM antibody 16–88. Shown is the ability of unmodified antibody (open circles) and derivatized antibody (filled symbols) to bind antigen in a competition binding assay with ^{125}I-labeled antibody. SPDP-derivatized antibody has 2 (filled circles), 8.6 (filled squares), or 16 (filled triangles) pyridyldithio groups per antibody molecule. TPCH-derivatized antibody has 1.8 (filled circles), 9.2 (filled squares), or 16.6 (filled triangles) pyridyldithio groups per antibody molecule. Modified from **ref. 10**.

4. Notes

1. **Figure 2** shows the effect of derivatization of human monoclonal IgM antibody 16–88 with the carbohydrate-directed crosslinker TPCH compared to the amino group-directed crosslinker SPDP. An increasing degree of antibody derivatization with SPDP results in increased inactivation of the antigen-binding function. Derivatization of the antibody with 16 SPDP molecules causes almost a complete loss of the antigen-binding activity (**Fig. 2**, left panel). In contrast, derivatization of the antibody with as many as 16.6 carbohydrate-directed TPCH molecules does not cause any measurable change in the antigen binding activity of the antibody (**Fig. 2**, right panel).

   Figure 3 shows the effect of coupling of CVF, a protein with an M_r of approx 150,000, to the SPDP-derivatized (left panel) or TPCH-derivatized antibody (right panel) on the antigen-binding activity. CVF conjugates prepared with SPDP-derivatized antibody exhibit an increased impairment of the antigen-binding function with increasing coupling ratios of CVF per antibody molecule. Taking into consideration the data from **Fig. 2**, it is evident that both SPDP derivatization and CVF coupling contribute to the compromise in antigen binding activity. At a low-coupling ratio (one to three CVF molecules per antibody)

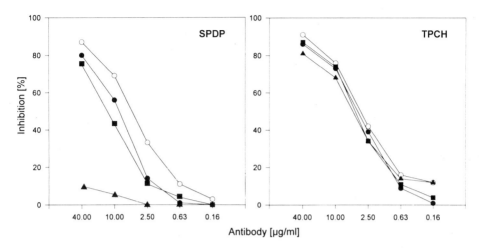

Fig. 3. Effect of CVF conjugation to SPDP-derivatized (left panel) and TPCH-derivatized IgM antibody 16–88 (right panel) on the antigen-binding activity of the resulting immunoconjugates. Shown is the ability of unmodified antibody (open circles) and antibody-CVF conjugates (filled symbols) to bind antigen in a competition binding assay with ^{125}I-labeled antibody. Left panel, antibody-CVF conjugates prepared with SPDP containing 1.2 mol of CVF per mol of antibody (derivatized with 2.4 pyridyldithio groups/antibody) (filled circles) 3 mol of CVF per mol of antibody (derivatized with 5 pyridyldithio groups/antibody) (filled squares) or 5.6 mol of CVF per mol of antibody (derivatized with 8.6 pyridyldithio groups/antibody) (filled triangles). Right panel, antibody-CVF conjugates prepared with TPCH containing 0.5 (filled circles), 2 (filled squares), or 3.2 mol of CVF (filled triangles) per mol of antibody (in all cases derivatized with 8 pyridyldithio groups/antibody). Modified from **ref. 10**.

the additional compromise as a result of CVF coupling is moderate, but it increases to more than 80% at a coupling ratio of five to six CVF molecules per antibody. In contrast, as shown in the right panel of **Fig. 3**, virtually no compromise in antigen-binding activity is observed when CVF is coupled to the TPCH-derivatized antibody. Even immunoconjugates containing three to four CVF molecules per antibody are almost indistinguishable in their antigen-binding activity compared to unmodified antibody.

Similar results are obtained with other effector proteins (e.g., ricin A-chain, barley toxin) coupled to the 16–88 IgM antibody *(9)*. Collectively, these data demonstrate that the carbohydrate moieties of antibody molecules can serve as attachment sites for multiple numbers of even large effector molecules such as CVF without impairment of the antigen binding function of the antibody.

2. **Figure 4** demonstrates that an increasing degree of derivatization of the 16–88 IgM antibody with TPCH causes an increased inactivation of the antibody's ability to activate complement. Because the carbohydrate moieties in the hinge region

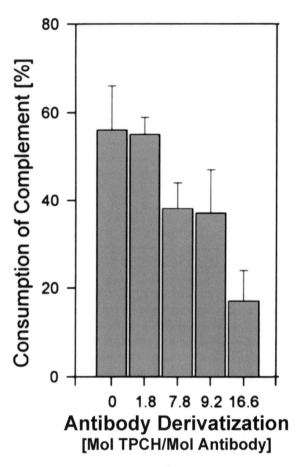

Fig. 4. Effect of TPCH derivatization of IgM antibody 16–88 on the complement-activating activity of the antibody as measured by a modified complement fixation assay. Modified from **ref. *10***.

of antibodies have been implicated in the process of complement activation by binding of complement component C1q, it is not surprising that chemical changes in the carbohydrate moieties of an antibody could cause a decrease in its ability to activate complement. In immunoconjugates the antibody component serves as the targeting moiety, which is intended to deliver the effector protein to a specific target. Accordingly, conservation of the antigen binding activity is of utmost importance for effective immunoconjugates. Other antibody functions, such as its ability to activate complement or to induce antibody-dependent cellular cytotoxicity, the so-called biological effector functions of antibodies, are usually not considered important for immunoconjugates. However, situations may exist where the biological effector functions of an antibody may contribute to the desired effect of an immunoconjugate; and it is important to be aware of the fact that derivatization of the oligosaccharide portion of antibody molecules may affect the biological effector functions.

3. Sialic acid residues are the target structure for the periodate oxidation of cis-diol groups and TPCH crosslinker incorporation into the antibody molecule as pre-treatment of the antibody with neuraminidase eliminates virtually all detectible sialic acid residues (approx 60 per 16–88 IgM antibody), resulting in no measurable incorporation of TPCH *(9)*.

It is important to perform the periodate oxidation in the presence of the TPCH crosslinker. The periodate oxidation in the absence of TPCH causes covalent intrachain and interchain crosslinks in the antibody molecule because of antibody amino groups reacting with the aldehydes. This effect is completely abolished if the periodate oxidation is performed in the presence of TPCH *(9)*.

The TPCH incorporation into the antibody molecule proceeds in a time-dependent manner during periodate oxidation. However, prolonged incubation (more than 15 min in the presence of 1 mM periodate) does not result in a further increase in the number of incorporated TPCH molecules but actually causes a decrease, indicating that the crosslinker molecule as well as the incorporated pyridyldithio groups are unstable in the presence of periodate *(9)*. Consistent with this observation, increasing the periodate concentration from 1 mM to 20 mM results in lower incorporation rates of the TPCH crosslinker *(9)*.

To increase the number of incorporated TPCH molecules, a two-step derivatization protocol can be used. After 15 min of oxidation with 1 mM periodate in the presence of TPCH crosslinker, the periodate is removed from the antibody by gel filtration. The eluted antibody is subsequently incubated in the presence of fresh crosslinker for up to 2.5 h. Using this two-step derivatization protocol, the number of incorporated TPCH molecules can be significantly increased *(9)*. It should be noted, that in the case of the 16–88 IgM antibody with its relatively large number of sialic acid residues, the two-step derivatization protocol is usually not necessary. The single 15-min incubation with the crosslinker in the presence of the periodate results in sufficient numbers of incorporated crosslinker molecules per antibody molecule and allows for successful immunoconjugate formation. However, the two-step protocol may be valuable in those cases where the glycoprotein to be derivatized with TPCH has only a relatively small number of sialic acid residues.

4. For amino group-directed derivatization of proteins with SPDP a concentration of 10 µM is sufficient for large molecules like IgM. The concentration of SPDP needs to be increased for the derivatization of smaller proteins (e.g., 100–150 µM for immunoglobulin G or CVF) and needs to be increased even further for relatively small proteins, such as barley toxin (e.g., 400 µM). The degree of derivatization can be controlled by decreasing or increasing the SDPD concentration for the derivatization of a given protein *(10,20)*.

Derivatization of glycoproteins with the carbohydrate-directed crosslinkers TPCH or TPMPH requires a somewhat higher concentration of the crosslinker (15 mM) during the oxidation. The degree of crosslinker incorporation can be controlled by the concentration of the crosslinker in the second step of the two-step derivatization procedure as well as by the length of incubation with the crosslinker during the first and second steps (compare with **Subheading 3.3.1.; ref. 9**).

The protein concentration for derivatization with a heterobifunctional cross-linking reagent should be above 1 mg/mL (1–3 mg/mL work very well). The protein concentration of the two protein partners to be coupled should also be at or above 1 mg/mL for each of the two partners. Depending on the molecular weight difference of the two coupling partners, the number of sulfhydryl-reactive groups in the antibody, the number of free sulfhydryl groups in the effector molecule, and the desired average ratio of effector molecule bound per antibody molecule, the effector molecule to be coupled may have to be present in several molar access over the antibody molecule.

5. All experiments reported in this manuscript use TPCH, a carbohydrate-directed heterobifunctional crosslinking reagent that introduces a pyridyldithio group. A structurally very similar carbohydrate-directed crosslinker molecule is TPMPH (**Fig. 1**). TPMPH uses mercaptopropionic acid rather than cysteine as a building block and therefore lacks the additional amino group present in TPCH. TPCH exhibits better incorporation into the glycoprotein to be derivatized compared to TPMPH (**9**). However, the somewhat better incorporation is offset by the more complex chemical synthesis of TPCH.

Other heterobifunctional crosslinking reagents with a hydrazide function for carbohydrate modification have also been synthesized, containing a maleimide group instead of a pyridyldithio group as their sulfhydryl-reactive function. No heterobifunctional crosslinking reagents have been reported combining a hydrazide function for oligosaccharide modification with a halide function as sulfhydryl-reactive group. Our laboratory has no direct experience with carbohydrate-directed heterobifunctional crosslinking reagents containing a maleimide function as their sulfhydryl-reactive group. However, we have used amino group-directed heterobifunctional crosslinking reagents containing the maleimide group as their sulfhydryl-reactive function (**20,23,24**). As several experimental findings relate to the maleimide group as well as the chemical nature of the crosslink between the two proteins, our pertinent findings for amino group-directed heterobifunctional crosslinking reagents are likely to apply to the corresponding carbohydrate-directed heterobifunctional crosslinking reagents as well.

The lower panel of **Fig. 1** shows the chemical structures of corresponding carbohydrate-directed and amino group-directed crosslinking reagents introducing a maleimide group. *N*-(β-maleimidopropionic acid) hydrazide (BMPH) and two related compounds differing only in the length of the aliphatic chain between the two reactive groups* are the equivalent of *N*-(γ-maleimidobutyryl-*N*-

---

N-(ε-maleimidocaproic acid) hydrazide (EMCH) and *N*-(κ-maleimidoundecanoic acid) hydrazide (KMUH) are carbohydrate-directed heterobifunctional crosslinking reagents where the hydrazide group and maleimide group are separated by 5 (EMCH) or 10 (KMUH) methylene groups, respectively.

hydroxysuccinimide ester (GMBS). 4-(*N*-Maleimidomethyl)-cyclohexane-1-carboxyl hydrazide (M_2C_2H) is the equivalent of the amino group-directed crosslinker succinimidyl-4-(*N*-maleimidomethyl)-cyclohexane-1-carboxylate (SMCC). Like BMPH and GMBS, M_2C_2H and SMCC also result in the formation of an aliphatic crosslink; however, the aliphatic crosslink contains a cyclohexane group. In contrast, the two corresponding crosslinkers 4-(4-*N*-maleimidophenyl)-butyric acid hydrazide (MPBH) and succinimidyl-4-(*p*-maleimidophenyl)-butyrate (SMPB) contain an aromatic ring that becomes part of the resulting intermolecular crosslink (**Fig. 1**).

The maleimide group exhibits a lower specificity for sulfhydryl groups compared with the pyridyldithio group. Heterobifunctional crosslinking reagents containing a maleimide group cause the formation of intramolecular as well as intermolecular covalent crosslinks in the derivatized proteins. These covalent crosslinks are nonreducable and result from maleimide groups reacting with amino groups in the same protein because of the high apparent concentration of amino groups in the proximity to where the crosslinker molecules are incorporated into a protein. Not surprisingly, derivatization of proteins with maleimide group-containing crosslinkers results in greater functional inactivation of the derivatized proteins *(20,23)*. Furthermore, because of the reaction of maleimide groups with amino groups, an increase in the crosslinker concentration during derivatization of the protein does not result in a corresponding increase in detectible sulfhydryl-reactive maleimide groups *(20)*.

Another important observation is that proteins derivatized with a heterobifunctional reagent containing an aromatic structure (such as SMPB) as well as immunoconjugates prepared with SMPB, which results in an aromatic structure in the intermolecular crosslink, exhibit significantly shorter plasma half-lives in mice *(23,24)*. Apparently, the aromatic structure causes more rapid elimination from the blood stream.

6. Pyridyldithio groups introduced into proteins with either TPCH or TPMPH are stable for a minimum of several days during storage at 4°C. A similar stability is observed when a maleimide group is introduced using a crosslinker with an aliphatic chain such as SMCC. In contrast, the number of detectible maleimide groups decreases within 3–5 d of storage at 4°C by approx 70% when the protein is derivatized with a heterobifunctional crosslinker containing an aromatic ring structure *(20)*.

After reduction of the pyridyldithio groups with DTT to introduce free sulfhydryl groups the proteins need to be immediately subjected to coupling as oxidation of the sulfhydryl groups will occur, leading to homopolymer formation. To prevent oxidation of the free sulfhydryl groups, the gel filtration columns used to separate the sulfhydryl group-containing proteins from the DTT should be equilibrated in deaerated PBS, and the coupling reaction with the other protein should be performed in vials flushed with nitrogen.

Antibody conjugates prepared with SPDP (which results in the formation of a disulfide bond) were stable for at least 1.5 yr at 4–6°C *(20)*.

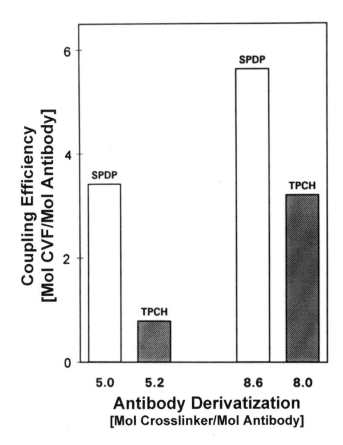

Fig. 5. Effect of the crosslinker attachment site on the coupling efficiency of CVF. The conjugation is performed with a 1.5-fold molar excess of sulfhydryl-derivatized CVF (2.8 mol of sulfhydryl groups per mol of CVF) over the number of IgM antibody-attached crosslinker molecules. Modified from **ref. *10***.

7. The coupling of effector molecules to the 16–88 IgM antibody derivatized with TPCH in its carbohydrate moieties is somewhat less efficient than the coupling of effector molecules to the antibody derivatized with SPDP at amino groups. Using the relatively large CVF molecule for coupling to the 16–88 IgM antibody derivatized with SPDP or TPCH, the number of CVF molecules bound to the TPCH-derivatized antibody is significantly less (**Fig. 5**). Not surprisingly, the lower coupling efficiency is worse at a lower degree of TPCH derivatization of the antibody, and improves with a higher degree of TPCH derivatization (**Fig. 5**). However, the coupling efficiency is always lower when compared with SPDP-derivatized antibody.

Significant differences in the coupling efficiency are observed for ricin A-chain and barley toxin (M_r approx 30,000 each), two significantly smaller proteins than

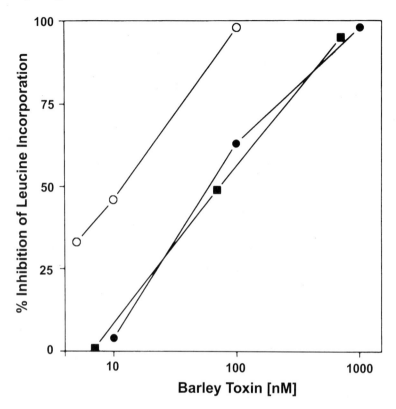

Fig. 6. Inhibition of protein synthesis by unmodified (open circles), SPDP-derivatized (closed circles), or IgM antibody-conjugated barley toxin (closed squares). Barley toxin is derivatized with 4.8 free sulfhydryl groups per toxin molecule and conjugated to IgM antibody 16-88 derivatized with 7.2 pyridyldithio groups per antibody molecule. Modified from **ref. 9**.

CVF. Using a fourfold molar excess of ricin A-chain or barley toxin over TPCH-derivatized 16–88 IgM antibody, the coupling efficiency of barley toxin (3.5 molecules per antibody molecule) is more than an order of magnitude better than the coupling efficiency of ricin A-chain (0.2 molecules per antibody molecule) *(9)*. The significantly lower degree of conjugation of the ricin A-chain is presumably a consequence of both, the accessibility of the natural free sulfhydryl group of ricin A-chain compared to the accessibility of the introduced free sulfhydryl groups of SPDP-derivatized barley toxin, and the fact that ricin A-chain has only one free sulfhydryl group whereas SPDP-derivatized barley toxin has several free sulfhydryl groups per toxin molecule.

Collectively, these results indicate that the coupling efficiency of effector molecules to carbohydrate-derivatized antibodies is lower because of the decreased accessibility of the carbohydrate moieties. Additional factors influencing the coupling efficiency are the size of the effector protein, the accessibility of its free sulfhydryl groups, and the total number of free sulfhydryl in the effector protein.

8. In addition to the lower coupling efficiency, CVF activity is somewhat compromised when coupled to TPCH-derivatized IgM antibody 16–88 compared with SPDP-derivatized antibody. To exert its hemolytic activity, CVF needs to interact with three proteins of the complement system (factor B, factor D, and complement component C5) *(19)*. Therefore, the reduced activity of CVF when coupled to the carbohydrate moieties of the antibody appears to be a steric constraint on the accessibility of the complement proteins to interact with CVF. This contention is supported by the observation that the activity of barley toxin coupled to TPCH-derivatized IgM antibody 16–88 is virtually indistinguishable from free (but SPDP-derivatized) barley toxin (**Fig. 6**). Please note that the derivatization of barley toxin with SPDP to introduce free sulfhydryl groups results in partial chemical inactivation of the barley toxin activity (**Fig. 6**; *see* also **Note 1**). However, no additional decrease in barley toxin activity is observed upon its coupling to TPCH-derivatized antibody (**Fig. 6**).

9. The Lowry method *(22)* is a reliable method to determine the protein concentration of crosslinker-derivatized proteins *(20)*. Absorbance at 280 nm is an equally reliable method unless the crosslinker contains a phenol group, leading to false high readings of the protein concentration *(20)*. The Bradford method *(25)* was found to be the least consistent method for protein determination of crosslinker-derivatized proteins *(20)*.

Acknowledgments

I would like to acknowledge the contributions of my coworkers and collaborators who were involved in this work, as is evident from the authorship of cited publications. Work performed in my laboratory at Georgetown University in Washington, D.C., was supported by NIH grants CA35525, CA45800, and CA01039.

References

1. Vogel, C.-W. (ed.) (1987) *Immunoconjugates. Antibody Conjugates in Radioimaging and Therapy of Cancer.* Oxford University Press, New York, NY.
2. Rodwell, J. D. (ed.) (1988) *Antibody-Mediated Delivery Systems.* Marcel Dekker, New York, NY.
3. Vogel, C.-W. and Bredehorst, R. (1997) Immunoconjugates, in *Encyclopedia of Human Biology, Vol. 4, 2nd Ed.* (Dulbecco, R. L., ed.), Academic Press, San Diego, CA, pp. 112.1–112.15.
4. Kreitman, R. J. (1999) Immunotoxins in cancer therapy. *Curr. Opin. Immunol.* **11,** 570–578.
5. Trail, P. A., King, H. D., and Dubowchik, G. M. (2003) Monoclonal antibody drug immunoconjugates for targeted treatment of cancer. *Cancer Immunol. Immunother.* **52,** 328–337.

6. Wong, S. S. (ed.) (1991) *Chemistry of Protein Conjugation and Cross-Linking.* CRC Press, Boca Raton, FL.

7. Hermanson, G. T. (ed.) (1996) *Bioconjugate Techniques.* Academic Press, San Diego, CA.

8. Carlsson, J., Drevin, H., and Axen, R. (1978) Protein thiolation and reversible protein-protein conjugation. *Biochem. J.* **173**, 723–737.

9. Zara, J., Wood, R., Boon, P., Kim, C.-H., Pomato, N., Bredehorst, R., et al. (1991) A carbohydrate-directed heterobifunctional cross-linking reagent for the synthesis of immunoconjugates. *Anal. Biochem.* **194**, 156–162.

10. Zara, J., Pomato, N., McCabe, R. P., Bredehorst, R., and Vogel, C.-W. (1995) Cobra venom factor immunoconjugates: Effects of carbohydrate-directed versus amino group-directed conjugation. *Bioconjug. Chem.* **6**, 367–372.

11. Ottenheijm, H. C. J., Liskamp, R. M. J., van Nispen, S. P. J. M., Boots, H. A., and Tijhuis, M. W. (1981) Total synthesis of the antibiotic sparsomycin, a modified uracil amino acid monoxodithioacetal. *J. Organic Chem.* **46**, 3273–3283.

12. Peters, L. C., Brandhorst, J. S., and Hanna, M. G., Jr. (1979) Preparation of immunotherapeutic autologous tumor cell vaccines from solid tumors. *Cancer Res.* **39**, 1353–1360.

13. Haspel, M. V., McCabe, R. P., Pomato, N., Janesch, J. J., Knowlton, J. V., Peters, L. C., et al. (1985) Generation of tumor cell-reactive human monoclonal antibodies using peripheral blood lymphocytes from actively immunized colorectal cancer patients. *Cancer Res.* **45**, 3951–3960.

14. McCabe, R. P., Peters, L. C., Haspel, M. V., Pomato, N., Carrasquillo, J. A., and Hanna, M. G., Jr. (1988) Preclinical studies on the pharmacokinetic properties of human monoclonal antibodies to colorectal cancer and their use for detection of tumors. *Cancer Res.* **48**, 4348–4353.

15. Vogel, C.-W. and Müller-Eberhard, H. J. (1984) Cobra venom factor: Improved method for purification and biochemical characterization. *J. Immunol. Methods* **73**, 203–220.

16. Roberts, W. K. and Selitrennikoff, C. P. (1986) Isolation and partial characterization of two antifungal proteins from barley. *Biochim. Biophys. Acta* **880**, 161–170.

17. Zara, J. J., Wood, R. D., Bredehorst, R., and Vogel, C.-W. (1992) S-(2-Thiopyridyl)-L-cysteine, a heterobifunctional crosslinking reagent. US Patent no. 5,157,123.

18. Vogel, C.-W. and Müller-Eberhard, H. J. (1981) Induction of immune cytolysis: Tumor-cell killing by complement is initiated by covalent complex of monoclonal antibody and stable C3/C5 convertase. *Proc. Natl. Acad. Sci. USA* **78**, 7707–7711.

19. Petrella, E. C., Wilkie, S. D., Smith, C. A., Morgan, A. C., Jr., and Vogel, C.-W. (1987) Antibody conjugates with cobra venom factor. Synthesis and biochemical characterization. *J. Immunol. Methods* **104**, 159–172.

20. Vogel, C.-W. (1988) Synthesis of antibody conjugates with cobra venom factor using heterobifunctional crosslinking reagents, in *Antibody-Mediated Delivery Systems* (Rodwell, J. D., ed.), Marcel Dekker, New York, NY, pp. 191–224.

21. Lee, D. S. C. and Griffiths, B. W. (1984) Comparative studies on Iodo-bead and Chloramine-T methods for radioiodination of human α-fetoprotein. *J. Immuol. Methods* **74,** 181–189.
22. Lowry, O. H., Rosebrough, N. J., Farr, A. L., and Randall, R. J. (1951) Protein measurement with the folin phenol reagent. *J. Biol. Chem.* **193,** 265–275.
23. Vogel, C.-W. (1987) Antibody conjugates without inherent toxicity: The targeting of cobra venom factor and other biological response modifiers, in *Immunoconjugates. Antibody Conjugates in Radioimaging and Therapy of Cancer* (Vogel, C.-W., ed.), Oxford University Press, New York, NY, pp. 170–188.
24. Vogel, C.-W. (1988) Antibody conjugates with cobra venom factor as selective agents for tumor cell killing, in *Cytolytic Lymphocytes and Complement: Effectors of the Immune System, Vol. 2* (Podack, E. R., ed.), CRC Press, Boca Raton, FL, pp. 135–151.
25. Bradford, M. M. (1976) A rapid and sensitive method for the quantitation of microgram quantities of protein utilizing the principle of protein-dye binding. *Anal. Miochem.* **72,** 248–254.

7

Synthesis of Hapten–Protein Conjugates Using Microbial Transglutaminase

Markus Meusel

Summary

Hapten–protein conjugates are essential in many immunochemical assays and in particular in assays using titration or competitive assay formats. By exploitation of the catalytic properties of the microbial transglutaminase from *Streptoverticillium mobarense* species (MTGase), that is, acyl transfer between γ-carboxamide groups and various primary amines, new techniques for the enzymatic modification of proteins were developed. One example of bioconjugation is the biotinylation of antibodies for immunochemical applications using two species of activated biotin. In this case, the activated biotin acts as the acyl acceptor and is coupled to the glutamine residues of a monoclonal antibody. Because of the substrate specificity of the MTGase with regard to the limited number of glutamine residues and the surrounding microenvironment, only a limited number of binding sites on the target protein are available; the proposed method is thus particularly suitable when only a few biotin molecules need to be attached. Another example for the modification of proteins is the synthesis of hapten–protein conjugates used in competitive-type immunoassays. Methods for the synthesis of 2,4-D-casein conjugates (2,4-dichlorophenoxyacetic acid, a herbicide) are presented. Various approaches, including a batch procedure and two *in situ* procedures, are described.

Key Words: Microbial transglutaminase; MTGase; bioconjugation; biotinylation; hapten–protein conjugates; coupling; immunoassay.

1. Introduction

Microbial transglutaminase (protein glutamine γ-glutamyltransferase, EC 2.3.2.13, MTGase) catalyses the acyl transfer reaction between the γ-carboxyamide group of glutaminyl residues and various primary amines *(1)*. When the ε-amino group of a lysine residue acts as the acyl acceptor, ε-(γ-Glu)Lys isopeptide bonds are formed. The reaction follows the scheme shown in **Fig. 1**.

From: *Methods in Molecular Biology, vol. 283: Bioconjugation Protocols: Strategies and Methods*
Edited by: C. M. Niemeyer © Humana Press Inc., Totowa, NJ

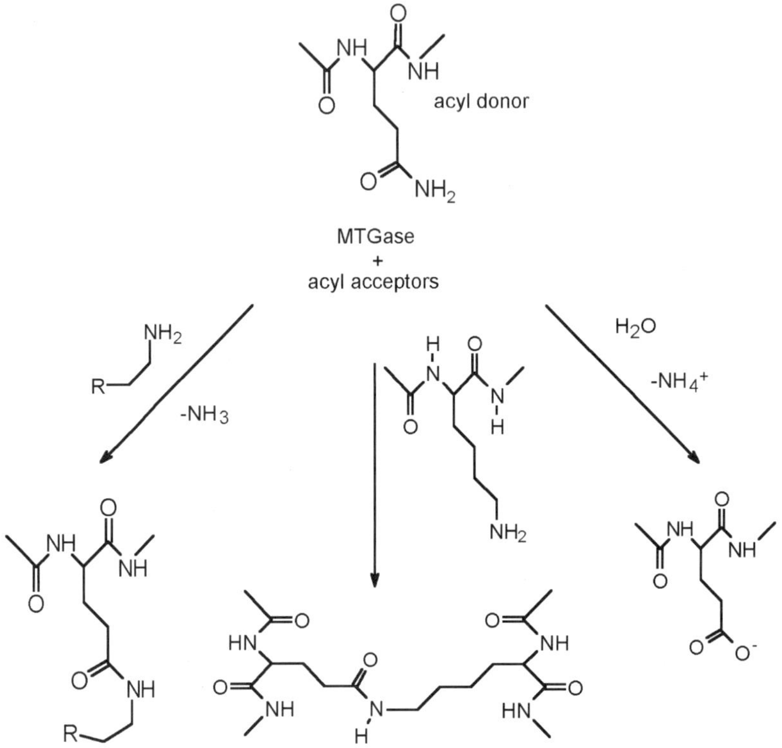

Fig. 1. Scheme of MTGase reaction.

Up to now, MTGase has been used mainly for food biotechnological applications, for example, for the alteration of the foaming and emulsifying properties of food protein (2) or to improve the nutritive value of food by crosslinking soy proteins or casein with lysine dipeptides (3) or other amino acids (4). Another important application is the gelification of soluble edible proteins, for example, casein (5,6).

Mammalian transglutaminases are found in many tissues and play an important role in the post translational modification of proteins or, as factor XIII, in blood coagulation (7). Until the end of the 1980s a mammalian transglutaminase from guinea pig (GTGase) was used for food and biotechnological applications. In 1989, however, a microbial transglutaminase from *Streptoverticillium* (MTGase) was isolated (8) and because of the low-cost availability of the enzyme in bulk, industrial applications of MTGase became feasible (9).

Compared with classical chemical methods, MTGase catalyzed crosslinking requires physiological conditions, which allow for reactions under mild conditions in buffered solutions. This makes MTGase a powerful tool in the field of analytical biotechnology, in particular, for the conjugation and modification of proteins.

In this chapter, the use of MTGase for bioconjugation will be described (*see* **Note 1**). One example will be the biotinylation of antibodies for immunochemical applications using two species of activated biotin. In this case, the activated biotin acts as the acyl acceptor and is coupled to the glutamine residues of the monoclonal antibody. Because only a limited number of binding sites on the target protein are available, the proposed method is particularly suitable when only a few biotin molecules need to be attached. Another example for use in bioconjugation will be the synthesis of hapten–protein conjugates used in competitive-type immunoassays. The herbicide 2,4-D-dichlorophenoxyacetic acid (2,4-D) is used as a model analyte to demonstrate the feasibility of MTGase for conjugate synthesis. Because 2,4-D is not a substrate for the MTGase reaction, it is aminofunctionalized before coupling to casein. Various approaches (batch procedures as well as *in situ* synthesis) will be described.

2. Materials

2.1. Chemicals

1. Microbial transglutaminase from *S. mobaraense* sp., MTGase, 1000 U/g (Ajinomoto Europe Sales GmbH, Hamburg, Germany).
2. Monoclonal anti-2,4-D-IgG (clone R2b E2/E5; Immunotech, Moscow, Russia).
3. Goat antimouse IgG-POD (Dianova, Hamburg, Germany).
4. Goat antimouse IgG (Sigma, St. Louis, MO).
5. Streptavidin–POD conjugate (Sigma).
6. Neutravidin (Pierce, Rockford, IL).
7. Biotinamido-5-pentylamine (BIAPA; Pierce).
8. *N*-ethylmaleimide (NEM; Sigma).
9. Bovine serum albumin (BSA; Sigma).
10. 2,4-Dichlorophenoxyacetic acid (Promochem, Wesel, Germany).
11. Biotin polyethylene oxide amine (biotin PEO–amine; Pierce).
12. Casein as sodium salt from bovine milk (Sigma).
13. Slide-A-Lyzer 0.1–0.5 mL or 0.5–3.0 mL, exclusion limit 3500 Daltons (Pierce).

Other chemicals were obtained as analytical reagent-grade products from Sigma, Aldrich (Milwaukee, WI), and Fluka (Buchs, Switzerland).

2.2. Buffers

Buffers and aqueous solutions should be prepared freshly using double deionized, bacterial filtered water. All ELISA-procedures were conducted on Nunc Maxi Sorp™-Microtiter plates (Nunc, Roskilde, Denmark).

1. Phosphate-buffered saline (PBS): 0.145 M sodium chloride, 8 mM sodium hydrogen phosphate, 2 mM potassium dihydrogen phosphate, pH 7.2.
2. Coating buffer: 50 mM sodium carbonate buffer, pH 9.6.
3. Blocking solution: 0.5% BSA in PBS.
4. Washing buffer solution, PBST: prepared from PBS by adding 0.05 % (v/v) polyoxyethylene sorbitan monolaurate (Tween-20).
5. Substrate solution for horseradish peroxidase, POD: must be prepared freshly by adding 200 μL of TMB stock solution (6 mg 3,3',5,5'-tetramethylbenzidine in 1 mL of dimethylsulfoxide) and 5 μL of hydrogen peroxide (30%) to 10 mL of 0.1 M acetate buffer, pH 5.5.
6. MTGase inhibitor solution, PBSN: prepared from PBS by adding 0.1% NEM.

3. Methods
3.1. Biotinylation of Antibodies

The methods described below in **Subheadings 3.1.1–3.1.4.** outline (1) the biotinylation of the antibody, (2) the enzyme-linked immunosorbent assay (ELISA) protocol for the test of the conjugates, (3) the investigation of the kinetics of the MTGase-catalyzed biotinylation, and (4) the corresponding ELISA procedure. Because only a limited number of binding sites on the antibody are available, the proposed method is particularly suitable when only a few biotin molecules need to be attached (*see* **Note 2**). For alternative biotinylation methods, refer to **Note 3**.

3.1.1. Biotinylation of Anti-2,4-D IgG

For biotinylation, two biotin derivatives can be used.

1. Add 78.5 μL of biotin–polyethylene oxide (biotin PEO–amine) or BIAPA (7.85 × 10^{-7} mol) and 20 μL of MTGase (0.1 U) in PBS to 1.57 × 10^{-8} mol anti-2,4-D IgG in 1000 μL of PBS (this corresponds to 0.3 mg/mL biotin PEO–amine or 0.26 mg/mL of BIAPA, 2.35 mg/mL of anti-2,4-D IgG, 0.2 mg/mL of MTGase; 50-fold molar excess of the amino-biotin). For optimization of the initial coupling step, *see* **Note 4**.
2. In parallel, control experiments in the absence of MTGase have to be performed (1.57 × 10^{-7} mol biotin PEO–amine or BIAPA, respectively, and 0.314 × 10^{-8} mol anti-2,4-D IgG in 200 μL of PBS).
3. Incubate overnight with gentle shaking at room temperature.
4. Dialyze the batches three times for 4 h in 2 L of deionized water (Slide-A-Lyzer).
5. Finally, each conjugate is diluted with PBS to a final volume of 2 mL (controls 300 μL).

To test the synthesized anti-2,4-D-IgG-biotin conjugates, different ELISA protocols must be applied. All washing steps are conducted with PBST, and the incubation steps are performed with gentle shaking at room temperature if not stated otherwise.

3.1.2. ELISA to Control Biotinylation

The success of the biotinylation is controlled in an ELISA format. The signals obtained from the biotinylated antibody should be significantly higher than those of the control values. As unspecific binding may also contribute to the measured signal it is recommended to perform a series of additional control experiments.

1. Coat the microtiter plate with neutravidin (10 μg/mL) in carbonate buffer (100 μL/well, 4°C overnight).
2. Wash the plate three times.
3. Block the plate for 1 h with 200 μL of blocking solution and wash three times afterwards.
4. Add the biotinylated anti-2,4-D antibody at a dilution of 1/10 in PBS (100 μL/ well) and incubate for 2 h at room temperature.
5. Wash five times and add the POD-labeled secondary antibody (goat antimouse IgG-POD) diluted in PBST (1/5000, 30 min at a volume of 100 μL per well). Wash the plate again five times.
6. For photometric determination add 100 μL of freshly prepared POD substrate solution to each well. Stop the substrate reaction after sufficient color development with 50 μL of 2 *M* sulfuric acid per well and measure the absorption values at 450 nm.

Recommended controls are as follows: (1) biotinylation batches without MTGase (incubation of the biotin derivative and the antibody only); and (2) the addition of goat-antimouse antibody to microtiter plate wells with unconjugated anti-2,4-D antibody and to a well blocked with BSA (showing unspecific binding of the monoclonal antibody and the POD conjugate).

To determine the conjugation ratio of biotin and monoclonal antibody, the use of matrix-assisted laser desorption ionization (MALDI)-mass spectrometry is recommended. The conjugation ratio can be calculated from the difference of the mass of the unconjugated antibody and the biotinylated species. Mass analysis was performed using an axial linear time of flight mass analyser. The system used uses a pulsed (3 ns) nitrogen laser emitting at 337 nm. An acceleration voltage of 20 kV was applied to the ion source. Ions were postaccelerated in front of the detector to a total ion energy of 38 keV. MALDI samples were prepared by mixing 2,5-dihydroxy benzoic acid and 2-hydroxy-5-methoxybenzoic acid 9:1 (v/v, each 20 g/L in 30% acetonitrile/water). Two microliters of this solution was mixed with 0.5 μL of the sample on the MALDI target and allowed to dry.

As an example, MALDI spectra of the unconjugated antibody and biotinylated species are shown in **Fig. 2**. From the differences in molecular weight of the antibody (the M^+ peak at 150,000 Daltons) and the conjugates, conjugation ratios of 1.87 ± 0.08 and 1.14 ± 0.11 were calculated for BIAPA (MW 328 Daltons) and biotin PEO–amine (MW 372 Daltons), respectively.

3.1.3. Kinetics of the Biotinylation Reaction

Because every single monoclonal antibody shows a different susceptibility to MTGase-catalyzed biotinylation, it is important to study the kinetics of the biotinylation reaction. To stop the biotinylation, NEM, a specific inhibitor of MTGase, is used.

1. Dilute a mixture of 10 µL of anti-2,4-D IgG (4×10^{-10} mol, in PBS), 0.4 µL of amino-functionalized biotin (biotin PEO–amine or BIAPA, 4×10^{-8} mol) and 4 µL of MTGase (2×10^{-3} U) with PBS to a final volume of 30 µL and allow it to react for time periods ranging from 0 to 34 h (2 mg/mL of anti-2,4-D IgG, 0.5 mg/mL of biotin PEO–amine, or 0.44 mg/mL of BIAPA, 0.13 mg/mL MTGase).
2. Take aliquots of 0.5 µL from the reaction mixture and dilute them immediately in 50 µL of MTGase inhibitor solution (PBSN).
3. Test the synthesized conjugates in ELISA as described in **Subheading 3.1.4.** Using this assay format, the removal of unconjugated biotin by dialysis is not required.

3.1.4. Investigation of Biotinylation Reaction Kinetics Using ELISA

1. Coat the microtiter plate with goat-anti-mouse IgG (2 µg/mL) in carbonate buffer, pH 9.6 (100 µL/well, 4°C overnight).
2. Wash the plate three times with PBS.
3. Block the plate for 1 h with 200 µL of blocking solution and wash again three times.
4. Add the biotinylated anti-2,4-D antibody diluted 1/1000 in PBSN (100 µL/well) and incubate for 2 h at room temperature.
5. Wash five times and add the POD-labeled streptavidin in PBST (1/5000, 1 h at a volume of 100 µL per well). Wash again five times.
6. For the photometric determination, add 100 µL of freshly prepared POD substrate solution to each well. After a sufficient color development, stop the enzymatic conversion of the substrate with 50 µL of 2 *M* sulfuric acid per well and measure the absorption at 450 nm.

To ascertain the specific nature of the interaction between (neutr)avidin and the biotinylated antibody, a competitive ELISA is described in **Note 5**. In general, the MTGase-catalyzed biotinylation proceeds slowly and is approx 80% complete after 2 d. For further information, *see* **Note 6**.

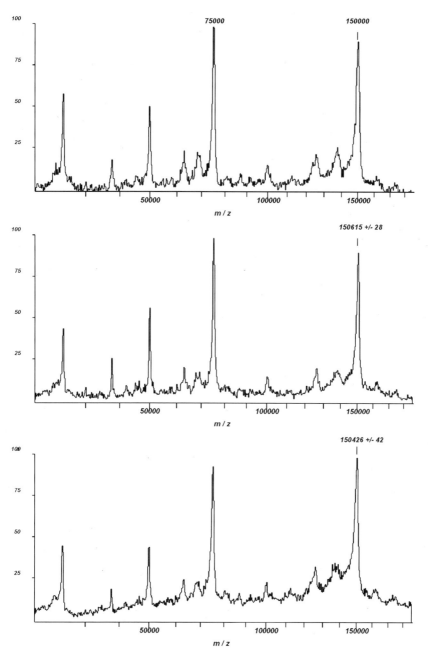

Fig. 2. Spectra of MALDI mass spectrometry. Top, spectrum of nonconjgated anti-2,4-D monoclonal antibody. M⁺ peak at 150 kDa. Middle, BIAPA modified antibody (MW of 150.615 Daltons). Bottom, biotin–PEO–amine-modified antibody (MW of 150,426 Daltons).

3.2. Synthesis of Hapten–Protein Conjugates for ELISA

The methods in **Subheading 3.2.1.** describe the aminofunctionalization of the herbicide 2,4-D, in **Subheading 3.2.2.** the synthesis of 2,4-D-casein conjugates with batch and *in situ* procedures, and in **Subheading 3.2.3.** a general ELISA protocol for testing the conjugates. It must be emphasized that the use of MTGase-generated conjugates is not limited to ELISA in microtiter plate formats (*see* **Note 7**).

3.2.1. Aminofunctionalization of 2,4-D

The herbicide 2,4-D serves as the model analyte to demonstrate the MTGase-mediated conjugate synthesis. Because 2,4-D itself is not a substrate for the MTGase reaction, it has to be aminofunctionalized before coupling to casein (*see* **Fig. 3**). Depending on their chemical structure, other haptens, however, may act directly as acyl acceptor in the MTGase reaction.

In a first step, the methyl ester of 2,4-D is synthesized (**steps 1–3**), and then the ester is amino-modified (**steps 4–7**):

1. Reflux a mixture of 25 mmol (5.53 g) 2,4-dichlorophenoxyacetic acid, 26 mmol (2.71 g) 2,2-dimethoxypropane, and 70 mg *p*-toluenesulfonic acid in 2.5 mL of absolute methanol overnight.
2. Add 30 mL of saturated sodium carbonate solution, extract the mixture with chloroform, dry with sodium sulfate, and evaporate.
3. Purify the crude product by vacuum distillation. A yield of approx 4.5 g of a white powder (melting point$_{1 \text{ mbar}}$: 107–108°C) can be expected.
4. Stir 5 mmol (1.17 g) of the methyl ester and 5 mL of 1,2-diaminoethane in 30 mL of absolute diethyl ether for 18 h at room temperature. In this step, a high molar excess of 1,2-diaminoethane is required to reduce crosslinking of the 2,4-D methyl ester.
5. After evaporation, remove excess diaminoethane in vacuo.
6. Take up the residue in methanol and pass it through a silica gel column with methanol as the eluent. Combine and reduce fractions containing the product by rotary evaporation.
7. Precipitate with diethyl ether and dry the product. A yield of approx 85 mg 2,4-dichlorophenoxy-*N*-(2-aminoethyl) acetamide, 2,4-D-A (a white crystalline powder) can be expected.

3.2.2. Synthesis of 2,4-D–Casein Conjugates

Conjugate synthesis is flexible and allows for various procedures depending on the application of the conjugates. A batch procedure as well as two *in situ* procedures are described.

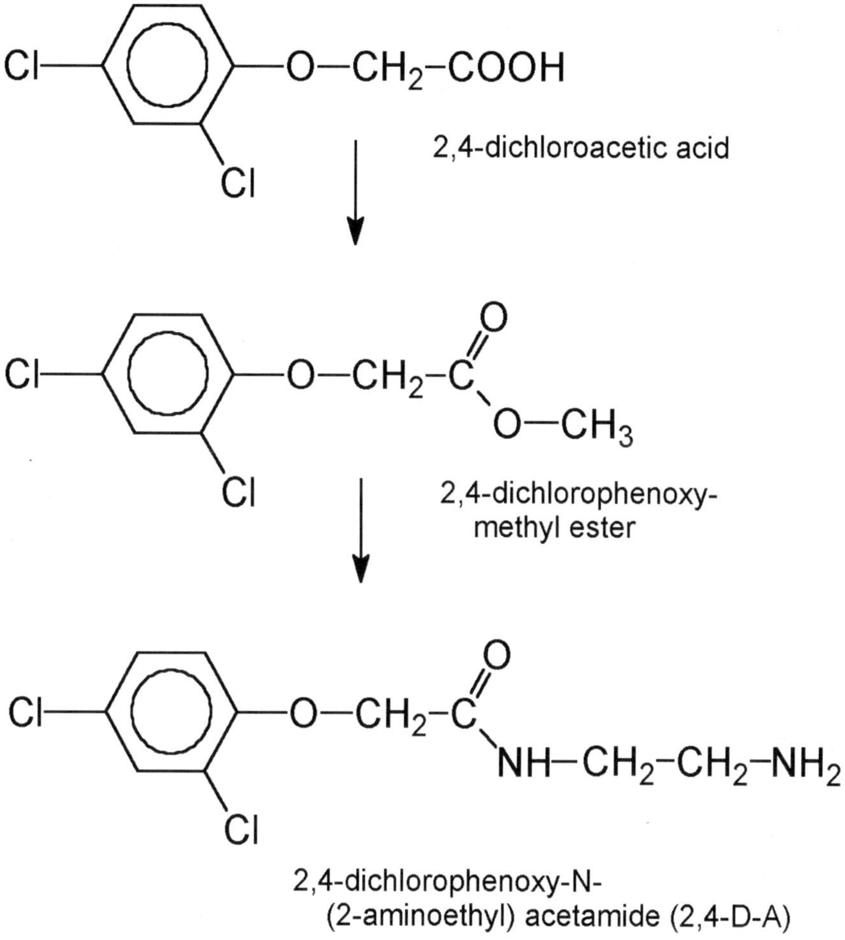

Fig. 3. Synthesis of 2,4-dichlorophenoxyacetic acid-*N*-(2-aminoethyl) acetamide.

3.2.2.1. Batch Procedure

The batch procedure follows the scheme shown in **Fig. 4**. 200 µL of a casein solution (1% bovine casein in PBS, pH 7.2, *see also* **Notes 8–10**), 20 µL of 2,4-dichlorophenoxy-*N*-(2-aminoethyl) acetamide (2,4-D-A) solution (0.1% in PBS, pH 7.2), and 20 µL of (0.1 U) MTGase solution (1% in PBS, pH 7.2) are added to 1760 µL of PBS, pH 7.2. After incubation overnight at room temperature under gentle shaking, the reaction is stopped with 100 µL of MTGase inhibitor solution (PBSN). The formation of the hapten-protein conjugates is monitored by ELISA according to the protocol described below (**Subheading 3.2.3.**). Conjugate concentrations refer to the casein concentration in the

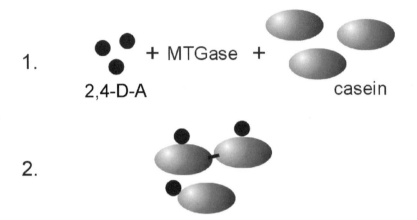

Fig. 4. Scheme of MTGase-catalysed immunoconjugate synthesis: batch procedure.

MTGase reaction. In the absence of free 2,4-D, the maximal absorption at a given 2,4-D–casein conjugate concentration should be obtained.

3.2.2.2. IN SITU PROCEDURES FOR CONJUGATE SYNTHESIS

Two alternatives are described. According to **Fig. 5**, in the first approach 2,4-D-A is coupled to a casein-layer already immobilized, whereas in the second approach immobilization and conjugation are conducted simultaneously (*see* also **Note 11**).

In the first step of the conjugate synthesis with a preimmobilized casein layer, casein is immobilized on the surface of the microtiter plate (0.1% casein in carbonate buffer, 100 μL per well, incubation for 2 h at room temperature). Subsequently, a solution of 20 μg/mL 2,4-D-A and 100 μg/mL MTGase in PBS, pH 7.4 (0.05 U/mL) is prepared and filled into the precoated wells of the microtiter plate at 100 μL per well to give a final 2,4-D-A concentration of 2 μg and a MTGase activity of 0.005 U per well. After an incubation of 2 h at room temperature, the microtiter plate is washed five times with PBST, pH 7.4, before 2,4-D calibration curves can be obtained according to the ELISA protocol described below. For the measurement of MTGase activity, refer to **Note 12**.

In the one-step synthesis, a solution of 20 μg/mL 2,4-D-A, 500 μg/mL casein, and 50 μg/mL (0.025 U/mL) MTGase in PBS, pH 7.4, is prepared and incubated in volumes of 100 μL per well for 2 h. After washing three times with PBST, pH 7.4, the microtiter plate is ready for the immunoassay. 2,4-D standard curves can be obtained following the ELISA procedure described below. If the kinetics of conjugate formation has to be investigated, refer to **Note 13**.

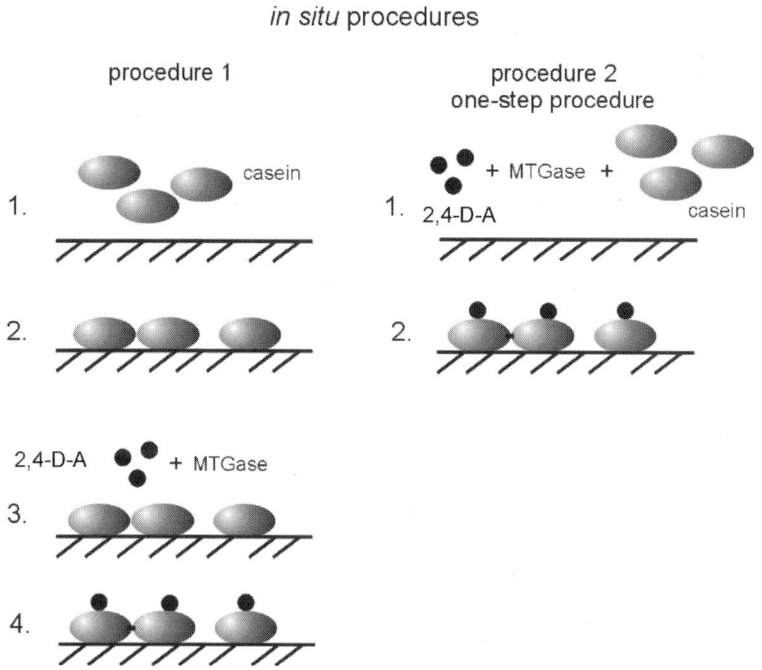

in situ procedures

Fig. 5. Scheme of MTGase-catalysed immunoconjugate synthesis: *in situ* procedures.

3.2.3. General ELISA Protocol for Testing the Hapten–Protein Conjugates

To test the hapten–protein conjugates synthesized by the MTGase-catalyzed reaction and to obtain 2,4-D standard curves, a protocol for a competitive ELISA is described. All washing steps are performed with PBST and the incubation steps take place under gentle shaking at room temperature.

1. Coat the microtiter plate with the hapten–protein conjugate diluted in carbonate buffer, pH 9.6 (0.01 to 100 μg/mL, 100 μL per well, 2 h).
2. Wash the microtiter plate three times.
3. In the batch procedure: block for 1 h with 200 μL of blocking solution. In the *in situ* procedures, blocking is optional.
4. Add preincubated analyte and primary monoclonal antibody (preincubation in vials for 10 min) and incubate for 10 min in the microtiter plate (100 μL per well).
5. Wash four times, add the POD-labeled secondary antibody (goat anti-mouse IgG-POD, 1:5000, diluted in PBST) and incubate for 10 min at a volume of 100 μL per well. Wash the plate again four times.
6. For photometric determination fill 100 μL of freshly prepared POD substrate solution into each well. After a sufficient color development, stop the enzymatic

conversion of the substrate with 50 µL of 2 *M* sulfuric acid per well. Measure the absorption at 450 nm against the reference wavelength at 630 nm.

4. Notes

1. It must be emphasized that the use of MTGase is not limited to the synthesis of hapten–protein conjugates. The crosslinking properties can, for example, also be exploited for protein–protein conjugation. Bechtold et al. *(10)*, for example, describe the enzymatic preparation of protein G-peroxidase conjugates, whereas Josten et al. *(11)* use MTGase for enzyme immobilization and the generation of stable-sensing surfaces.

2. A survey of the Swiss prot data bank reveals a glutamine to lysine ratio of about 1:1.4 for an average mouse immunoglobulin G. It is noteworthy, however, that in total about 90 lysine residues are present, most of them accessible on the protein surface. To avoid overlabeling, it is recommended to work with a low excess of label in a time-controlled experiment. Using glutamine residues of a protein as the substrate for MTGase, however, it has to be considered that the neighboring amino acids play a crucial role in determining the affinity of the substrate *(8)*. This is an important advantage with respect to a more selective reaction. First, the number of glutamine residues is smaller compared with the number of lysines, and second, only a small number of glutamine residues with the correct microenvironment are accessible to modification by the large MTGase. As a result, a moderate labeling ratio and a high retention of biological activity by the target protein can be expected. In a study of Josten et al. *(12)*, the conjugation ratio (biotin molecules per antibody) did not exceed 2.0.

3. Alternative methods to the use of MTGase for biotinylation include biotinylating reagents such as *N*-hydroxysuccinimide (NHS)–biotin, the sulfonated derivative S-NHS–biotin, or the long chain S-NHS-LC–biotin. For some applications site-specific biotinylation is mandatory, for example, if the entire peptide chain must remain accessible for antibody or receptor binding. Selo et al. *(13)* achieved specific biotinylation of the α-amino group using NHS-LC–biotin, defined experimental conditions, and a pH of 6.5 for the preferential incorporation of biotin into the N terminus. Furthermore, a highly specific method was described by Pavlinkova et al. *(14)*, in which biotin was coupled to 8-azidopurine nucleotides or nucleosides. These biotin derivatives exhibit a high affinity binding to a hydrophobic pocket formed at the V_H-V_L interface of immunoglobulins. Subsequent photolysis forms covalent bonds with surrounding side chains.

4. Please note that because of the accessibility of the glutamine residues and the different microenvironments, every single antibody reacts differently in a MTGase-catalyzed biotinylation. It is thus recommended to alter the concentrations of the individual compounds and the incubation times during the process of assay optimisation. To increase the coupling ratio, for example, the PBS buffer used for conjugation may be replaced by Tris-HCl, pH 7.5.

5. To ascertain the specific nature of the interaction between (neutr)avidin and biotinylated anti-2,4-D antibody, a competitive ELISA can be used. In this assay,

the biotinylated antibody and free biotin in solution compete for binding to the neutravidin. Based on the assay scheme in **Subheading 3.1.4., step 5**, the POD-labeled streptavidin together with biotin at concentrations ranging from 0 to 10^{-4} mol in PBST (100 μL/well) can be added. If the biotinylation of the antibody was successful, a competition with free biotin will be observed.

6. It is helpful to follow the progress of biotinylation by ELISA. In general, biotinylation by MTGase is a slow process which may require incubation times of up to one day. As the MTGase-mediated conjugate synthesis with proteins of known and good substrate properties (casein, gluten) requires no more than 1–2 h *(4,15,16)* the slow biotinylation of antibodies may be attributed to unfavorable substrate properties. Furthermore, the nature of the biotin derivative also has a significant impact on the coupling efficiency. Josten et al. *(12)* showed that the best results were obtained with biotin PEO–amine, which was attributed to the longer and more polar spacer arm.

7. The application of MTGase-generated conjugates (**Subheading 3.2.**) is not only limited to ELISA in microtiter plates. The high stability of the conjugate layers was exploited by Jülicher et al. *(17)* applying the one-step immobilization procedure described here to screen-printed immunoelectrodes for 2,4-D determination in methanolic soil extracts. In these experiments, hapten–protein conjugates had to be immobilized on a carbon working electrode. Simple adsorption of the conjugates was not possible because the electrodes showed neither sufficient reproducibility nor stability. The problem of immobilization could be circumvented by using the transglutaminase-catalyzed *in situ* procedure. The authors attributed the high stability of the conjugate layer to the crosslinked casein, an effect also described by Motoki et al. *(18)*.

8. Although the procedures described here (**Subheading 3.2.**) are based on casein as the acyl donor in the transglutaminase reaction other proteins may also be readily used. Casein was chosen as carrier protein because it provides several glutamine residues that are known to be easily crosslinked by transglutaminase *(19)*. Besides casein, fibrinogen and gelatin can readily be crosslinked by MTGase, although they are not of general use in immunochemistry. The widely used proteins BSA and keyhole limpet hemocyanin could not be used as carriers in MTGase-mediated conjugate synthesis *(15)*.

9. If BSA has to be used as the carrier protein, the addition of dithiothreitol in the coupling reaction may be investigated as BSA is a substrate for MTGase only under reductive conditions *(6)*.

10. According to Christensen et al. *(19)*, casein provides four to five glutamine residues suitable for MTGase coupling (α_{S1}-, α_{S2}-, and κ-casein with four accessible residues, β-casein with five residues). Thus, a moderate conjugation ratio is expected, and overconjugation very unlikely.

11. In the *in situ* coupling procedure (**Subheading 3.2.2.**), additional blocking steps can be avoided as the 0.1% casein solution saturates all binding sites on the microtiter plate. However, if the background signal of the ELISA is too high, an additional blocking step with BSA is recommended.

12. The activity of MTGase can be measured by a colorimetric hydroxamate procedure according to Folk and Cole *(20)*. For a comprehensive discussion of transglutaminase activity assays, *see* Jeitner et al. *(21)*.

13. As for the biotinylation, it may be helpful to investigate the kinetics of 2,4-D-casein conjugate formation: a mixture of 200 µL of casein solution (1% in PBS, pH 7.4), 100 µL of 2,4-D-A solution (0.1% in PBS, pH 7.4) and 50 µL of (0.24 U) MTGase solution (1% in PBS, pH 7.4) in 1650 µL of PBS, pH 7.4, is prepared and allowed to react from 0 to 280 min as described before (*see* "batch procedure" in **Subheading 3.2.2.**). 5-µL aliquots from the reacted mixture are diluted immediately in 995 µL of 0.1% NEM-containing carbonate buffer, pH 9.6. The conjugates synthesised are then tested in the competitive ELISA.

References

1. Wilhelm, B., Meinhardt, A., and Seitz, J. (1996) Transglutaminases: purification and activity assays. *J. Chromatogr. B.* **684,** 163–177.

2. Kato, A., Wada, T., Kobayashi, K., Seguro, K., and Motoki, M. (1991) Ovomucin-food protein conjugates prepared through the transglutaminase reaction. *Agric. Biol. Chem.* **55,** 1027–1031.

3. Ikura, K., Okumura, K., Yoshikawa, M., Sasaki, R., and Chiba, H. (1985) Incorporation of lysyldipeptides into food proteins by transglutaminase. *Agric. Biol. Chem.* **49,** 1877–1878.

4. Ikura, K., Yoshikawa, M., Sasaki, R., and Chiba, H. (1981) Incorporation of amino acids into food proteins by transgluatminase. *Agric. Biol. Chem.* **45,** 2587–2592.

5. Ikura, K., Kometani, T., Yoshikawa, M., Sasaki, R., and Chiba, H. (1980) Crosslinking of casein components by transglutaminase. *Agric. Biol. Chem.* **44,** 1567–1573.

6. Nonaka, M., Tanaka, H., Okiyama, A., Motoki, M., Ando, H., Umeda, K., and Matsuura, A. (1989) Polymerisation of several proteins by Ca^{2+}-independent transglutaminase derived from microorganisms. *Agric. Biol. Chem.* **53,** 2619–2623.

7. Greenberg, C. S., Birckbichler, P. J., and Rice, R. H. (1991) Transglutaminases: multifunctional cross-linking enzymes that stabilise tissues. *FASEB J.* **5,** 3071–3077.

8. Ando, H., Adachi, M., Umeda, K., Matsuura, A., Nonaka, M., Uchio, R., et al. (1989) Purification and characterisation of a novel transglutaminase derived from micro-organisms. *Agric. Biol. Chem.* **53,** 2613–2617.

9. Dickinson E. (1997). Enzymic cross-linking as a tool for food colloid rheology control and interfacial stabilisation. *Trends Food Sci. Technol.* **8,** 334–339.

10. Bechtold, U., Otterbach, J. T., Pasternack, R., and Fuchsbauer, H. L. (2000) Enzymic preparation of protein G-peroxidase conjugates catalysed by transglutaminase. *J. Biochem. (Tokyo)* **127,** 239–245.

11. Josten, A., Meusel, M., Spener, F., and Haalck, L. (1999) Enzyme immobilisation via microbial transglutaminase: a method for the generation of stable sensing surfaces. *J. Mol. Catal. B* **7,** 57–66.

12. Josten, A., Haalck, L., Spener, F., and Meusel, M. (2000) Use of microbial transglutaminase for the enzymatic biotinylation of antibodies. *J. Immunol. Meth.* **240,** 47–54.

13. Sélo, I., Négroni, L., Créminon, C. Grassi, J., and Wal, J. M. (1996) Preferential labelling of N-terminal groups in peptides by biotin: application to the detection of specific anti-peptide antibodies by enzyme immunoassays. *J. Immunol. Meth.* **199,** 127–138.

14. Pavlinkova, G., Rajagopalan, K., Muller, S., Chavan, A., Sievert, G., Lou, D., et al. (1997) Site-specific photobiotinylation of immunoglobulins, fragments and light chain dimers. *J. Immunol. Meth.* **201,** 77–88.

15. Josten, A., Meusel, M., and Spener, F. (1998) Microbial transglutaminase-mediated synthesis of hapten-protein conjugates for immunoassays. *Anal. Biochem.* **258,** 202–208.

16. Yoshikawa, M, Goto, M, Ikura, K., Sasaki, R., and Chiba, H. (1982) Transglutaminase-catalysed formation of coenzymatically active NAD^+ analog casein conjugates. *Agric. Biol. Chem.* **46,** 207–213.

17. Jülicher, P., Haalck, L., Meusel, M., Cammann, K., and Spener, F. (1998) *In situ* antigen immobilisation for stable organic-phase immunoelectrodes. *Anal. Chem.* **70,** 3362–3367.

18. Motoki, M., Aso, H., Seguro, K., and Nio, N. (1987) α_{S1}-Casein film preparation using transglutaminase. *Agric. Biol. Chem.* **51,** 993–996.

19. Christensen, B. M., Sorensen, E., Hojrup, P., Petersen, T. E., and Rasmussen, L. K. (1996) Localisation of potential transglutaminase cross-linking sites in bovine caseins. *J. Agric. Food Chem.* **44,** 1943–1947.

20. Folk, J. E. and Cole, P. W. (1965) Structural requirements of specific substrates for guinea pig liver transglutaminase. *J. Biol. Chem.* **240,** 2951–2960.

21. Jeitner, T. M., Fuchsbauer, H. L., Blass, J. P., and Cooper, A. J. (2001) A sensitive fluorometric assay for tissue transglutaminase. *Anal. Biochem.* **292,** 198–206.

II

NUCLEIC ACID CONJUGATES

8

Fluorescent Sample Labeling for DNA Microarray Analyses

Verena Beier, Andrea Bauer, Michael Baum, and Jörg D. Hoheisel

Summary

Three fluorophor-labeling methods for gene expression profiling on deoxyribonucleic acid (DNA) microarrays are described. All three techniques start from total ribonucleic acid (RNA) samples. Two procedures are based on first-strand complementary DNA synthesis by reverse transcription. Label is introduced either by direct incorporation of fluorescently labeled nucleotides or indirectly by incorporation of aminoally-dUTP and subsequent coupling of fluorescent dyes. The third method is based on an amplification of antisense RNA by in vitro transcription subsequent to first- and second-strand complementary DNA synthesis. While the first two methods are applied mainly in analyses on microarrays made from spotted polymerase chain reaction products or long oligonucleotides, the last procedure is mostly used for experiments on *in situ* synthesized oligonucleotide arrays.

Key Words: DNA microarray; oligonucleotide array; hybridization; fluorescent labeling.

1. Introduction

Deoxyribonucleic acid (DNA) microarrays of all formats, whether made by spotting prefabricated molecules or by *in situ* synthesis, have become a popular and versatile technique to analyse gene expression profiles on a genome-wide basis *(1,2)*. All ribonucleic acids (total RNA) are extracted from the samples, labeled appropriately, and analyzed on arrays of gene-specific DNA fragments. A very common analysis mode is the labelling of two related samples with a pair of fluorescent dyes, Cy3 and Cy5 for example, followed by their simultaneous hybridization to a single microarray *(3)*. Upon binding, the relative fluorescence intensities produced at the microarray spots are determined, instantly indicating differences in the amount of each transcript (**Fig. 1**).

Although quite a number of labeling protocols exist, there are three procedures that are currently used most frequently. All three rely on oligo(dT)-priming so that there is no need to isolate mRNA before labeling. Direct labeling

From: *Methods in Molecular Biology, vol. 283: Bioconjugation Protocols: Strategies and Methods*
Edited by: C. M. Niemeyer © Humana Press Inc., Totowa, NJ

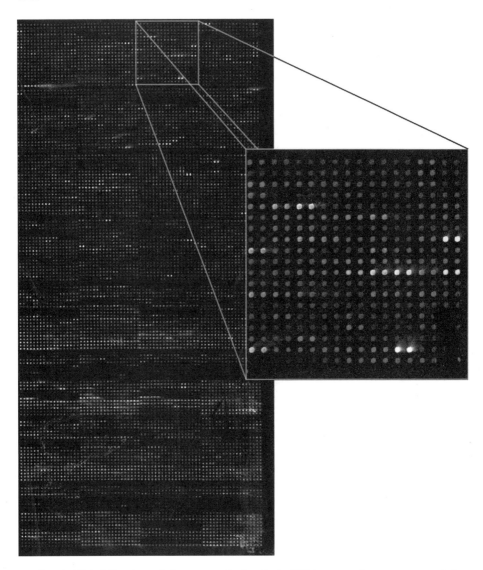

Fig. 1. Hybridization of fluorescently labeled cDNA to a microarray containing 4700 gene-specific PCR products of *Neurospora crassa* (*4*). Using the aminoallyl-labeling protocol, 15 µg of total RNA isolated from *N. crassa*-mycelium grown on minimal medium and 15 µg of total RNA isolated from mycelium grown on complete medium were labeled with Cy3 and Cy5, respectively, and hybridized simultaneously.

during reverse transcription of the RNA is performed by incorporation of nucle-otides, to which a fluorophor is bound. Alternatively, the dye is added in a two-step process: initially, aminoallyl-dUTP is incorporated, and the fluorophor is only attached upon completion of complementary DNA (cDNA) synthesis by a chemical coupling of reactive NHS esters of the respective dye. With the latter method, usually better labeling efficiency and consistency is achieved because the nucleotides with the relatively small aminoallyl-group have a much better incorporation rate compared to those, which have the rather bulky fluorescence molecules attached.

A third method is based on the production of antisense RNA (aRNA) by in vitro transcription amplification *(5)*. First-strand cDNA synthesis is performed using an oligo(dT)-primer that is linked to the promoter sequence of T7 RNA polymerase. After second-strand cDNA synthesis, aRNA is synthesized in rather large quantities via in vitro transcription. The labeling of the aRNA can be performed during the in vitro transcription by incorporation of fluorescence-labeled or biotinylated nucleotides. Alternatively, unlabeled aRNA can be produced and used as template in yet another round of reverse transcription as described previously. When biotinylated nucleotides are used, a reaction with streptavidine–phycoerythrine conjugate is necessary for staining the target after hybridization.

2. Materials

2.1. Reverse Transcription With Direct Incorporation of Fluorescently Labeled Nucleotides

2.1.1. Reagents

1. Oligo(dT)$_{12-18}$ primer (0.5 µg/µL), Invitrogen (Karlsruhe, Germany).
2. SuperScript™II RNase H- reverse transcriptase (200 U/µL), 5X first-strand buffer, 0.1 M DTT.
3. Ribonuclease H (2 U/µL), Invitrogen.
4. RNaseOUT™ (40 U/µL), Invitrogen.
5. Deoxyribonucleotide-5'-triphosphates dATP, dCTP, dGTP, and dTTP (100 mM), MBI Fermentas (St. Leon-Roth, Germany).
6. Cy3-dCTP, Cy5-dCTP (1 mM), Amersham Biosciences (Freiburg, Germany).
7. QIAquick® PCR purification kit, Qiagen (Hilden, Germany).

2.2. Aminoallyl Labeling

2.2.1. Reagents

1. Oligo(dT)$_{12-18}$ primer (0.5 µg/µL), Invitrogen.
2. SuperScript™II RNase H-reverse transcriptase (200 U/µL), 5X first-strand buffer, 0.1 M DTT, Invitrogen.
3. Deoxyribonucleotide-5'-triphosphates dATP, dCTP, dGTP, and dTTP (100 mM), MBI Fermentas.
4. 5-(3-Aminoallyl)-dUTP (aminoallyl-dUTP) Sigma-Aldrich (Deisenhofen, Germany).

5. FluoroLink™ Cy3 monofunctional dye (PA23001) Amersham Biosciences.
6. FluoroLink™ Cy5 monofunctional dye (PA25001), Amersham Biosciences.

2.2.2. Buffers (see **Note 1**)

1. 100 m*M* Aminoallyl-dUTP.
2. Cy3-/Cy5-monofunctional dye: dissolve the content of one tube in 72 µL of DMSO; aliquot 4.5 µL of this mixture in 16 individual tubes; dry the aliquots immediately in vacuo and store them frozen at –80°C in the dark (*see* **Note 2**).
3. 50X dNTP-mix: 10 µL each of 100 m*M* dATP, 100 m*M* dGTP and 100 m*M* dCTP, 4 µL of 100 m*M* aminoallyl-dUTP and 6 µL of 100 m*M* dTTP.
4. 1 *M* NaOH.
5. 0.5 *M* Ethylenediamine tetraacetic acid (EDTA).
6. 1 *M* N-hydroxyethylpiperazine-*N*'-2-ethanesulfonate (HEPES), pH 7.5.
7. 0.1 *M* NaHCO$_3$, pH 9.0.
8. 4 *M* Hydroxylamine.

2.3. Antisense RNA Amplification by In Vitro Transcription

2.3.1. Reagents

2.3.1.1. First-Strand Synthesis

1. Deoxyribonucleotide-5'-triphosphates dATP, dCTP, dGTP, and dTTP (100 m*M* each), MBI Fermentas.
2. RNaseOUT™ (40 U/µL).
3. SuperScript™II RNase H-reverse transcriptase (200 U/µL), 5X first-strand buffer, Invitrogen.
4. T7-T$_{(24)}$-Primer (5'-GGC CAG TGA ATT GTA ATA CGA CTC ACT ATA GGG AGG CGG T$_{(24)}$-3'; 100 pmol/µL), Biospring (Frankfurt, Germany).

2.3.1.2. Second-Strand Synthesis

1. DNA polymerase I (*Escherichia coli*; 10 U/µL), Invitrogen.
2. *E. coli* DNA ligase (10 U/µL), Invitrogen.
3. Ribonuclease H (2 U/µL), Invitrogen.
4. T4 DNA polymerase (5 U/µL), Invitrogen.
5. 5X second-strand buffer, Invitrogen.
6. Rotiphenol, Carl Roth GmbH & Co. KG (Karlsruhe, Germany).

2.3.1.3. In Vitro Transcription

1. Megascript T7 kit–inclusive nucleoside-triphosphates (75 m*M*), 10X reaction buffer, enzyme-mix, and DNAse I (2 U/µL), Ambion (Huntington, UK).
2. Biotin-11-CTP, NEN (Rodgau-Jügesheim, Germany).
3. Biotin-16-UTP, Roche Diagnostics (Mannheim, Germany).
4. RNeasy® mini-kit of Qiagen.

2.3.2. Buffers

1. 10 mM dNTP-mix.
2. 7.5 M Ammonium acetate.
3. 24:1 (v/v) Chloroform/isoamyl alcohol.
4. 80% Ethyl alcohol.

3. Methods

3.1. Reverse Transcription With Direct Incorporation of Fluorescently Labeled Nucleotides

Starting from total RNA, a first-strand cDNA synthesis is performed, during which fluorescently labeled nucleotides are incorporated. Subsequently, the RNA template is hydrolyzed, and the single-stranded cDNA is purified. It can be used directly for hybridization onto microarrays made of spotted polymerase chain reaction (PCR) products or long oligonucleotides (*see* also **Note 3**).

3.1.1. First-Strand cDNA Synthesis

1. Dissolve 10–15 µg of total RNA in 15 µL of water, add 5 µL of oligo(dT)$_{12-18}$ primer and mix by pipetting. After an incubation at 70°C for 10 min, place the mixture on ice.
2. On ice, add 8.5 µL of first-strand buffer, 3.5 µL of DTT; 10 µL each of dATP, dGTP, and dTTP; 2 µL of dCTP; 2 µL of Cy3-dCTP or Cy5-dCTP, respectively, 1 µL of RNaseOUT™ and 2 µL of SuperScript™II transcriptase; mix well.
3. Incubate the reaction for 1 h at 42°C, add another 2 µL of SuperScript™II transcriptase and incubate again at 42°C for at least 3 h (*see also* **Note 4**).
4. Incubate at 70°C for 10 min, add 1 µL of RNase H and incubate 20 min at 37°C. Continue immediately with cDNA purification or store overnight at –20°C.

3.1.2. Purification of Labeled cDNA

The purification of the labeled cDNA is done with the QIAquick® PCR purification kit according to the manufacturer's instructions. All centrifugation steps are performed at room temperature at 16,000g.

1. Add 225 µL of buffer PB to the sample and mix carefully by pipetting.
2. Apply the sample to the QIAquick column and centrifuge for 1 min.
3. After discarding the flow-through, add 750 µL of buffer PE to the column for washing and centrifuge for 1 min. Discard the flow-through and centrifuge again for 1 min to remove the buffer completely.
4. For elution, place the QIAquick column in a clean tube, add 50 µL of water, incubate 1 min at room temperature, and centrifuge for 1 min (*see* also **Note 5**). Repeat this elution step with another 50 µL of water.

3.2. Aminoallyl Labeling

This method is also based on reverse transcription of total RNA. However, the incorporation of fluorescence-labeled dCTP is replaced by the use of aminoallyl-dUTP. After reverse transcription, fluorescent dyes in form of monoreactive NHS-esters are coupled to the amino residues of the single stranded cDNA. After removal of unbound dye, the sample is ready for hybridization to spotted microarrays (*see* **Note 3**).

3.2.1. First-Strand cDNA Synthesis

1. Dissolve 10–15 µg of total RNA in 9.5 µL of water; add 5 µL of oligo(dT)$_{12–18}$ primer and mix by pipetting. After incubation at 70°C for 10 min, place the mixture on ice.
2. On ice, add 6 µL of 5X first-strand buffer, 0.6 µL of 50X dNTP-mix, 3 µL of DTT, 3 µL of water, and 1.9 µL of SuperScript II transcriptase.
3. Incubate at 42°C for at least 3 h (*see* **Note 4**).
4. For hydrolysis of the RNA, add 10 µL of NaOH and 10 µL of ethylenediamine tetraacetic acid and incubate at 65°C for 10 min. Subsequently, neutralize with 25 µL of *N*-Hydroxyethylpiperazine-*N*'-2-ethanesulfonate.

3.2.2. Purification of First-Strand Synthesis Reaction

Purification of the first-strand synthesis product is performed with the QIAquick PCR purification kit according to the manufacturer's instructions. All centrifugation steps are conducted at room temperature at 16,000*g*.

1. Add 25 µL of water and 500 µL of buffer PB to the sample and mix carefully by pipetting.
2. Apply the sample to the QIAquick column and centrifuge for 1 min.
3. After discarding the flow-through, add 750 µL of buffer PE for washing and centrifuge for 1 min. Discard the flow-through and centrifuge again for 1 min to remove the buffer completely.
4. For elution, place the QIAquick column in a clean tube, add 30 µL of water, incubate 1 min at room temperature, and centrifuge for 1 min (*see* **Note 5**). Repeat this elution step with another 30 µL of water.

3.2.3. Coupling of Monofunctional NHS Esters of the Cy-Dyes

1. Dry the cDNA sample *in vacuo*.
2. Dissolve dried cDNA pellet in 9 µL of NaHCO$_3$.
3. Add the dissolved cDNA to one aliquot of either Cy3- or Cy5-monofunctional dye and incubate at room temperature in the dark for 1 h.
4. For quenching unbound Cy-dye, add 4.5 µL of hydroxylamine and incubate 15 min at room temperature in the dark.

3.2.4. Purification of Labeled cDNA

The purification of the labeled cDNA was done with the QIAquick® PCR purification kit according to the manufacturer's instructions:

1. Add 70 µL of water and 500 µL of buffer PB to the sample and mix carefully by pipetting.
2. Apply the sample to the QIAquick column and centrifuge for 1 min.
3. After discarding the flow-through, add 750 µL of buffer PE for washing and centrifuge for 1 min. Discard the flow-through and centrifuge again for 1 min to remove the buffer completely.
4. For elution, place the QIAquick column in a clean tube, add 30 µL of buffer EB, incubate 1 min at room temperature and centrifuge for 1 min. Repeat this elution step with another 30 µL of buffer EB.

3.3. Antisense RNA Amplification by In Vitro Transcription

This labeling method was extensively tested for working on *in situ* synthesized oligonucleotide arrays using the Geniom technology of febit (*see* **Note 6**). In a first step, total RNA is reverse transcribed into double-stranded cDNA. After purification, the cDNA is used as template for an in vitro transcription with concomitant incorporation of biotinylated nucleotides (*see* **Note 3**). The fluorescence labeling of the target is performed after hybridization to an oligonucleotide array by an incubation with streptavidine–phycoerythrine conjugate.

3.3.1. First-Strand Synthesis

1. 15 µg of total RNA is dissolved in 10 µL of water and placed on ice.
2. Add 1 µL of T7-T(24)-primer, mix well and incubate at 70°C for 10 min by shaking gently (500 rpm; *see* also **Note 7**).
3. During this incubation, mix in a second tube the reverse transcription mix, made of 4 µL of 5X first-strand buffer, 2 µL of DTT, 1 µL of dNTP-mix, and 0.5 µL of RNaseOUT. Preheat this mix to 50°C.
4. Place the primer mix also at 50°C, incubate briefly and add the reverse transcription mix.
5. Incubate at 50°C for 2 min. Then add 1.5 µL of SuperScript II transcriptase.
6. Incubate at 50°C for 1 h, then place immediately on ice and proceed directly to second-strand synthesis.

3.3.2. Second-Strand Synthesis

1. Prepare the second-strand synthesis mix on ice: mix 30 µL of 5X second-strand buffer, 4 µL of DNA polymerase I, 3 µL of dNTP-mix, 1 µL of DNA ligase, 1 µL of RNase H, and 91 µL of water.

2. Add the second-strand synthesis mix to the first-strand synthesis reaction and incubate at 16°C for 2 h.
3. Add 2 μL of T4 DNA polymerase and incubate at 16°C for 5 min.
4. Place the reaction on ice and proceed immediately to cDNA purification.

3.3.3. cDNA Purification

1. Add 1 volume rotiphenol to the second-strand synthesis reaction, mix by vortexing at maximum speed for 1 min and centrifuge at 16,000g for 5 min.
2. Transfer the aqueous phase carefully to a new tube and place the sample on ice.
3. Repeat the two steps with 1 vol chloroform/isoamyl alcohol.
4. For DNA precipitation, add 0.5 vol 7.5 *M* ammonium acetate and 2.5 vol 100% ethyl alcohol and centrifuge at 4°C at 16,000g for 20 min.
5. After removing the supernatant, wash the pellet twice with 375 μL of 80% ethyl alcohol. In between, centrifuge at 4°C at 16,000g for 7 min.
6. Dry the cDNA pellet *in vacuo* and dissolve it in 1.5 μL of water. The cDNA can either be used directly for in vitro transcription or stored overnight at –20°C.

3.3.4. In Vitro Transcription

1. Prepare on ice the nucleotide mix, made up of 2 μL of ATP, 2 μL of GTP, 1.5 μL of CTP, 1.5 μL of UTP, 3.75 μL of Biotin-11-CTP, 3.75 μL of Biotin-16-UTP, and 2 μL of 10X reaction buffer.
2. Add the nucleotide mix and 1.5 μL of of the kit's enzyme mix to the purified cDNA.
3. Seal the tube with parafilm and incubate the reaction at 37°C for 6 h shaking gently.
4. For hydrolysis of the DNA, add 1 μL of DNase I and incubate at 37°C for 15 min.
5. Place the tube on ice and proceed immediately to aRNA purification.

3.3.5. aRNA Purification

For purification of the aRNA, the RNeasy Mini Kit (Qiagen) is used as recommended by the manufacturer. All centrifugation steps are performed at room temperature at 16,000g.

1. Adjust the volume to 100 μL by adding 20 μL of water to the aRNA. Then, add 350 μL of buffer RLT and mix by vortexing.
2. Add 250 μL of ethyl alcohol and mix carefully by pipetting.
3. Transfer the sample gently to an RNeasy column and centrifuge for 1 min.
4. Place the RNeasy column into a fresh 2-mL collection tube and add 500 μL of RPE buffer. Centrifuge for 1 min and discard flow-through. Repeat this washing step a second time.
5. For elution, place the RNeasy column into a fresh tube. Add 40 μL of water, incubate at room temperature for 4 min and centrifuge the column for 1 min. Repeat this elution step with another 40 μL of water.

4. Notes

1. When working with RNA, all buffers must be prepared with DEPC-treated water and be autoclaved.
2. When aliquoting the Cy-dyes, be sure to hide them from light. The DMSO used for aliquoting needs to be completely water-free and stored over molecular sieve.
3. Before hybridization, target is dried *in vacuo* and dissolved in an appropriate volume of hybridization buffer.
4. For a higher yield, the incubation of the reverse transcription reaction must be at least for 3 h; we recommend incubation times up to 12 h.
5. The elution from the QIAquick columns is very pH sensitive. Therefore, take care that the water used really has at least pH 7. This results in a much higher yield. If the water has a pH below 7.0, adjust the pH with 1 N NaOH. Using buffer EB as recommended by the manufacturer is not possible, because this buffer contains Tris, whose amino groups would react with the NHS-esters of the Cy-dyes.
6. Using biotinylated aRNA for hybridisation to spotted cDNA microarrays is not recommended. The subsequent reaction with streptavidin-phycoerythrine conjugate would produce a high background.
7. All incubation steps during this protocol are performed by gentle shaking at 500 rpm.

Acknowledgments

We wish to thank our colleagues for helpful discussions and suggestions. This work was funded by the German Federal Ministry of Education and Research (BMBF).

References

1. Brown, P. O. and Botstein, D. (1999) Exploring the new world of the genome with DNA microarrays. *Nature Genet.* **21,** 33–37.
2. Lockhart, D. J. and Winzeler, E. A. (2000) Genomics, gene expression and DNA arrays. *Nature* **405,** 827–836.
3. Shalon, D., Smith, S. J., and Brown, P. O. (1996) A DNA microarray system for analyzing complex DNA samples using two-color fluorescent probe hybridisation. *Genome Res.* **6,** 639–645.
4. Aign, V. and Hoheisel, J. (2003) Analysis of nutrient-dependent transcript variations in Neurospora crassa. *Fungal Genet. Biol.* **40,** 225–233.
5. Van Gelder, R. N., von Zastrow, M. E., Yool, A., Dement, W. C., Barchas, J. D., and Eberwine, J. H. (1990) Amplified RNA synthesised from limited quantities of heterogeneous cDNA. *Proc. Natl. Acad. Sci. USA* **87,** 1663–1667.

9

High-Density Labeling of DNA for Single Molecule Sequencing

Susanne Brakmann

Summary

Two unusual enzymatic activities are required for the realization of a single molecule sequencing: a polymerase for copying a deoxyribonuclease (DNA) target into complementary flurophore-labeled DNA, and an exonuclease for the successive hydrolysis of the completely dye-labeled DNA. Recently, we found that the wild-type Klenow fragment of *Escherichia coli* DNA polymerase I is well-suited for the synthesis of DNA in a reaction set-up that contains exclusively specific rhodamine-labeled analogs of the natural pyrimidine nucleotides (dCTP and dTTP). This protocol describes the procedure used for the preparation of DNA that is labeled at all pyrimidine bases of one strand, as well as an example of enzymatic downstream processing of the DNA product.

Key Words: DNA sequencing; single molecule detection; DNA polymerase; fluorescence.

1. Introduction

The international race to sequence the human genome as well as the genomes of other model organisms has encouraged efforts to realize a "single molecule sequencing," an idea that nourishes the hope to simplify and speed up the task of sequencing DNA segments as long as 50,000 bp and joining the sequence information of genome fragments *(1–5)*. The different strategies to realize single molecule sequencing are based on the fact that single fluorescent molecules can be identified within milliseconds *(6)*, and they combine sequential enzymatic hydrolysis of individual DNA molecules with subsequent identification of released monomers by their fluorescence characteristics, either wavelengths, or fluorescence lifetimes, or both *(5,7)*.

Two unusual enzymatic activities are required for this technique: (1) the complete and faithful synthesis of DNA copies exclusively from fluorescently labeled analogs of the four types of bases (A, G, C, and T) and

From: *Methods in Molecular Biology, vol. 283: Bioconjugation Protocols: Strategies and Methods*
Edited by: C. M. Niemeyer © Humana Press Inc., Totowa, NJ

(2) the exonucleolytic degradation of the completely labeled DNA. Recently, we presented a solution to the problem of complete enzymatic labeling by identifying the wild-type Klenow fragment of *Escherichia coli* DNA polymerase I (KF) as the first natural polymerase that retains full activity and fidelity in the sole presence of rhodamine-labeled deoxynucleoside triphosphates *(8)*. We demonstrated that the complete substitution of all pyrimidine bases of one DNA strand and thus, complete labeling of every base pair by the respective fluorescent analogs can be achieved using KF in a primer-extension reaction *(9)*. This protocol illustrates the techniques used for labeling of any DNA fragment with commercially available nucleotide analogs and for purification and handling of the labeled DNA product (*see* **Note 1**).

2. Materials

1. Double-stranded DNA template, such as a linearized plasmid or a polymerase chain reaction (PCR)-generated fragment. Example used here: PCR fragment comprising the coding sequence of bacteriophage T7 RNA polymerase (2700 bp).
2. Oligonucleotide primer that binds to the respective plasmid or fragment. Example used here: 5'-GGC GTT AGT GAT GGT GAT GGT GAT GCG CGA ACG CGA AGT CCG ACT CTA AG-3'. For immobilization of the dye-labeled DNA product, the oligonucleotide primer is linked to a 5'-biotin.
3. 1 mM dNTP stock solutions.
4. 1 mM Tetramethylrhodamine-(TAMRA)-dUTP, e.g., FluoroRed™ (Amersham Biosciences).
5. 1 mM RhodamineGreen-(R110)-dCTP, e.g., R110-dCTP (Applied Biosystems).
6. Klenow polymerase, exonuclease-deficient (New England Biolabs).
7. 10X Buffer for Klenow polymerase: 100 mM Tris-HCl, pH 7.5, 50 mM MgCl$_2$, 75 mM dithiothreitol.
8. 0.5 M Ethylenediamine tetraacetic acid (EDTA), pH 8.0.
9. Dimethyl sulfoxide (DMSO; p.A.).
10. 1,4-Dioxane (p.A.).
11. *i*-Propanol (p.A.).
12. Ethanol (p.A.).
13. Water (high-performance liquid chromatography grade).
14. Streptavidin-coated beads, e.g., Dyna beads (Dynal), and magnetic separator.
15. 2X Binding and washing buffer (BW buffer): 10 mM Tris-HCl, pH 7.5, 1 mM EDTA, 2 M NaCl.
16. Agarose, for example, Seakem GTG (FMC Bioproducts), and electrophoresis buffer (1X TAE), loading buffer (e.g., with bromophenol blue as the tracking dye), as well as electrophoresis equipment.
17. "Salt trap" buffer for electroelution: 3 M sodium acetate, 0.03 % Bromophenol blue.
18. 100X Electroelution buffer: 1 M Tris-acetate, pH 8.0, 10 mM EDTA.
19. Electroelution equipment (*see* **Fig. 2**).
20. Exonuclease III (United States Biochemicals).

21. 10X Buffer for exonuclease III (United States Biochemicals): 0.5 M Tris-HCl, pH 7.6, 50 mM MgCl$_2$, 50 mM dithiothreitol, 0.5 mg/mL bovine serum albumin.
22. 0.5 M EDTA.

3. Methods

The methods described below comprise (1) the preparation of a dye-labeled copy of a defined DNA fragment, (2) its purification using immobilization on streptavidin-coated beads, and (3) the alternative purification without immobilization. Furthermore, an exemplary enzymatic digestion of dye-labeled DNA will be presented.

3.1. Labeling of DNA With Two Fluorescent Dyes

1. Dilute 2.5 pmol double-stranded DNA template (2700 bp) and 10 pmol biotinylated primer in 80 μL of Tris-HCl (50 mM, pH 7.5) and place the tube in a heating block (e.g., PCR cycler).
2. Heat to 100°C for 5 min, then switch machine off. After approx 30 min, the mixture is cooled to approx 30°C (*see* **Note 2**).
3. Add 5 nmol of each dATP, dGTP, TAMRA-dUTP, R110-dCTP.
4. 50 μL 10X Klenow buffer.
5. 50 μL DMSO.
6. 50 U Klenow polymerase.
7. Add water to 500 μL.
8. Incubate 1 h at 37°C.
9. Terminate the reaction by addition of 20 μL EDTA.
10. For immobilization, proceed with **Subheading 3.2.**; for other purposes, proceed with **Subheading 3.3.**

DNA consisting of one normal strand and a complementary strand that is substituted with rhodamine-labeled nucleotide analogs at all pyrimidine positions differs substantially from natural DNA (*see* **Note 3**). The predominant observations include increased tendency to form aggregates upon solution in aqueous media (i.e., low solubility!), and lack of cooperative melting behavior in aqueous solution.

Using transmission electron microscopy, we were able to visualize the TAMRA-dUTP/R110-dCTP-labeled reaction product (**Fig. 1**; in cooperation with Dr. D. Czerny, Göttingen, Germany). The dye-labeled DNA forms an even distribution of individual molecules with shape and length similar to the expected features of a natural 2700-bp DNA fragment.

3.2. Immobilization and Purification of Dye-Labeled DNA Using Streptavidin-Coated Beads

Beads may be loaded with DNA at a maximal ratio of 10^5:1 (DNA:beads). Magnetic Dyna beads used by us were supplied at a concentration of approx 10^9 beads/mL.

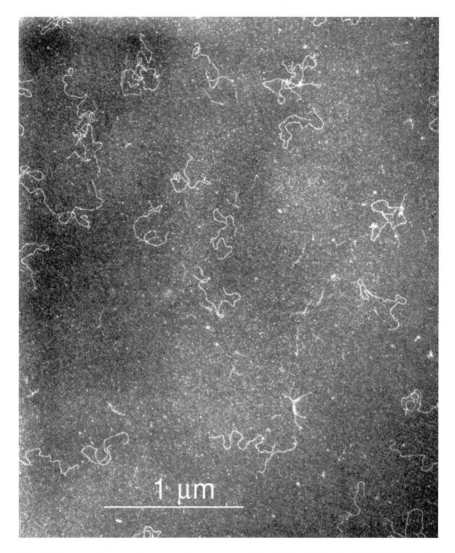

Fig. 1. Transmission electron micrograph of ds-DNA labeled with TAMRA-dUTP and R110-dCTP (graphite surface; staining with uranyl acetate). Reprinted from **ref. 9** with the permission of Wiley–VCH.

1. Take 200 μL bead suspension (approx 2×10^8 beads) and wash the beads with 200 μL of 1X BW buffer using a magnetic separator.
2. Repeat the washing step twice.
3. Resuspend beads in 100 μL of 2X BW buffer.
4. Add 100 μL of solution containing biotinylated DNA (approx 10^{11} molecules) from the labeling reaction (*see* **Subheading 3.1.**).

5. Shake at least 1 h at room temperature, thereby keeping all beads in suspension.
6. Purification from excess fluorescent nucleotides and other reagents is performed with a series of washing steps that should be followed exactly. Start washing the beads once with 100 μL of 80 % *i*-propanol.
7. Wash once with 100 μL of 80% ethanol.
8. Wash once with 100 μL of 80% *i*-propanol.
9. Wash once with 100 μL of 80% ethanol.
10. Wash once with 100 μL water.
11. Resuspend beads in 100 μL of 10 m*M* Tris-HCl.

DNA purified using this procedure is essentially free of fluorescent monomer and thus, suited for single-molecule detection, for example, by fluorescence correlation spectroscopy.

3.3. Purification Without Immobilization

Dye-labeled DNA may be purified by agarose gel electrophoresis. We stain these agarose gels with ethidium bromide because DNA with one strand labeled at every pyrimidine usually is nonfluorescent and thus, is invisible (*see* **Note 3**). Routinely, we recover the dye-labeled DNA by electroelution instead of using commercially available kits because our method yields highest amounts of purified DNA. A schematic diagram showing an electroelution chamber is given below (**Fig. 2**).

1. Prepare an agarose gel (e.g., 1.2% agarose for resolving the 2700 bp product).
2. Mix DNA solution from the labeling reaction (**Subheading 3.1.**) with an appropriate amount of loading buffer, load mixture onto the agarose gel, run the electrophoresis at approx 50 mA until the tracking dye (Bromophenolblue) has reached approximately two thirds of the migration distance.
3. Stain the gel with ethidium bromide solution (0.5 μg/mL in water; approx 20 min), visualize the DNA using a UV transilluminator (if possible, with reduced intensity).
4. Using a sharp scalpel, excise the DNA band, cut it into small pieces, and place them in an electroelution chamber that is filled with 1X electroelution buffer. Fill the V-channels with salt trap buffer (approx 75 μL).
5. Apply 150 V (9 mA) for approx 60–90 min, avoiding gas bubbles at the V channels. During this time, DNA migrates from the agarose piece into the salt trap where an excess of counter ions prevents further migration.
6. Collect the solutions from V-channels using a pipet with a long pipet tip (e.g., those used for loading samples onto sequencing gels). Precipitate the DNA by addition of 2 vol ethanol (100%) and incubate at –20°C for at least 1–2 h.
7. Recover the DNA by centrifugation at 13,000 rpm; wash pellets with 70% ethanol and repeat the centrifugation step.
8. Dissolve the DNA in 10 m*M* Tris-HCl, pH 7.5, containing 50% 1,4-dioxane.

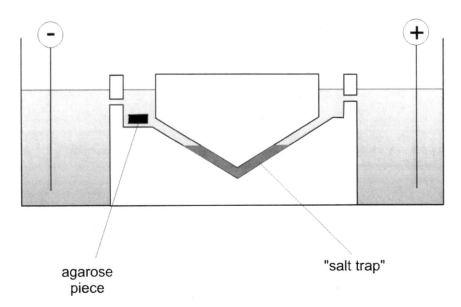

Fig. 2. Schematic representation of an electroelution chamber. The V-channel is filled with 3 *M* sodium acetate buffer that represents a "salt trap," where migrating DNA molecules are stopped. The DNA can be recovered by precipitation from the salt trap solution.

3.4. Enzymatic Manipulation of Dye-Labeled DNA: Sequential Degradation With Exonuclease III

Despite its reduced solubility in aqueous media, dye-labeled DNA may serve as a substrate in further enzymatic reactions. Therefore, organic cosolvents are necessary for keeping the DNA in solution; many enzymes, however, will be inactive under these conditions. The sequential digestion of TAMRA-dUTP/ R110-dCTP-labeled DNA by *E. coli* exonuclease III may serve here as a successful example *(10)*: This enzyme exhibited sufficient stability and activity upon addition of 25% 1,4-dioxane that mediated the solution of dye-labeled DNA. The reaction may, for example, be performed in a fluorescence microcuvet (volume: approx 70 μL) and followed fluorometrically: Because of the liberation of fluorescent monomers, fluorescence will increase steadily. Alternatively, analysis of the reaction may be performed using polyacrylamide gel electrophoresis. We analyzed samples using sequencing gels and resolution of the fluorescent products on an ABI 373A sequencer:

1. Take 40 fmol dye-labeled DNA from a solution in 10 m*M* Tris-HCl, pH 7.5, containing 50% (v/v) 1,4-dioxane.
2. Heat 5 min to 50°C and then let the sample cool to room temperature.

3. Add 10X exonuclease buffer and H_2O to reach a final concentration of 1X reaction buffer and 25% (v/v) 1,4-dioxane (total reaction volume, e.g., 70 μL).
4. Add 50 U exonuclease III and mix.
5. Incubate at 37°C.
6. At the same time, take samples (e.g., 7 μL) after 2, 5, 10, 15, 20, 30, and 60 min. Each sample must be stopped with 1 μL of 0.5 M EDTA and then kept on ice.
7. Alternative A: Analyze all samples using polyacrylamide gel electrophoresis and detect the products after staining with ethidium bromide and UV illumination. Alternative B: Analyze all samples using fluorescence detection by an automatic sequencing system.

4. Notes

1. The complete incorporation of two dye-labeled nucleotides by exonuclease-deficient Klenow fragment proceeds independent of sequence and template and yields high amounts of double-stranded DNA product. In principle, the primer extension reaction can be applied to templates as long as 10,000 bp; we also were successful in copying bacteriophage λ DNA with complete substitution of both pyrimidine nucleotides. With longer templates, the error rate of Klenow polymerase should be taken into account. However, the rate of misincorporation did not exceed a maximum of 10^{-4} with our experiments.
2. We performed the primer annealing with a slow cooling step (within approx 30 min). During this time period, a partial rehybridization of the complementary single DNA strands will also occur. This has not been a problem with templates up to 9000 bp—the DNA polymerase will perform the "nick translation" by all means, starting at the 3' terminus of the annealed primer. However, if extensive smearing is observed with the product DNA after agarose gel electrophoresis, rapid cooling is recommended (similar to PCR procedures, that is, within seconds to few minutes).
3. DNA that is labeled at high density with rhodamine dyes exhibits some unusual properties: (1) The close proximity of incorporated dyes promotes fluorescence quenching. As a result, the product DNA is usually nonfluorescent. (2) DNA labeled with a high amount of rhodamine dyes tends to form aggregates at polar surfaces and in aqueous solution. Upon addition of organic solvents like DMSO or 1,4-dioxane, formation of aggregates is reduced significantly. Furthermore, recording the absorbance is possible only in the presence of one of these cosolvents. (3) Dye-labeled double-stranded DNA may be converted into single-stranded DNA by heating in 100% DMSO; the DNA does not melt in aqueous media! This may be essential if annealing of a primer for another round of primer extension or amplification is desired.

References

1. Nguyen, D. C., Keller, R. A., Jett, J. H., and Martin, J. C. (1987) Detection of single molecules of phycoerythrin in hydrodynamically focused flows by laser-induced fluorescence. *Anal. Chem.* **59,** 2158–2161.

2. Ambrose, W. P., Goodwin, P. M., Jett, J. H., Johnson, M. E., Martin, J. C., Marrone, B. L., et al. (1993) Application of single molecule detection to DNA sequencing and sizing. *Ber. Bunsenges. Phys. Chem.* **97,** 1535–1542.

3. Eigen, M. and Rigler, R. (1994) Sorting single molecules: Applications to diagnostics and evolutionary biotechnology. *Proc. Natl. Acad. Sci. USA* **91,** 5740–5747.

4. Service, R. F. (1999) Deconstructing DNA for faster sequencing. *Science* **283,** 1669.

5. Stephan, J., Dörre, K., Brakmann, S., Winkler, T., Wetzel, T., Lapczyna, M., et al. (2001) Towards a general procedure for sequencing single DNA molecules. *J. Biotechnol.* **86,** 255–267.

6. Rigler, R., Mets, Ü. Widengren, J., and Kask, P. (1993) Fluorescence correlation spectroscopy with high count rate and low background: analysis of translational diffusion, *Eur. Biophys. J.* **22,** 169–175.

7. Dörre, K., Brakmann, S., Brinkmeier, M., Han, K.-T., Riebeseel, K., Schwille, P., et al. (1997) Techniques for single molecule sequencing. *Bioimaging* **5,** 139–152.

8. Brakmann, S. and Nieckchen, P. (2001) The large fragment of Escherichia coli DNA polymerase I synthesizes DNA exclusively from fluorescently labeled nucleotides. *ChemBiochem* **2,** 773–777.

9. Brakmann, S. and Löbermann, S. (2001) High-density labeling of DNA: Preparation and characterization of the target for single molecule sequencing. *Angew. Chemie Int. Ed.* **40,** 1427–1429.

10. Brakmann, S. and Löbermann, S. (2002) A further step towards single molecule sequencing: *Escherichia coli* exonuclease III degrades DNA that is fluorescently labeled at each basepair. *Angew. Chemie Int. Ed.* **41,** 3215–3217.

10

Sequence-Specific DNA Labeling
Using Methyltransferases

Goran Pljevaljčić, Falk Schmidt, Alexander Peschlow, and Elmar Weinhold

Summary

Sequence-specific labeling of native deoxyribonucleic acid (DNA) still represents a more-or-less unsolved problem. Difficulties mainly arise from the necessity to combine two different functions: sequence-specific recognition of DNA and covalent bond formation between the label and DNA. DNA methyltransferases (MTases) naturally possess these two functions and transfer a methyl group from the cofactor S-adenosyl-L-methionine (AdoMet) to adenine or cytosine residues within specific DNA sequences, typically ranging from two to eight base pairs. Unfortunately, the methyl group itself is a very limited reporter group and it would be desirable to transfer larger chemical entities with DNA MTases. Replacement of the methionine side chain of the natural cofactor AdoMet by an aziridinyl residue leads to the synthetic cofactor N-adenosylaziridine, which is quantitatively, base- and sequence-specifically coupled with DNA in a DNA MTase-catalyzed reaction. By attaching interesting reporter groups to a suitable position of N-adenosylaziridine a large variety of new synthetic cofactors are obtained for sequence-specific labeling of DNA. This method is illustrated by coupling primary amino groups and biotin to short duplex oligodeoxynucleotides or plasmid DNA using the DNA MTase M·TaqI.

Key Words: Sequence-specific DNA labeling; enzymatic DNA labeling; DNA modification; DNA methyltransferase; DNA modifying enzyme; DNA methylation; S-adenosyl-L-methionine; AdoMet; aziridine cofactor; cofactor engineering; N-adenosylaziridine; primary amino group; biotin.

1. Introduction

Deoxyribonucleic acid (DNA) methyltransferases (MTases) transfer the activated methyl group of the cofactor S-adenosyl-L-methionine (AdoMet) to the amino groups of adenine and cytosine or the 5-position of cytosine within specific double-stranded DNA sequences *(1,2)*. In addition to AdoMet (**Fig. 1,**

From: *Methods in Molecular Biology, vol. 283: Bioconjugation Protocols: Strategies and Methods*
Edited by: C. M. Niemeyer © Humana Press Inc., Totowa, NJ

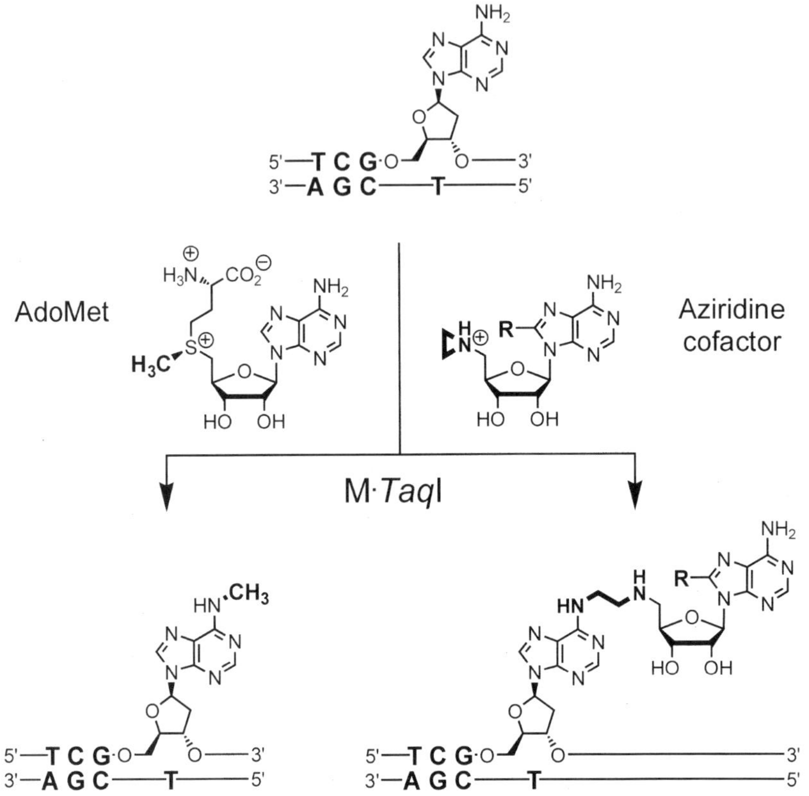

Fig. 1. Reactions catalyzed by the DNA methyltransferase from *Thermus aquaticus* (M·*Taq*I). Naturally, M·*Taq*I catalyzes the transfer of the activated methyl group from the cofactor *S*-adenosyl-L-methionine (AdoMet) to the amino group of adenine within the double-stranded 5'-TCGA-3' sequence (left). With *N*-adenosylaziridine or derivatives carrying a reporter group at the 8-position the M·*Taq*I-catalyzed opening of the aziridine ring leads to coupling of the whole cofactor to the amino group of adenine within the double-stranded 5'-TCGA-3' sequence and hence to sequence-specific labeling of DNA (right).

left) the synthetic nucleoside *N*-adenosylaziridine (R = H) is a cofactor for the adenine-specific DNA MTase from *Thermus aquaticus* (M·*Taq*I; **Fig. 1**, right) and is quantitatively, base- and sequence-specifically coupled with the 5'-TCGA-3' recognition sequence within a short duplex oligodeoxynucleotide *(3)*. We have extended this work and used the aziridine cofactor as coupling reagent for a reporter group. An *N*-adenosylaziridine derivative containing a fluorophor attached via a flexible linker to the 8-position of the adenine ring (R=NH[CH$_2$]$_4$NH-dansyl) was sequence-specifically coupled with a short duplex oligodeoxynucleotide and plasmid DNA by M·*Taq*I *(4)*. This novel technique called sequence-specific methyltransferase-induced labeling of DNA (SMILing DNA) is very versatile and can be used to sequence-specifically couple a wide variety of chemical groups to DNA. Sequence-specific labeling of short synthetic and long native DNA with either a primary amino group or biotin will be used to illustrate the method which could find interesting applications in functional studies of DNA modifying enzymes, molecular biology, DNA-based medical diagnosis and nanobiotechnology.

2. Materials

1. TLC plates Silca gel 60 F_{254} (Merck, Darmstadt, Germany).
2. 2-Bromoethylamine hydrobromide (Fluka, Taufkirchen, Germany).
3. 2',3'-*O*-Isopropylideneadenosine (Sigma, Taufkirchen, Germany).
4. Bromine (Aldrich, Taufkirchen, Germany).
5. Sodium hydrogensulfite (Fluka).
6. 1,4-Diaminobutane (Merck).
7. Triethylamine (Fluka).
8. Dowex 50 W × 4, 100–200 mesh, H^+-form (Fluka).
9. Silca gel 60, 230–400 mesh, particle size 0.040–0.063 mm (Merck).
10. 6-Nitroveratrylchloroformate (Fluka).
11. 4-Dimethylaminopyridine (Merck).
12. Mesyl chloride (Merck).
13. *N*-Ethyldiisopropyl amine (Fluka).
14. Reversed-phase high-performance liquid chromatography (HPLC) columns Prontosil, 5 µm, 120 Å, 250 × 4.6 mm and Prontosil, 5 µm, 120 Å, 250 × 8 mm (Bischoff, Leonberg, Germany).
15. Triethanolamine (Merck).
16. Succinimidyl biotin (Fluka).
17. M·*Taq*I expression vector pAGL15-M13 and *Escherichia coli* host cells ER 2267 (New England Biolabs, Beverly, MA).
18. Luria Bertani (LB) medium.
19. Antibiotics ampicillin and kanamycin.
20. Isopropyl-β-D-thio-galactopyranoside.
21. Phenylmethylsulfonyl fluoride.
22. Buffer A: Mes/*N*-hydroxyethylpiperazine-*N*'-2-ethanesulfonate (HEPES)/sodium acetate buffer mix, 6.7 mM each, pH 7.5, 1 mM ethylenediamine tetraacetic acid, 10 mM β-mercaptoethanol and 10% glycerol.
23. Buffer B: Mes/HEPES/sodium acetate buffer mix, 6.7 mM each, pH 6.0, 300 mM potassium chloride, 0.2 mM 1,4-dithiothreitol, and 10% glycerol.
24. Buffer C: 40 mM Tris acetate, pH 7.9; 20 mM magnesium acetate; 100 mM potassium acetate; 2 mM 1,4-dithiothreitol; and 10% glycerol.
25. Cation exchange material Poros HS/M, 50 mL (Applied Biosystems, Darmstadt, Germany).

26. Heparin Sepharose CL 6B, 50 mL (Amersham Biosciences, Freiburg, Germany).
27. Gel filtration column Superdex 75, 26 × 600 mm (Amersham Biosciences).
28. Ultrafiltration devices Jumbosep, 10 kDa cut-off (Pall, Dreieich, Germany) and Centriprep YM-10 (Millipore, Eschborn, Germany).
29. Coomassie Protein Assay Reagent (Pierce, KMF, St. Augustin, Germany).
30. Complementary oligodeoxynucleotides containing the 5'-TCGA-3' sequence.
31. 10X labeling buffer: 200 mM Tris acetate, pH 6.0, 100 mM magnesium acetate and 500 mM potassium acetate.
32. Triton X-100.
33. Anion exchange column Poros 10 HQ, 10 μm, 4.6 × 10 mm (Applied Biosystems).
34. Proteinase K (Qiagen, Heiden, Germany).
35. pUC19 plasmid DNA.
36. R·*Eco*RI and R·*Taq*I restriction endonuclease (MBI Fermentas, St. Leon-Rot, Germany).
37. Qiagen PCR purification Kit.

3. Methods

The methods described below outline (1) the chemical synthesis of aziridine cofactors for DNA labeling with primary amino groups and biotin, (2) the expression and purification of the M·*Taq*I DNA MTase, (3) the sequence-specific labeling of short duplex oligodeoxynucleotides and (4) the sequence-specific labeling of plasmid DNA.

3.1. Synthesis of Aziridine Cofactors

Water-sensitive chemical reactions are carried out in oven-dried glassware under argon atmosphere and solvents are dried using standard techniques *(5)*. Thin-layer chromatography (TLC) is performed with DC alumina or glass plates and compounds are visualized by UV light (254 nm). Column chromatographic separations of compounds are achieved by using 50 to 100 times more silica gel than sample material. NMR spectra are recorded with CDCl$_3$ (δ_H = 7.24) or [d_6]DMSO (δ_H = 2.49) as solvent. Electrospray ionization mass spectra (ESI-MS) are obtained in positive ion mode and samples are dissolved in a mixture of methanol and water (1:1 v/v) containing 5% formic acid.

3.1.1. Aziridine

The synthesis is based on literature procedures (**refs. *6* and *7*; *see* Note 1**).

1. Dissolve potassium hydroxide (13.3 g, 237 mmol) in water (25 mL) at 0°C and add 2-bromoethylamine hydrobromide (12.0 g, 58.6 mmol).
2. Stir the solution at 10°C for 1 h.
3. Distill aziridine onto potassium hydroxide plates twice and store the compound at 4°C over potassium hydroxide plates.

Aziridine (1.3 mL, 43%) is obtained as a colorless liquid (head temperature 55°C); ^{1}H NMR (500 MHz, CDCl$_3$): δ = 1.24 (s, 4H).

3.1.2. 8-Bromo-2',3'-O-Isopropylideneadenosine (2)

2',3'-O-Isopropylideneadenosine (**1**; **Fig. 2**) is brominated using a modified literature procedure (**8**).

1. Add bromine (0.62 mL, 12.1 mmol) in potassium acetate buffer (20 mL, 1 *M*, pH 3.9) to a solution of nucleoside **1** (2.31 g, 7.53 mmol) in potassium acetate buffer (300 mL, 1 *M*, pH 3.9) within 15 min at 0°C.
2. Stir the orange solution at room temperature for 15 h until no starting material is detected by TLC.
3. Add a saturated solution of sodium hydrogensulfite to reduce excess of bromine, which will lead to a white precipitate.
4. Adjust the pH of the resulting suspension to 7 using a sodium hydroxide solution (220 mL, 10 *M*).
5. Collect the white precipitate by filtration, wash the precipitate with water (350 mL), and freeze-dry the precipitate.

Nucleoside **2** (2.40 g, 82%) is obtained as a white solid (R_f 0.70, CH$_2$Cl$_2$/ CH$_3$OH 9:1); ^{1}H NMR (500 MHz, [d_6]DMSO): δ = 1.32 (s, 3H; acetonide–CH$_3$), 1.54 (s, 3H; acetonide–CH$_3$), 3.40–3.52 (m, 2H; 5'-H), 4.14-4.17 (m, 1H; 4'-H), 5.02 (dd, ^3J = 2.8, 6.0 Hz, 1H; 3'-H), 5.10 (t, ^3J = 5.8 Hz, 1H; 5'-OH), 5.65 (dd, ^3J = 2.8, 6.0 Hz, 1H; 2'-H), 6.01 (d, ^3J = 2.8 Hz, 1H; 1'-H), 7.53 (s, br., 2H; 6-NH$_2$), 8.14 (s, 1H, 2-H); ESI-MS m/z (%): 387.9 (100) [M + H]$^+$, 216.3 (75) [8-bromoadenine + H]$^+$.

3.1.3. 8-Amino[1''-(4''-Aminobutyl)]-2',3'-O-Isopropylideneadenosine (3)

1. Add 1,4-diaminobutane (0.82 mL, 8.1 mmol) and dry triethylamine (TEA; 2.26 mL, 16.3 mmol) to a solution of nucleoside **2** (628 mg, 1.62 mmol) in dry DMSO (10 mL) under argon atmosphere.
2. Stir the solution at 110°C for 4 h and monitor the reaction progress by TLC.
3. Remove the solvent under reduced pressure and use the crude product directly in the next step.

For analytical purposes, the residue is dissolved in water (50 mL), the pH adjusted to 5.3 with acetic acid (0.1 *M*) and the crude product purified by cation exchange chromatography on Dowex 50 W × 4, H$^+$-form (100 g, elution with 0.6 L water followed by 1 L of 1 *M* potassium hydroxide). Fractions containing the product are extracted with chloroform. The organic layers are combined and the solvent is removed under reduced pressure to yield nucleoside **3** (639 mg, 100%) as a white solid (R_f 0.44, butanol/acetic acid/water 12:3:5); ^{1}H NMR (500 MHz, CDCl$_3$): δ = 1.33 (s, 3H; acetonide-H), 1.48–1.55 (m, 2H; linker-H), 1.61 (s, 3H; acetonide-H), 1.64–1.70 (m, 2H; linker-H), 2.66–2.73

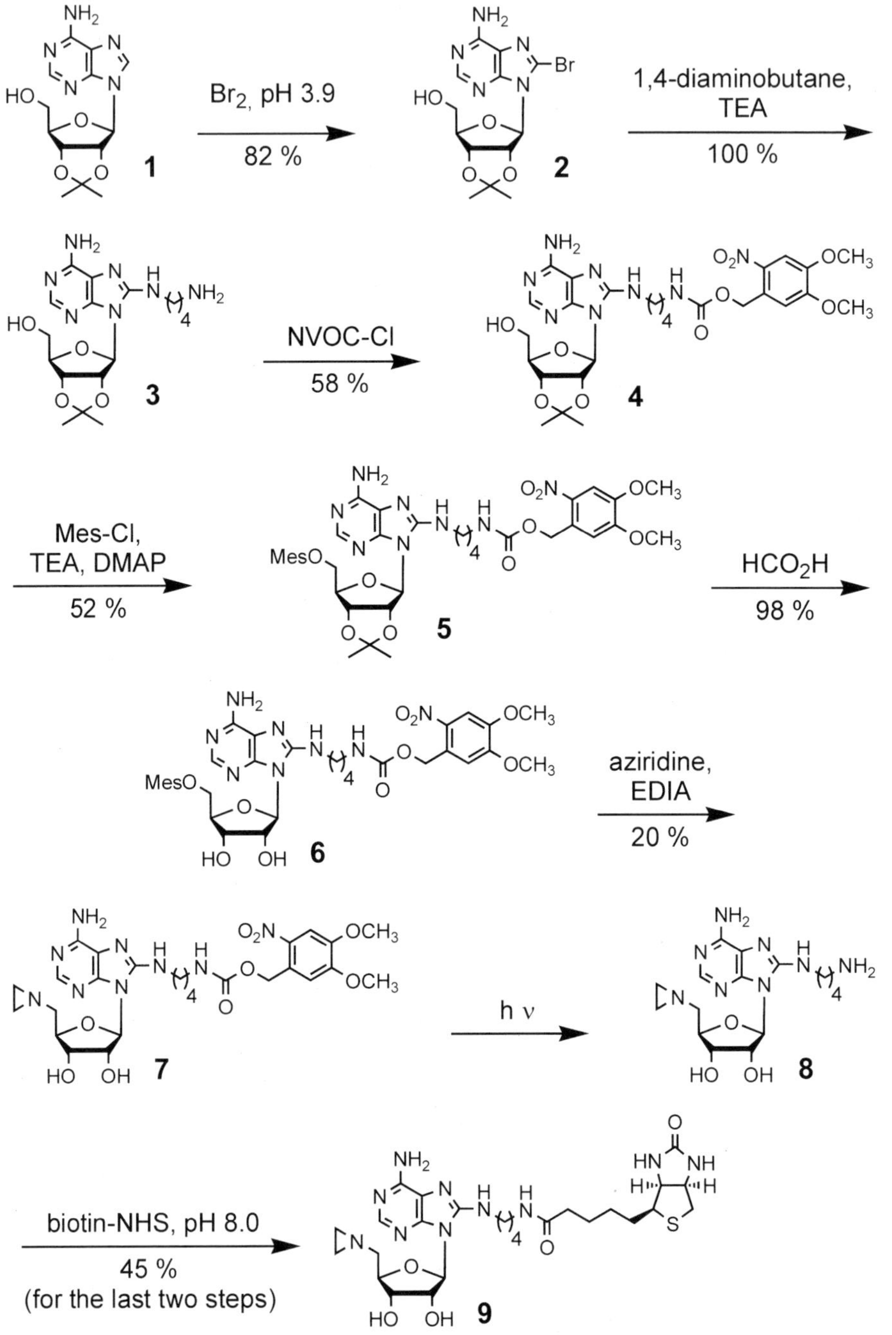

Fig. 2. Chemical synthesis of aziridine cofactors **7**, **8**, and **9** for sequence-specific labeling of DNA by DNA MTases.

(m, 2H; linker-H), 3.33–3.42 (m, 2H; linker-H), 3.77–3.91 (m, 2H; 5'-H), 4.28–4.30 (m, 1H; 4'-H), 4.99 (dd, ^3J = 2.7, 6.3 Hz, 1H; 3'-H), 5.08 (dd, ^3J = 4.8, 6.3 Hz, 1H; 2'-H), 5.39 (s, br., 2H; 6-NH$_2$), 6.15 (d, ^3J = 4.5 Hz, 1H; 1'-H), 6.55-6.60 (m, 1H; linker-NH), 8.10 (s, 1H; 2-H); ESI-MS m/z (%): 394.3 (25) [M + H]$^+$, 222.3 (100) [8-Amino[1'-(4'-aminobutyl)]-adenine + H]$^+$.

3.1.4. 8-Amino[1''-(N''-6-Nitroveratryl-Oxocarbonyl)-4''-Aminobutyl]-2',3'-O-Isopropylideneadenosine *(4)*

1. Add 6-nitroveratrylchloroformate (NVOC-Cl; 875 mg, 3.18 mmol) to a solution of nucleoside **3** (1.00 g, 2.54 mmol) in dry pyridine (25 mL).
2. Stir the resulting solution under argon atmosphere at room temperature for 3 h and monitor the progress of the reaction by TLC.
3. After complete conversion treat the solution with water (25 mL) at 0°C and remove the solvent under reduced pressure.
4. Purify the crude product by column chromatography (silica gel, 25 g, elution with CH$_2$Cl$_2$/CH$_3$OH 93:7).

Nucleoside **4** (928 mg, 58%) is obtained as a yellow solid (R_f 0.35, CH$_2$Cl$_2$/CH$_3$OH 9:1); ^{1}H NMR (500 MHz, [d_6]DMSO): δ = 1.29 (s, 3H; acetonide-H), 1.47–1.49 (m, 2H; linker-H), 1.53 (s, 3H; acetonide-H), 1.58–1.61 (m, 2H; linker-H), 3.03–3.07 (m, 2H; linker-H), 3.29–3.30 (m, 2H; linker-H), 3.54–3.57 (m, 2H; 5'-H), 3.85 (s, 3H; methoxy-H), 3.87 (s, 3H; methoxy-H), 4.14–4.15 (m, 1H; 4'-H), 4.95 (dd, ^3J = 2.9, 6.1 Hz, 1H; 3'-H), 5.31 (s, 2H; NVOC-CH$_2$), 5.35 (dd, ^3J = 3.7, 6.5 Hz, 1H; 2'-H), 5.47 (br. s, 1H; 5'-OH), 6.03 (d, ^3J = 3.7 Hz, 1H; 1'-H), 6.51 (s, 2H; 6-NH$_2$), 6.93 (t, ^3J = 5.2 Hz, 1H; linker-NH), 7.16 (s, 1H; NVOC-arom. H), 7.46 (t, ^3J = 5.8 Hz, 1H; linker-NH), 7.68 (s, 1H; NVOC-arom. H), 7.90 (s, 1H; 2-H); ESI-MS m/z (%): 633.2 (100) [M + H]$^+$.

3.1.5. 8-Amino[1''-(N''-6-Nitroveratryl-Oxocarbonyl)-4''-Aminobutyl]-2',3'-O-Isopropylidene-5'-O-Mesyladenosine *(5)*

1. Prepare a solution of nucleoside **4** (204 mg, 322 μmol), 4-dimethylaminopyridine (DMAP; 40 mg, 0.327 mmol) and dry triethylamine (TEA; 1.2 mL, 8.6 mmol) in dry methylene chloride (10 mL) and cool the solution under argon atmosphere to 0°C.
2. Add mesyl chloride (Mes-Cl; 55.5 μL, 0.707 mmol) and stir the solution for 90 min.
3. Quench the reaction by adding a cold, saturated sodium hydrogencarbonate solution (3 mL).
4. Extract the solution with ice-cold chloroform (3 × 5 mL), combine the organic phases, and remove the solvent under reduced pressure.
5. Purify the crude product by column chromatography (silica gel, 11 g, elution with CH$_2$Cl$_2$/CH$_3$OH 95:5).

Nucleoside **5** (118 mg, 52%; *see* **Note 2**) is obtained as an orange solid (R_f 0.47, CH$_2$Cl$_2$/CH$_3$OH 9:1); ^{1}H NMR (400 MHz, [d_6]DMSO): δ = 1.29 (s, 3H;

acetonide-H), 1.42–1.48 (m, 2H; linker-H), 1.50 (s, 3H; acetonide-H), 1.53–1.61 (m, 2H; linker-H), 3.01 (s, 3H; mesyl-H), 3.02–3.05 (m, 2H; linker-H), 3.26–3.32 (m, 2H; linker-H), 3.83 (s, 3H; methoxy-H), 3.84 (s, 3H; methoxy-H), 4.19 (dd, ^2J = 10.2, ^3J = 7.2 Hz, 1H; 5'-Ha), 4.24–4.31 (m, 1H; 4'-H), 4.33–4.38 (m, 1H; 5'-Hb), 5.09 (dd, ^3J = 6.3, 3.0 Hz, 1H; 3'-H), 5.28 (s, 2H; NVOC-CH$_2$), 5.60 (dd, ^3J = 6.3, 1.7 Hz, 1H; 2'-H), 6.10 (d, ^3J = 1.7 Hz, 1H; 1'-H), 6.52 (s, 2H; 6-NH$_2$), 7.05 (t, ^3J = 5.2 Hz, 1H; linker-NH), 7.13 (s, 1H; NVOC-arom. H), 7.45 (t, ^3J = 5.5 Hz, 1H; linker-NH), 7.64 (s, 1H; NVOC-arom. H), 7.88 (s, 1H; 2-H); ESI-MS m/z (%): 711.3 (71) [M + H]$^+$, 615.5 (100) [cyclonucleoside]$^+$.

3.1.6. 8-Amino[1''-(N''-6-Nitroveratryl-Oxocarbonyl)-4''-Aminobutyl]-5'-O-Mesyladenosine (6)

1. Dissolve nucleoside **5** (144 mg, 202 μmol) in aqueous formic acid (49%, 20 mL).
2. Stir the resulting solution at room temperature for 5 d.
3. After complete conversion remove the solvent under reduced pressure, and co-evaporate remaining solvent with a mixture of water and methanol (1:1, 3 × 5 mL) under reduced pressure.

Nucleoside **6** (132.2 mg, 98%; see **Note 2**) is obtained as a yellow solid (R_f 0.22, CH$_2$Cl$_2$/CH$_3$OH 9:1); ESI-MS m/z (%): 671.2 (45) [M + H]$^+$, 575.2 (100) [cyclonucleoside]$^+$.

3.1.7. 8-Amino[1''-(N''-6-Nitroveratryl-Oxocarbonyl)-4''-Aminobutyl]-5'-(1-Aziridinyl)-5'-Deoxyadenosine (7)

1. Dissolve nucleoside **6** (44.8 mg, 66.8 μmol) in aziridine (1.1 mL; see **Subheading 3.1.1.**) and N-ethyldiisopropylamine (EDIA; 400 μL).
2. Stir the solution under argon atmosphere at room temperature for 3 d and monitor the reaction by analytical reversed-phase HPLC (Prontosil, 5 μm, 120 Å, 250 × 4.6 mm, detection at 280 and 350 nm). Compounds are eluted with acetonitrile (14% for 10 min, followed by a linear gradient to 42% in 40 min and to 70% in 10 min) in triethylammonium acetate buffer (0.1 M, pH 7.0) and a flow rate of 1 mL/min.
3. Remove the solvent under reduced pressure after completeness of the reaction.
4. Purify the crude product by column chromatography (silica gel, 2.8 g, elution with CH$_2$Cl$_2$/CH$_3$OH 9:1).

Nucleoside **7** (8.0 mg, 20%; see **Note 3**) is obtained as a yellow solid (R_f 0.22, CH$_2$Cl$_2$/CH$_3$OH 9:1; HPLC retention time 21.3 min); ^{1}H NMR (500 MHz, [d_6]DMSO): δ = 1.25–1.27 (m, 2H; linker-H), 1.37–1.39 (m, 2H; linker-H), 1.43–1.45 (m, 2H; aziridine-H), 1.46–1.50 (m, 2H; aziridine-H), 1.67–1.69 (m, 2H; linker-H), 1.98 (dd, ^2J = 13.0 Hz, ^3J = 3.1 Hz, 1H; 5'-Ha), 2.97 (dd, ^2J = 13.0 Hz, ^3J = 3.4 Hz, 1H; 5'-Hb), 3.00–3.04 (m, 2H; linker-H), 3.85 (s, 3H; methoxy-H), 3.86 (s, 3H; methoxy-H), 3.97 (m, 1H; 4'-H), 4.23 (m, 1H; 3'-H),

4.71 (m, 1H; 2'-H), 5.18–5.19 (m, 1H; OH), 5.31 (s, 2H; NVOC-CH$_2$), 5.92 (d, ^3J = 7.2 Hz, 1H; 1'-H), 6.41 (s, 2H; 6-NH$_2$), 7.16 (s, 1H; NVOC-arom. H), 7.49 (t, ^3J = 5.7 Hz, 1H; linker-NH), 7.61 (t, ^3J = 5.7 Hz, 1H; linker-NH), 7.68 (s, 1H; NVOC-arom. H), 7.88 (s, 1H; 2-H); ESI-MS m/z (%): 618.3 (100) [M + H]$^+$.

3.1.8. 8-Amino[1''-(4''-Aminobutyl)]-5'-(1-Aziridinyl)-5'-Deoxyadenosine (8)

1. Add a solution of nucleoside **7** (0.38 mg, 0.62 (μmol; determined by UV spectroscopy; *see* **Subheading 3.3.** for extinction coefficient) in DMSO (200 μL) to triethanolamine hydrochloride buffer (750 μL, 100 mM, pH 8.0).
2. Irradiate the solution in a quartz cuvette with a mercury lamp at room temperature for 40 min (*see* **Note 4**) and monitor the reaction progress by analytical reversed-phase HPLC (Prontosil, 5 μm, 120 Å, 250 × 4.6 mm, detection at 280 nm). Compounds are eluted with acetonitrile (4.9% for 10 min, followed by a linear gradient to 28% in 30 min and to 70% in 10 min) in triethylammonium acetate buffer (0.1 M, pH 7.0) and a flow rate of 1 mL/min. Because complete conversion occurs, the solution can be used directly in the next step.

Nucleoside **8** (*see* **Note 3**; HPLC retention time = 4.6 min); ESI-MS m/z (%): 379.3 (100) [M + H]$^+$.

3.1.9. 8-Amino[1''-(N''-Biotinyl)-4''-Aminobutyl]-5'-(1-Aziridinyl)-5'-Deoxyadenosine (9)

1. Add a solution of succinimidyl biotin (biotin-NHS; 0.66 mg, 1.93 μmol) in DMSO (500 μL) directly to the solution of deprotected nucleoside **8** (*see* **Subheading 3.1.8.**).
2. Stir the mixture at room temperature for 40 min.
3. Purify the crude product by preparative reversed-phase HPLC (Prontosil, 5 μm, 120 Å, 250 × 8 mm, detection at 280 nm). Compounds are eluted with acetonitrile (4.9% for 10 min, followed by a linear gradient to 28% in 30 min and to 70% in 10 min) in triethylammonium hydrogencarbonate buffer (0.01 M, pH 8.6) and a flow rate of 3 mL/min (*see* **Note 5**).
4. Freeze-dry product containing fractions.

Nucleoside **9** (0.17 mg, 45%; determined by UV spectroscopy; *see* **Subheading 3.3.** for extinction coefficient and **Note 3**) is obtained as a white solid; (HPLC retention time 13.5 min); ^{1}H NMR (500 MHz, [d_6]DMSO): δ = 1.28–1.30 (m, 2H; aliphat. H), 1.31–1.32 (m, 2H; aliphat. H), 1.36–1.48 (m, 8H; 4 × aliphat. H, 2 × aziridine-H), 1.55–1.59 (m, 2H; aliphat. H), 1.69–1.71 (m, 2H; aliphat. H), 1.98 (dd, ^3J = 3.0 Hz, ^2J = 13.7 Hz, 1H; 5'-Hb), 1.99–2.06 (m, 2H; aliphat. H), 2.56 (d, ^3J = 12.8 Hz, 1H; SCH$_2$, Ha), 2.80 (dd, ^3J = 4.9, 12.8 Hz, 1H; SCH$_2$, Hb), 2.98 (dd, ^3J = 3.1 Hz, ^2J = 13.1 Hz, 1H; 5'-Ha), 3.01–3.03 (m, 1H; SCH), 3.15–3.26 (m, 2H; aliphat. H), 3.97 (dt, ^3J = 2.1, 3.7 Hz, 1H; 4'-H), 4.10 (dd, ^3J = 4.5, 7.9 Hz, 1H; SCHRC<u>H</u>), 4.23 (dd, ^3J = 2.1, 5.3 Hz, 1H; 3'-H), 4.28 (dd, ^3J = 4.7,

7.7 Hz, 1H; SCH$_2$CH), 4.69 (dd, ^3J = 5.1, 7.3 Hz, 1H; 2'-H), 5.92–5.93 (d, ^3J = 7.3 Hz, 1H; 1'-H), 6.34 (s, br., 1H; biotin-NH), 6.39 (s, 2H; 6-NH$_2$), 6.40 (s, 1H; biotin-NH), 7.57 (t, ^3J = 5.5 Hz, 1H; linker-NH), 7.78 (t, ^3J = 5.6 Hz, 1H; linker-NH), 7.88 (s, 1H; 2-H); ESI-MS m/z (%): 605.3 (100) [M + H]$^+$.

3.2. Expression and Purification of the M·TaqI DNA MTase

This protocol for expression and purification of M·TaqI is based on literature procedures *(9,10)*. M·TaqI is expressed in ER 2267 *E. coli* cells harboring the pAGL15-M13 plasmid, which carries the gene for M·TaqI under the inducible tac promotor.

1. Grow cells in flasks (6 × 5 L) containing LB medium (6 × 2 L) supplemented with ampicillin (100 mg/L) and kanamycin (35 mg/L) at 37°C to an optical density of 0.6 at 600 nm.

2. Induce protein expression by adding isopropyl-β-D-thio-galactopyranoside (0.1 m*M* final concentration) and continue cell growth for 4 h.

3. Harvest cells by centrifugation (20 min at 2800*g* and 4°C) and store the cell paste (~50 g wet weight) at –20°C.

4. Lyse cells by sonication in buffer A (200 mL, Mes/HEPES/sodium acetate buffer mix, 6.7 m*M* each, pH 7.5, 1 m*M* ethylenediamine tetraacetic acid, 10 m*M* β-mercaptoethanol, and 10% glycerol) containing sodium chloride (200 m*M*) and phenylmethylsulfonyl fluoride (0.2 m*M*) and centrifuge the suspension (1 h at 35,000*g* and 4°C).

5. Load the cleared supernatant onto a cation exchange column (Poros HS/M, 26 × 95 mm) and elute with a linear gradient of sodium chloride (0.2 to 1.0 *M*) in buffer A and pool fractions containing M·TaqI.

6. Dilute the combined fractions twofold with buffer A, load the solution onto a heparin column (heparin Sepharose CL 6B, 26 × 95 mm), and elute with a linear gradient of sodium chloride (0.2 to 1.0 *M*) in buffer A. Pool fractions containing M·TaqI and concentrate the protein solution by ultrafiltration (Jumbosep, 10 kDa cut off).

7. Load the concentrated protein solution (5–10 mL) onto a gel filtration column (Superdex 75, 26 × 600 mm), elute with buffer A containing sodium chloride (0.6 *M*) and pool fractions containing M·TaqI.

8. Dilute the combined fractions threefold with buffer A and reload the protein solution onto the cation exchange column (Poros HS/M, 26 × 95 mm) for removal of any natural cofactor bound to M·TaqI. Extensively wash the column with buffer B (10 L, Mes/HEPES/sodium acetate buffer mix, 6.7 m*M* each, pH 6.0, 300 m*M* potassium chloride, 0.2 m*M* 1,4-dithiothreitol, and 10% glycerol, *see* **Note 6**) and elute M·TaqI with a linear gradient of potassium chloride (0.1 to 1 *M*) in buffer C (40 m*M* Tris acetate, pH 7.9, 20 m*M* magnesium acetate, 100 m*M* potassium acetate, 2 m*M* 1,4-dithiothreitol, and 10% glycerol).

9. Pool fractions containing M·TaqI, concentrate the protein solution by ultrafiltration (Centriprep YM-10) to 2–3 mL, dilute the protein solution twofold with glycerol, and store the protein solution at –20°C.

The concentration of M·*Taq*I can be estimated using Coomassie Protein Assay Reagent and bovine serum albumin as standard. Typically, approx 50 mg of M·*Taq*I are obtained. Molar concentrations are calculated using a molecular weight of 47,850 g/mol for M·*Taq*I.

3.3. Labeling of Short Duplex Oligodeoxynucleotides

Labeling of short duplex oligodeoxynucleotides is illustrated with the duplex **10·11** (**Fig. 3A** and **Fig. 4A**) containing the 5'-TCGA-3' recognition sequence of M·*Taq*I (*see* **Note 7**). Extinction coefficients for the single strands at 260 nm (**10**: 129,700 L mol^{-1} cm^{-1}; **11**: = 122,700 L mol^{-1} cm^{-1}) are calculated according to the nearest-neighbor method *(11)*. Hybridization of the complementary strands is achieved by mixing equal molar amounts (measured by UV absorption at 260 nm) in Tris acetate buffer (20 m*M*, pH 6.0) containing magnesium acetate (10 m*M*) and potassium acetate (50 m*M*), heating to 95°C for 2 min and slow cooling to room temperature in a heat block over a period of 2–3 h. Concentrations of aziridine cofactors **7**, **8**, and **9** are determined by UV spectroscopy using the published extinction coefficient of 8-amino[1"-(6"-aminohexyl)]-adenosine-5'-phosphate at 278 nm of 20,450 L mol^{-1} cm^{-1} *(12)*.

3.3.1. Labeling With Primary Amino Groups

Labeling of duplex **10·11** with a primary amino group is achieved with aziridine cofactor **8** *(***Fig. 3A***)*, which is freshly prepared by UV light-induced deprotection of nucleoside **7** (*see* **Note 8**).

1. Add a solution of nucleoside **7** (80 nmol dissolved in 6.5 µL DMSO) to triethanolamine hydrochloride buffer (93.5 µL, 10 m*M*, pH 8.0) in an Eppendorf tube (0.5 mL) and attach the tube to an irradiation glass apparatus with a mercury lamp using adhesive tape.
2. Irradiate the solution at room temperature for 30 min (complete deprotection can be verified by analytical reversed-phase HPLC; *see* **Subheading 3.1.8.**). The resulting solution of deprotected nucleoside **8** (800 µ*M*) is then directly used in a 10-fold dilution for the DNA labeling reaction.
3. Prepare a solution of aziridine cofactor **8** (80 µ*M*), duplex **10·11** (10 µ*M*) and M·*Taq*I (11 µ*M*) in Tris acetate buffer (20 m*M*, pH 6.0) containing magnesium acetate (10 m*M*), potassium acetate (50 m*M*), and Triton X-100 (0.01%) and incubate at 37°C. The progress of the reaction can be monitored by anion exchange HPLC (Poros 10 HQ, 10 µm, 4.6 × 10 mm, detection 260 nm). Compounds are eluted with potassium chloride (0.2 *M* for 5 min, followed by a linear gradient to 0.4 *M* in 5 min, to 0.6 *M* in 20 min, and to 1 *M* in 5 min) in Tris hydrochloride buffer (10 m*M*, pH 7.6) and a flow rate of 4 mL/min (**Fig. 3B**). The duplex **10·11** elutes with a retention time of 22.3 min (**Fig. 3B**, trace a) and after 22 h incubation almost all starting material is converted to a fast eluting complex between M·*Taq*I and the product duplex **10·12** (**Fig. 3B**, trace b).

A

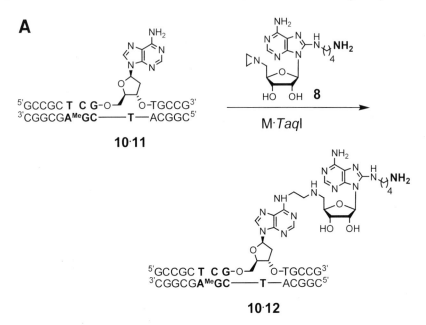

10·11

M·TaqI

10·12

B

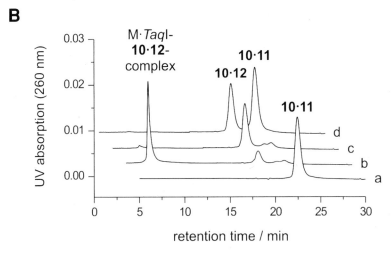

Fig. 3. Labeling of duplex oligodeoxynucleotide **10·11** with a primary amino group. Reaction scheme for the M·TaqI-catalyzed nucleophilic attack of the exocyclic amino group of adenine within the 5'-TCGA-3' sequence (A^{Me} = N-6-methyl-2'-deoxy-adenosine; *see* **Note 7**) leads to opening of the aziridine ring of cofactor **8** and covalent coupling of the whole cofactor carrying a free primary amino group with the duplex (**A**). The reaction is monitored by anion exchange HPLC and the chromatograms (**B**) show the following: Starting duplex **10·11** (trace a), reaction after 22 h (trace b), additional treatment with Proteinase K (trace c) and coinjection of product duplex **10·12** with starting duplex **10·11** (trace d).

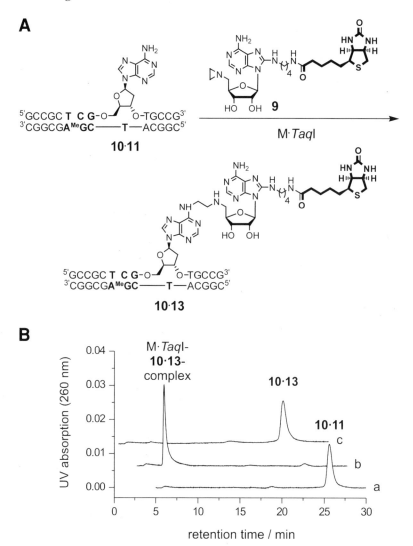

Fig. 4. Labeling of duplex oligodeoxynucleotide **10·11** with biotin. Reaction scheme for the M·*Taq*I-catalyzed coupling of the biotinylated aziridine cofactor **9** with the exocyclic amino group of adenine within the 5'-TCGA-3' sequence (A^{Me} = *N*6-methyl-2'-deoxyadenosine; *see* **Note 7**; **A**). The reaction is monitored by anion exchange HPLC and the chromatograms (**B**) show the following: Starting duplex **10·11** (trace a), reaction after 22 h (trace b), and additional treatment with Proteinase K (trace c).

4. Dilute the labeling solution twofold with Tris hydrochloride buffer (50 mM, pH 8.0) containing calcium chloride (1 mM), add a solution of Proteinase K (0.2 mg per nmol duplex, 20 mg/mL) and incubate at 65°C for 1 h. This step

serves to release the product duplex **10·12** from the complex and the product duplex **10·12** elutes after 19.6 min (**Fig. 3B**, trace c). A coinjection of starting duplex **10·11** and product duplex **10·12** is also shown in **Fig. 3B** (trace d). This anion exchange HPLC method can also be used to isolate the labeled duplex and further modification of duplex **10·12** with amine-reactive probes should be possible (*see* **Note 9**).

3.3.2. Labeling With Biotin

Although the product duplex **10·12** with a primary amino group could be further modified with succinimidyl biotin to yield biotinylated DNA, it appears more convenient to first attach the biotin group to the cofactor and then use the aziridine cofactor **9** for labeling. The labeling reaction with cofactor **9** (**Fig. 4A**) is performed as described for the labeling reaction with cofactor **8** (*see* **Subheading 3.3.1.**).

1. Prepare a solution of aziridine cofactor **9** (80 μ*M*, predissolved in DMSO), duplex **10·11** (10 μ*M*), and M·*Taq*I (11 μ*M*) in Tris acetate buffer (20 m*M*, pH 6.0) containing magnesium acetate (10 m*M*), potassium acetate (50 m*M*), and Triton X-100 (0.01%) and incubate at 37°C for 22 h. The progress of the reaction can be monitored by anion exchange HPLC (**Fig. 4B**). At the end of the reaction the starting duplex **10·11** has disappeared and a fast eluting complex between M·*Taq*I and the product duplex **10·13** is formed.
2. Release the product duplex **10·13** from the complex by treatment with Proteinase K as described above (*see* **Subheading 3.3.1.**). The free product duplex **10·13** elutes about 1 min earlier during anion exchange HPLC than the starting duplex **10·11**.

3.4. Labeling of Plasmid DNA With Biotin

Labeling of plasmid DNA is illustrated with linearized pUC19 DNA (*see* **Note 10**).

1. Linearize pUC19 DNA (0.25 μg/μL) by treatment with R·*Eco*RI (10 U per μg of plasmid DNA) in Tris hydrochloride buffer (50 m*M*, pH 7.5) containing magnesium chloride (10 m*M*), sodium chloride (100 m*M*), Triton X-100 (0.02%), and bovine serum albumin (0.1 mg/mL; the recommended buffer supplied by the manufacturer) and incubate at 37°C for 1 h.
2. Prepare a solution of R·*Eco*RI-linerized pUC19 (0.025 μg/μL, 14.1 n*M*, three recognition sequences for M·*Taq*I), M·*Taq*I (46.5 n*M*), and aziridine cofactor **9** (80 μ*M*, predissolved in DMSO) in Tris acetate buffer (20 m*M*, pH 6.0) containing magnesium acetate (10 m*M*), potassium acetate (50 m*M*), and Triton X-100 (0.01%) and incubate at 60°C for 3 h. The progress of the labeling reaction can be monitored in a DNA protection assay (**Fig. 5**).
3. Remove aliquots (8 μL, 0.2 μg linearized pUC19) after different incubation times, add restriction endonuclease R·*Taq*I (1 μL, 5 U) and 10X reaction buffer (1 μL) supplied by the manufacturer to each aliquot and incubate the mixtures at 65°C

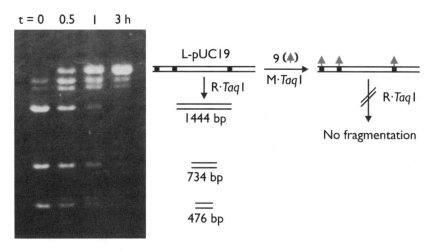

Fig. 5. M·*Taq*I-catalyzed labeling of plasmid DNA with biotin. The progress of the labeling reaction with the aziridine cofactor **9** is analyzed in a DNA protection assay. At the beginning of the labeling reaction (t = 0 h) fragmentation of R·*Eco*RI-linearized pUC19 DNA (L-pUC19) with the restriction endonuclease R·*Taq*I (one of the four recognition sequences of R·*Taq*I in pUC19 overlaps with the R·*Eco*RI recognition sequence) leads to three major bands (1444, 734, and 476 bp) on an agarose gel (an additional 32-bp fragment is too small to be observed on the gel). With increasing reaction times (t = 0.5 and 1 h), these bands disappear and bands corresponding to longer intermediates or the full-length linearized plasmid (2686 bp) appear. After 3 h the DNA is almost completely protected against fragmentation by R·*Taq*I, indicating that the three recognition sequences of R·*Taq*I are blocked by covalent modification. No DNA protection against cleavage by R·*Taq*I is observed in the absence of either M·*Taq*I or the aziridine cofactor **9** (not shown).

for 1 h. Afterward, add a solution (2 µL) of glycerol (30%) containing bromophenol blue (0.25%) to each aliquot and analyze the samples by standard agarose gel electrophoreses.

4. Remove M·*Taq*I from the plasmid DNA by adjusting the pH of the solution to 8.0, add Proteinase K (10 µg per µg DNA) and incubate the solution at 65°C for 1 h.
5. Purify the labeled plasmid DNA with the QiagenPCR purification Kit according to the instructions given by the manufacturer.

4. Notes

1. Caution: Aziridine is hazardous and should be handled with care in a fume hood.
2. The mesylated nucleosides **5** and **6** have a strong tendency to form cyclonucleosides by nucleophilic attack of the nitrogen at the 3-position of the adenine ring on the activated 5'-carbon. Thus, it is best to use nucleosides **5** and **6** as quickly as possible in the next step.

3. Aziridines are sensitive to acids and rapidly polymerize under acidic conditions. Thus, never expose aziridine or the aziridine cofactors **7**, **8**, and **9** to acidic conditions below pH 6.0.

4. For larger scale syntheses an illumination glass apparatus (max. 1 L) with an inner mercury lamp (125 W) was used.

5. HPLC purification is performed with triethylammonium hydrogencarbonate buffer to avoid acidification (*see* **Note 3**) upon freeze-drying.

6. This step is necessary to remove any natural cofactor from M·*Taq*I.

7. M·*Taq*I recognizes the double-stranded palindromic 5'-TCGA-3' sequence and can alkylate the adenine residues in both strands leading to two different hemialkylated products in the first turnover. In order to facilitate product analysis we used the hemimethylated duplex **10·11** in which the lower strand **10** contains *N*6-methyl-2'-deoxyadenosine (A^{Me}) instead of 2'-deoxyadenosine within the recognition sequence, which can not be modified by M·*Taq*I. Therefore, alkylation is directed to the upper strand **11**. However, modification by M·*Taq*I is not restricted to hemimethylated substrates and nonmethylated recognition sequences can be labeled as well (*see* **Subheading 3.4.**).

8. Alternatively, nucleoside **7** can be used to label the duplex in a M·*Taq*I-catalyzed reaction. Afterwards the primary amino group can be deprotected by irradiation with UV light.

9. The amino groups of the nucleobases adenine, guanine, and cytosine in DNA are not nucleophilic because the nitrogen lone pairs are conjugated with the ring systems.

10. Labeling of plasmid DNA can also be performed with circular plasmid DNA.

Acknowledgments

We thank Nathalie Bleimling and Kerstin Glensk for the preparation of M·*Taq*I and Roger Goody for his continuous support. This work was supported by the QIAGEN GmbH, Hilden, Germany.

References

1. Cheng, X. (1995) Structure and function of DNA methyltransferases. *Annu. Rev. Biophys. Biomol. Struct.* **24**, 293–318.

2. Jeltsch, A. (2002) Beyond Watson and Crick: DNA methylation and molecular enzymology of DNA methyltransferases. *ChemBioChem* **3**, 275–293.

3. Pignot, M., Siethoff, C., Linscheid, M., and Weinhold, E. (1998) Coupling of a nucleoside with DNA by a methyltransferase. *Angew. Chem. Int. Ed.* **37**, 2888–2891.

4. Pljevaljcic, G., Pignot, M., and Weinhold, E. (2003) Design of a new fluorescent cofactor for DNA methyltransferases and sequence-specific labeling of DNA. *J. Am. Chem. Soc.* **125**, 3486–3492.

5. Perrin, D. D. and Armarego, W. L. F. (eds.) (1988) *Purification of Laboratory Chemicals*. Pergamon Press, Oxford, England.

6. Gabriel, S. (1888) Ueber Vinylamin und Bromäthylamin. *Chem. Ber.* **21,** 2664–2669.
7. Gabriel, S. and Stelzner, R. (1895) Ueber Vinylamin. *Chem. Ber.* **28,** 2929–2938.
8. Ikehara, M., Tada, H., and Kaneko, M. (1968) Studies of nucleosides and nucleotides—XXXV. Purine cyclonucleosides—5. Synthesis of purine cyclonucleoside having 8,2'-O-anhydro linkage and its cleavage reactions. *Tetrahedron* **24,** 3489–3498.
9. Holz, B., Klimasauskas, S., Serva, S., and Weinhold, E. (1998) 2-Aminopurine as a fluorescent probe for DNA base flipping by methyltransferases. *Nucleic Acids Res.* **26,** 1076–1083.
10. Goedecke, K., Pignot, M., Goody, R. S., Scheidig, A. J., and Weinhold, E. (2001) Structure of the *N*6-adenine DNA methyltransferase M·*Taq*I in complex with DNA and a cofactor analog. *Nat. Struct. Biol.* **8,** 121–125.
11. Cantor, C. R., Warshaw, M. M., and Shapiro, H. (1970) Oligonucleotide interactions. 3. Circular dichroism studies of the conformation of deoxyoligonucleotides. *Biopolymers* **9,** 1059–1077.
12. Barker, R., Trayer, I. P., and Hill, R. L. (1974) Nucleoside phosphates attached to agarose. *Methods Enzymol.* **34,** 479–491.

11

Hapten Labeling of Nucleic Acids for Immuno-Polymerase Chain Reaction Applications

Michael Adler

Summary

A method for the ultrasensitive protein detection in the range of 0.01 to 10,000 amol of the model antibody anti-mouse-IgG from rabbit is described, using a combination of Immuno–polymerase chain reaction (PCR) and PCR–enzyme-linked immunosorbent assay (PCR-ELISA). The antibody was first immobilized on antigen-coated microplates; in a second step, a commercially available DNA-labeled species-specific antibody was added; and finally the deoxyribonucleic marker was amplified in a PCR step, including twofold labeling with biotinylated primer and a hapten-coupled nucleotide during PCR. Subsequently, the labeled PCR product was immobilized on streptavidin-coated microplates and detected with an antibody–enzyme conjugate. The protocol could easily be adapted to the detection of other antibodies or antigens by exchanging the antigen-specific antibody. Several modifications of the method as well as optimization steps, potential error sources, and countermeasures are discussed.

Key Words: Immuno-PCR; ELISA; PCR-ELISA; indirect assay; secondary antibody.

1. Introduction

The polymerase chain reaction (PCR; **ref. *1***) is a powerful tool both for the modification and amplification of nucleic acids. The combination of the exponential amplification power of PCR with antibody-based immunoassays, such as the enzyme-linked immunosorbent assay (ELISA, e.g., **ref. *2***), allows for the detection of proteins at a level of a few hundred molecules. This Immuno-PCR (IPCR) method, first introduced in 1992 by Sano et al. *(3)*, is because of its enormous sensitivity very demanding for the experimentation and therefore not in common use, even though there is a high interest in ultrasensitive protein detection and many successful applications of this technique in research studies *(3–46)*.

Modifications of the method, especially the development of preconjugated deoxyribonucleic acid (DNA)–antibody reagents *(30)* and the combination with

From: *Methods in Molecular Biology, vol. 283: Bioconjugation Protocols: Strategies and Methods*
Edited by: C. M. Niemeyer © Humana Press Inc., Totowa, NJ

PCR–ELISA *(18,27)* detection methods for the amplificated product, nowadays gives even the nonspecialized researcher the opportunity to use this technique for his or her experiments.

In this chapter an example of the detection of the model protein anti-mouse IgG from rabbit in an indirect IPCR-assay, the application of the IPCR, as well as the usage of PCR–ELISA for the quantification of (I)PCR amplificates is shown. The complete method described could easily be adapted to detect other antibodies. The PCR–ELISA technique could also be combined with standard PCR assays for the quantitative detection of nucleic acids alone.

In the protocol introduced in this chapter, a typical indirect ELISA set-up (*see* **Fig. 1A**) for the detection of a primary antibody was modified for IPCR. The antigen was immobilized on microplates and in a first step coupled with an antigen-specific primary antibody from rabbit. Subsequently, a commercially available DNA-labeled species-specific secondary antibody was added and, after a washing step for the removal of nonspecifically bound reagents, the DNA marker was amplified in a PCR (*see* **Fig. 1B**). The amplificate was labeled during amplification with a biotinylated primer and a hapten-coupled nucleotide. After the PCR, the labeled PCR product was immobilized on streptavidin-coated microplates and detected with an antibody–enzyme conjugate (**Fig. 1C**). In the **Notes** section, several modifications of the method, as well as optimization steps, potential error sources, and countermeasures, are discussed.

1.1. Abbreviations

mIgG: Mouse-IgG
αm-rIgG: Rabbit-anti-mouse IgG
αr-gIgG(b): Biotinylated goat-anti-rabbit IgG
AP: Alkaline phosphatase
ELISA: Enzyme-linked immunosorbent assay
IPCR: Immuno-PCR
HP: Horseredish peroxidase
PCR: Polymerase chain reaction
RSR: "Anti-rabbit IgG secondary reagent," goat-anti rabbit-IgG coupled wih DNA
STV: Streptavidin

2. Materials
2.1. Equipment

1. Digital pipet 1–10, 10–100, and 100–1000 µL (Eppendorf, Hamburg, Germany).
2. Multichannel pipet 5–50 µL, 50–300 µL (Finnpipette, Thermo Electron, Dreieich, Germany).
3. Microtiter-plate compatible PCR thermocycler (e.g., MWG, Ebersberg, Germany, MJ Research, Perkin-Elmer, Boston, MA).

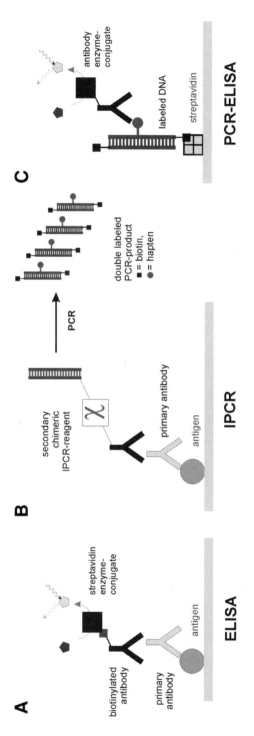

Fig. 1. Comparison of indirect ELISA, IPCR, and PCR-ELISA. **A**, ELISA: The antigen mouse-IgG is immobilized on a microplate surface, coupled in a first step with an antigen specific primary rabbit-antimouse antibody, and subsequently with a biotinylated species-specific secondary goat-antirabbit antibody. The detection is conducted with a STV–enzyme conjugate that binds to the biotinylated antibody and changes, for instance, a colorless substrate into a colored product. The assay could be used for the detection of either the immobilized antigen or the primary antibody. **B**, IPCR: The assay is conducted similar to the previously described ELISA, but instead of a species-specific biotinylated antibody, a species-specific antibody-DNA conjugate is used for detection. During PCR amplification of the DNA marker, biotin and hapten labels are incorporated in the amplification product. **C**, PCR–ELISA: The biotin- and hapten-labeled PCR amplificate is immobilized on a STV-coated microplate surface and coupled with an antibody–enzyme conjugate, which is subsequently detected by the enzymatic reaction.

4. Microtiter-plate reader for fluorescence detection (e.g., Victor, Perkin-Elmer).
5. Orbital shaker for microtiter plates (Heidolph, Schwabach, GmbH).

2.2. Consumables

1. TopYield™ starter kit (no. 248917 NUNC, Wiesbaden, Germany).
2. Polypropylen microtiter plates (no. 267245 NUNC).
3. Siliconisated tubes (1.5 mL., e.g., Biozym, Hess, Oldendorf, Germany, Starlab, Ahrensburg, Germany).
4. Black polystyrol microtiter plates (no. 475515, NUNC).
5. Filter tips for the pipets (e.g., Eppendorf, Biozym).

2.3. Reagents and Buffers

1. 50 mM Borate buffer, pH 9.5.
2. Mouse-IgG (I-5381, Sigma, Taufkirchen, Germany).
3. Rabbit-anti-mouse IgG (B-7264, Sigma).
4. Biotinylated goat-anti-rabbit IgG (B-7264, Sigma).
5. Anti-rabbit IgG DNA conjugate (RSR, cat. no. 42–02, Chimera Biotec GmbH, Dortmund, Germany).
6. Conjugate dilution buffer (CDB, cat. no. 30–04, Chimera Biotec).
7. Assay-Buffer A (cat. no. 30–02, Chimera Biotec).
8. Assay-Buffer B (cat. no. 30–03, Chimera Biotec).
9. Blocking solution (cat. no. 30–01, Chimera Biotec).
10. Primer mix (cat. no. 30–05, Chimera Biotec).
11. PCR supplement mix (cat.-no. 30–07, Chimera Biotec).
12. *Taq* polymerase (Biomaster GmbH, Cologne, Germany, including 10X buffer and MgCl$_2$ stock solution).
13. Digoxigenin dUTP (1835289, Roche, Mannheim, Germany).
14. Amplificate buffer (cat. no. 32–01, Chimera Biotec).
15. Anti-digoxigenin alkaline phosphatase conjugate (1093274, Roche).
16. Streptavidin (STV) alkaline phosphatase conjugate (1093266, Roche).
17. AttoPhos® substrate (1681982, Roche).
18. STV, recombinant (1721666, Roche).

3. Methods

The methods described in the following outline include the performance of the IPCR assay (**Subheading 3.1.**), the labeling of the nucleic acid marker DNA during amplification (**Subheading 3.2.**), PCR–ELISA detection of the labeled DNA (**Subheading 3.3.**), and quantitative analysis of the data read-out (**Subheading 3.4.**).

3.1. Immuno-PCR

Before PCR amplification, the antibody to be detected is coupled with the DNA marker.

	A	B	C	D	E	F	G	H
1	0.01 amol	0.1 amol	1 amol	10 amol	100 amol	1000 amol	NC IPCR	10,000 amol
2	0.01 amol	0.1 amol	1 amol	10 amol	100 amol	1000 amol	NC IPCR	10,000 amol
3	0.3 amol	30 amol	300 amol	NC PCR	0.3 amol	30 amol	300 amol	NC PCR

Fig. 2. Typical IPCR sample set-up on microplate: Rows 1 and 2, calibration curve in double determination; row 3, "unknown" samples.

3.1.1. Preparation of Microplate Surfaces

1. Immobilize a uniform concentration of 10 µg/mL of mIgG in borate buffer in a volume of 30 µL/well on TopYield microplates for 12–48 h at 4°C (*see* **Notes 1** and **3**).
2. Subsequently, wash the IgG-coated modules three times for three minutes each with 240 µL/well buffer A.
3. Incubate the modules with 240 µL/well blocking solution for at least 12 h at 4°C (*see* **Note 2**). The blocked modules are stable for approx 1 wk at 4°C.

Parallel to the preparation of the antibody-coated modules for immuno-PCR, streptavidin-coated microplates for the PCR–ELISA should be prepared (*see* **Subheading 3.3.1.**).

3.1.2. Antigen–Antibody Coupling

1. Wash the coated modules two times for 30 s and two times for 4 min under orbital shaking with 240 µL/well buffer B.
2. Incubate with αm-rIgG in a 10-fold dilution series ranging from 0.01 to 10,000 amol/30 µL well volume (approx 50 fg/mL to 50 ng/mL) in buffer B containing 10% blocking solution (**Fig. 2**). Include one well containing only buffer B/10% blocking solution as an IPCR-negative control. Also prepare three "spiking" samples with 0.3, 30, and 300 amol/30 µL. All dilution steps should be conducted in siliconisated cups. All samples should be conducted in double determination (*see* **Notes 3** and **12**).

3.1.3. Coupling With IPCR Reagent

1. After 25 min incubation at room temperature under orbital shaking wash four times with buffer B as described previously.
2. Add 30 µL/well RSR in a dilution of 1:300 in conjugate dilution buffer (Chimera Biotec, *see* **Note 4**).
3. Incubate again for 25 min as described previously.
4. Wash seven times with buffer B (4 × 30 s, 3 × 3 min) and two times for 1 min with buffer A.

Table 1
PCR Mastermix for Eight Wells

PCR supplement mix	203 µL
10X Buffer	25µL
50 mM MgCl$_2$	15 µL
0.3 mM Dig-dUTP	2 µL
100 µM PrimerMix	5 µL
Taq polymerase	1.25 µL
Total	250 µL

3.1.4. Control–ELISA

Parallel to the IPCR, a conventional ELISA is conducted as a control experiment (**Fig. 1**). Perform the assay steps in **Subheadings 3.1.1.** and **3.1.2.** as described previously.

1. After 25 min incubation at room temperature under orbital shaking wash four times with buffer B as described (**Subheading 3.1.2., step 1**).
2. Add 30 µ/well "αr-gIgG(b)" in a concentration of 5 nM in conjugate dilution buffer (Chimera Biotec).
3. Incubate again for 25 min as described.
4. Wash four times with buffer B as described.
5. Add 30 µ/well streptavidin alkaline phosphatase conjugate in a dilution of 1:5000 in reagent dilution buffer.
6. Incubate again for 25 min as described.
7. Wash four times with buffer B as described and three times for 1 min with buffer A.
8. Add 50 µL of AttoPhos™ substrate (Roche) to each well.
9. Incubate for 15 min at room temperature under orbital shaking.
10. Measure fluorescence intensity at 550 nm.

For quantification of the data, *see* **Subheading 3.5.**

3.2. Labeling and Amplification of the Marker DNA

During PCR, a twofold labeling of the DNA amplificate with a hapten-coupled nucleotide and biotinylated primers is conducted (*see* **Note 5**).

1. Prepare a PCR mastermix according to the recipe in **Table 1**, calculated for one row of eight wells. For a larger amount of wells, use multiples of the volumes given in **Table 1** (e.g., for the above described calibration curve and spiking probes, 3 × 8 wells of PCR mastermix are needed, including a PCR-negative control in double determination) The preparation of the PCR mix should be performed either during the last washing steps or in advance, using frozen aliquots (*see* **Note 6**).
2. Pipet 30 µL of the PCR mastermix in each well.

Table 2
IPCR Method

Time	Temperature	Repeats
5 min	95°C	1×
1 min	50°C	
1 min	72°C	28×
12 s	95°C	
5 min	50°C	1×
5 min	72°C	1×

3. Seal the wells with adhesive foil (*see* **Note 7**).
4. Perform a PCR according to the method in **Table 2**.
5. After completion of the PCR, add 30 µL of amplificate buffer to each well.
6. Transfer the amplificate in a polypropylene microplate. The plate could be sealed with adhesive foil and the amplificate stored for up to 4 wk at 4°C.
7. Dilute the amplificate 1:800 in buffer A by adding 5 µL of amplificate in 200 µL of buffer A and subsequently 10 µL of this dilution to 200 µL of buffer A. The dilution should be carried out in a polypropylene microtiter plate, using a multi-channel pipet (*see* **Note 8**).

The plate could be sealed with adhesive foil and the dilution stored for up to 2 d at 4°C.

3.3. Detection of the Amplified Product With PCR–ELISA

The double-labeled PCR amplificate is immobilized on microtiter plates and coupled with an antibody-enzyme conjugate (*see* **Note 9**). Subsequently, the detection is carried out using a fluorescence-generating substrate.

3.3.1. Preparation of Microplate Surfaces

STV-coated microplates were prepared in house similar to IgG-coated plates (*see* **Note 10**). The microplates should be prepared in advance of the steps in **Subheadings 3.1.** and **3.2.**

1. Immobilize a uniform concentration of 10 µg/mL of STV (Roche) in a volume of 50 µL/well on polystyrol-microtiter modules (e.g., black MaxiSorp F16 for fluorescence assays, *see also* **Note 9**) for 72 h at 4°C.
2. Subsequently, wash the STV-coated plates three times for 3 min each with 240 µL/well buffer A.
3. Incubate the modules with 150 µL/well blocking solution for at least 12 h at 4°C.

The blocked modules were stored with the blocking solution, sealed with adhesive foil and are stable for approx 4 wk at 4°C.

	A	B	C	D	E	F	G	H
IPCR 1	0.01 amol	0.1 amol	1 amol	10 amol	100 amol	1000 amol	NC IPCR	10,000 amol
IPCR 2	0.01 amol	0.1 amol	1 amol	10 amol	100 amol	1000 amol	NC IPCR	10,000 amol
IPCR 3	0.3 amol	30 amol	300 amol	NC PCR	0.3 amol	30 amol	300 amol	NC PCR
IPCR 1	0.01 amol	0.1 amol	1 amol	10 amol	100 amol	1000 amol	NC IPCR	10,000 amol
IPCR 2	0.01 amol	0.1 amol	1 amol	10 amol	100 amol	1000 amol	NC IPCR	10,000 amol
IPCR 3	0.3 amol	30 amol	300 amol	NC PCR	0.3 amol	30 amol	300 amol	NC PCR
Control	ELISA-NC	ELISA-NC						

Fig. 3. Typical PCR–ELISA sample set-up on microplate, corresponding to the IPCR shown in **Fig. 2**.

3.3.2. Immobilization of the IPCR Product

1. Apply 50 µL of the diluted amplificate (**Subheading 3.2., step 7**) to each well in duplicate (**Fig. 3**).
2. Incubate for 45 min at room temperature under orbital shaking.

3.3.3. Coupling With the Detection Enzyme

1. Wash four times with buffer B as described (**Subheading 3.1.2., Note 1**).
2. Add 50 µL of a 1:5000 dilution of anti-digoxigenin alkaline phosphatase conjugate (Roche) in buffer A to each well.
3. Incubate for 45 min at room temperature under orbital shaking.

3.4. Quantitative Analysis of the Data

After detection of the fluorescence intensities, mean values and error will be determined, the data will be normalized against the negative controls without αm-rIgG antibody and the calibration curve will be used for quantification of the amount of αm-rIgG in the "unknown" samples.

1. Wash four times with buffer B as described above (**Subheading 3.1.2., Note 1**) and three times for 1 min with buffer A.
2. Add 50 µL of AttoPhos™ substrate (Roche) to each well.
3. Incubate for 15 min at room temperature under orbital shaking (*see* **Note 11**).
4. Measure fluorescence intensity at 550 nm.

File the raw data in an data sheet and proceed as follows:

Determine the average value of the PCR–ELISA double determination	$\rightarrow$	M_E
Determine the standard deviation of the PCR–ELISA duplicates	$\rightarrow$	S_E
Determine the average value of the IPCR double determination	$\rightarrow$	M_I
Determine the standard deviation of the IPCR duplicates	$\rightarrow$	S_I
Determine the average value of the negative controls without αm-rIgG	$\rightarrow$	M_N
Determine the quotient from M_I and M_N	$\rightarrow$	Q_N

Example:

Sample "X"

IPCR-double determination: A_X and B_X

First value of a ELISA-double determination for IPCR "A_X":	M_{1A}
Second value of a ELISA-double determination for IPCR "A_X":	M_{2A}
First value of a ELISA-double determination for IPCR "B_X":	M_{1B}
Second value of a ELISA-double determination for IPCR "B_X":	M_{2B}

$$M_{E-A} = (M_{1A} + M_{2A})/2, \qquad M_{E-B} = (M_{1B} + M_{2B})/2,$$

$$S_E = \sqrt{\Sigma(M - M_E)^2/n}$$

$$M_I = (M_{E-A} + M_{E-B})/2$$

and so on.

For quantification, plot log(Q_N) of the calibration curve samples against the log (spiked antibody concentration) and carry out a linear regression. The resulting equitation will be used for the determination of the concentration of the three "unknown" samples (*see* **Note 12**). Results of a typical experiment are shown in **Fig. 4**.

4. Notes

1. For IPCR amplification, microplate surfaces combining protein binding ability and compatibility to PCR thermocylers are necessary. Polypropylene modules as common for PCR tubes have insufficient binding capacity, whereas most ELISA plates are not compatible to PCR cyclers. In comparison experiments (*see also* **ref. 28**), TopYield™ modules has shown superior performance as the material of choice for IPCR. However, especially in competitive assays, the choice of plate materials could be limited because of commercially available kits, for example, with solid-phase immobilized antigenes that could not easily be substituted by other plate materials. In this case, either a waterbath PCR cycler could be applied, which is not recommended because of prolonged PCR time and inhomogeneities in amplification caused by air bubbles, or the plates could be handled normally

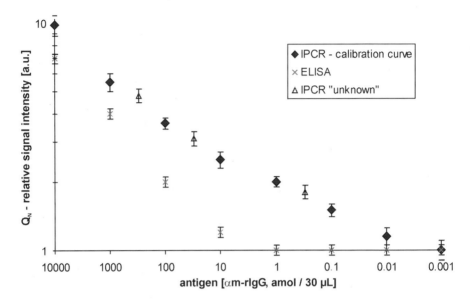

Fig. 4. Plot of IPCR calibration curve (triangles), "unknown" samples (squares), and control-ELISA (X) for the detection of an anti-mouse IgG from rabbit (αm-rIgG). The detection limit of the assay is between 0.1 and 0.01 amol αm-rIgG in 30 μL of buffer volume or approx 50–500 fg/mL. Plotted are relative signal intensities of the signal/negative control ration (Q_N, *see* **Subheading 3.4.**) against the spiked concentrations; the negative control without αm-rIgG gives in this plot a signal of 1.

until the addition of the PCR mix and then denatured for 5 min at 95°C, followed by the transfer of the PCR mix in cycler-compatible modules. Neither method could be compared in sensitivity and homogeneity to an IPCR assay performed in proper handled TopYield™ modules, but they allow for a first test of the method.

2. Regarding the signal amplification potential of the PCR amplification, it is of uttermost importance to reduce nonspecific binding, which results in false-positive results. To accomplish this, three strategies have to be followed simultaneously:

   a. Blocking of unspecific interactions. Either antigen or DNA marker used in IPCR could be nonspecifically bound to the microplate surface. Therefore, blocking of the modules with proteins and DNA with a specific IPCR buffer is needed. BSA alone as used in many ELISAs is not sufficient.

   b. Quality of the antibodies. It has been repeatedly proven *(4,27,41)* that the detection limit of the IPCR is strongly dependent on the performance of the antibodies. Generally, polyclonal antibodies have shown superior performance compared with monoclonal antibodies, so that, whenever possible, polyclonal antibodies has to be preferred for application. Especially in sandwich IPCR, the use of an identical polyclonal antibody for capture and detection is well-suited for the reduction of unspecific interaction.

c. Avoidance of contamination. Because of the sensitivity of the IPCR method against contamination, filter tips should be used for pipetting in each step. Great care has to be given to the avoidance of cross-contamination. It is recommended to store reagents in small aliquots, sufficient for one assay each.

In case of high nonspecific background signals, all reagents, especially the PCR mix, should be exchanged with new aliquots. The concentration of the antibodies immobilized on the surface and/or used as primary detection antibodies could be lowered in threefold increments. However, it must be considered that because of the set-up of the assay, in contrast, for example, to a standard PCR reaction, a small signal will be obtained for the negative control of the IPCR regardless of all washing steps. Therefore, it is recommended to use signal-to-noise ratios for quantification purposes and comparison of different experiments (*see below*).

3. The indirect IPCR method described above could be easily adapted to a variety of typical ELISA set-ups. The given protocol is well-suited for the quantification of a sample containing an unknown amount of the antigen-specific antibody in solution. For example, by replacing the concentration gradient of the primary antibody with a fixed amount of the antibody, the assay could be used for the quantification of the immobilized antigen. This is a typical set-up if no functionalized primary antibody for a given antigen is available. If the immobilized mouse IgG in the experiment described above is exchanged with a specific capture antibody, a sandwich assay could be established. Capture antibodies should be immobilized in concentrations ranging from 10–50 µg/well. A number of different IPCR protocols for model applications are given in (*5*). For detection from biological matrix (e.g., plasma or cell culture medium), the assay also has to be modified. The sample material should be diluted typically 1:3 in an appropriate buffer (e.g., sample dilution buffer; Chimera Biotec) to minimize matrix effects. Additionally, concentrations of the reagents used have to be adjusted in further optimization effects. It is recommended to prepare a calibration curve from a biological matrix that is as similar as possible to the matrix of the samples. If the samples to be analyzed are handled in a specific way (e.g., samples were taken from animals and frozen at –20°C), the spiked samples for the calibration curve necessary for quantification should be treated accordingly.

4. Using a DNA–antibody conjugate for IPCR greatly reduces the number of steps involved for coupling antigen and marker DNA and thus, potential error sources. As the binding efficiency even of the STV–biotin system, commonly applied for the stepwise build-up of a marker complex from biotinylated detection antibody, STV and biotinylated DNA (*8,9,14,27,39,46*) is only approx 10% for each step (*27*), the usage of ready-to-use antibody–DNA conjugates has to be preferred against three subsequent incubations. However, the exact amount of the conjugate needed for optimal IPCR performance is dependent on the binding efficiency to the target antigen and potential unspecific binding partners. Therefore, for each novel assay, different concentrations of the antibody–DNA conjugate should be tested in a range from 10-fold dilution to 10-fold higher concentration of the

suggested dilution. In case of real-time detection (*see* below, **Note 5**), uniformly an approx 10-fold higher concentration of the reagent is needed.

5. For detection with PCR–ELISA, labeling of the PCR product could be achieved with labeled nucleotides or primers. The amount of labeling nucleotide could be varied to introduce randomly a certain amount of label while the usage of one or two labeled primers will result in specific one- or two-timed labeled DNA. Alternatively, unlabeled primers could be used and the PCR product could be immobilized using complementary capture oligonucleotides (enzyme-linked immunosorbent oligonucleotide assay [ELOSA]). Using this technique, the coamplification of an internal competitor is possible, which allows for a further increase of the sensitivity and significance of the assay. (For a typical protocol, *see* **refs.** *4,18*) Besides PCR–ELISA- or –enzyme-linked immunosorbent oligonucleotide assay-based detection methods, a on-line detection of the DNA amplification is also possible with the combination of real-time PCR and IPCR. For application of this technique, which massively reduces hands-on time and overall duration of the IPCR, a real-time PCR thermocycler (e.g., ABIprism 7000, Applied Biosystems, Foster City, CA) and specific TaqMan™ probes are needed. (For a typical protocol, *see* **ref.** *5*) The PCR–ELISA protocol described above could also easily adapted to any conventional PCR method by adding biotinylated primer and Dig-dUTP to the PCR reagent, thus enabling a semiquantitative PCR detection of nucleic acids, highly superior to, for instance, signal intensity determination of gel electrophoresis *(47–55)*.

6. Nowadays, a broad range of companies offer different Taq polymerases for PCR amplification. Because different *Taq* polymerases show varying performance in combination with the reagents necessary for IPCR, several polymerases should be compared during optimization of the assay. Generally, the IPCR marker DNA is so chosen that no special abilities of the polymerase (e.g., extended proofreading, adaptation on extra long DNA) were needed. The protocol described previously was optimized for biomaster *Taq* (biomaster GmbH, Cologne, Germany). For different polymerases, buffer and $MgCl_2$, which are commonly supplemented with the *Taq*, have to be adjusted according to manufacturer's instruction. PCR mastermix as described in **Subheading 3.2.1.** could be prepared and tested in advance. Aliquots of 250 µL each, sufficient for eight wells of 30 µL of PCR, could be frozen and stored at –80°C and rethawn when needed for IPCR without loss in activity.

7. As important as the choice of plate material is the sealing of the modules in PCR. It is necessary to evaluate different adhesive foils in combination with the thermocycler applied in IPCR. Best results in avoiding cross-contamination during PCR were obtained with a thermocyler with active pressure from the lid, for example, the MWG thermocyler. The use of mineral oil supernatant is not recommended because of the difficulties in removing the oil for further processing in PCR–ELISA. If uniformly too high or too low signals were obtained, the number of cycles during PCR could be adjusted between 26 and 35 cycles, respectively. It is preferable to use a lower number of cycles to avoid the amplification of nonspecific background.

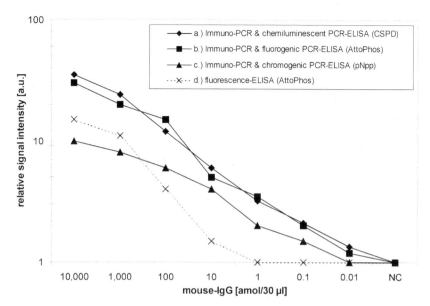

Fig. 5. Comparison of different detection methods for PCR–ELISA subsequent to direct IPCR detection of solid-phase immobilized rabbit-IgG (for more details, *see* **ref. 27**). Plotted are relative signal intensities (Q_N, *see* **Subheading 3.4.**) against the spiked concentrations, the negative control without antigen gives in this plot a signal of 1. Best sensitivities were obtained for a and b, enabling an 1000-fold improve of the ELISA detection limit without IPCR (d).

8. It is recommended to test different dilutions of the PCR amplificate ranging from 1:50 to 1:2000 in establishing a new test because this factor gives an easy method to enhance the performance of the PCR–ELISA. The product of one IPCR is sufficient for a number of PCR–ELISA assays with different dilutions. For photometric quantification, a higher concentration of the PCR amplificate is necessary, for example, 1:80 in the protocol described.

9. Besides alkaline phosphatase (AP), horseradish peroxidase is often applied in ELISA. For PCR–ELISA, alkaline phosphatase is recommended because of its simpler handling and the stability of the enzymatic reaction without auto-inactivation of the enzyme owing to peroxide as observed for horseradish peroxidase. Because AP requires a cofactor for function and binds to phosphate groups, it is necessary to avoid phosphate- or EDTA-containing buffers when handling alkaline phosphatase. If such buffers have to be used, a separate washing step with phosphate- and EDTA-free buffer previous to the application of the substrate should be included in the protocol for removing the inactivating components. Commercially available are photometric (pNpp, Sigma), fluorescence- (4-MUP, AttoPhos, Roche), or chemiluminescence- (CSPD®, Roche) generating substrates for AP (**Fig. 5**). Highest sensitivity is obtained using chemiluminescence techniques, cheapest detection is

possible with photometric substrates. Fluorescence gives the user a good compromise between handling, which is cumbersome for chemiluminescence as additional steps are required and the signals are not stable, and sensitivity, which is reduced for photometry. The choice of plate material for PCR–ELISA depends on the intended detection method: Transparent modules are well suited for photometry but promote crosstalk between wells. White modules should be chosen for chemiluminescence but are giving a high background in fluorescence, where black modules should be preferred. It is recommended to combine a low number of cycles in PCR with a sensitive PCR–ELISA detection method, e.g. fluorescence.

10. As an alternative to in-house prepared STV-coated plates, these plates are also commercially available from a variety of suppliers, for example, Roche. The use of STV is favorable compared with the biotin-binding protein avidin because of the generally lower background signal of STV owing to unspecific binding against avidin.

11. If the signals obtained for 15-min incubation at room temperature are too low, the plate should be remeasured in 10-min intervals and/or the incubation should be conducted at 37°C. For photometric quantification, using, for example, pNpp (*see* **Note 9**), measurement after 45-min incubation at 37°C is recommended.

12. Generally, whenever possible in research applications, IPCR assays should be performed in triplicates for the elimination of outlier signals. Well-established and validated assays could also be conducted in duplicates. Single determinations should be avoided due to the enzymatic amplification steps involved in the method. PCR negative controls containing pure PCR mastermix without DNA and IPCR negative controls with no DNA-antibody conjugate, no target antigen or no capture antibody should be included in each assay. *See also* **Note 2** for troubleshooting in case of positive negative controls. Additionally, as negative control for the PCR–ELISA, wells without labeled DNA, enzyme and/or substrate should be measured. The average error of the IPCR method should be below 30% between two IPCR double determinations (S_I) and below 15% between two ELISA double determinations (S_E).

Acknowledgments

This work was supported by Chimera Biotec GmbH, Dortmund, Germany.

References

1. Mullis, K. B. and Faloona, F. (1987) Specific synthesis of DNA in vitro via a polymerase-catalyzed chain reaction. *Methods Enzymol.* **155**, 335–350.
2. Crowther, J. R. *ELISA; Theory and Practice.* Methods in Molecular Biology (Walker, J. M., ed.), Humana Press, Totowa, NJ, 1995.
3. Sano, T., Smith, C. L., and Cantor, C. R. (1992) Immuno-PCR: very sensitive antigen detection by means of specific antibody-DNA conjugates. *Science* **258**, 120–122.
4. Adler, M., Langer, M., Witthohn, K., et al. (2003) Detection of rViscumin in plasma samples by immuno-PCR. *Biochem. Biophys. Res. Commun.* **300**, 757–763.

5. Adler, M., Wacker, R., and Niemeyer, C. M. (2003) A real-time immuno-PCR assay for routine ultrasensitive quantification of proteins. *Biochem. Biophys. Res. Commun.* **308,** 240–250.
6. Baumler, A. J., Heffron, F., and Reissbrodt, R. (1997) Rapid detection of Salmonella enterica with primers specific for iroB. *J. Clin. Microbiol.* **35,** 1224–1230.
7. Cao, Y. (2002) In-situ immuno-PCR. A newly developed method for highly sensitive antigen detection in situ. *Methods Mol. Biol.* **193,** 191–196.
8. Case, M., Major, G. N., Bassendine, M. F., Burt, A. D., et al. (1997) The universality of immuno-PCR for ultrasensitive antigen detection. *Biochem. Soc. Trans.* **25,** 374S.
9. Case, M. C., Burt, A. D., Hughes, J., Palmer, J. M., Collier, J. D., Bassendine, M. F., et al. (1999) Enhanced ultrasensitive detection of structurally diverse antigens using a single immuno-PCR assay protocol. *J. Immunol. Methods* **223,** 93–106.
10. Chang, T. C. and Huang, S. H. (1997) A modified immuno-polymerase chain reaction for the detection of beta-glucuronidase from *Escherichia coli. J. Immunol. Methods* **208,** 35–42.
11. Chye, S. M., Lin, S. R., Chen, Y. L., et al. (2004) Immuno-PCR for detection of antigen to Angiostrongylus cantonensis circulating fifth-stage worms. *Clin. Chem.* **50,** 51–57.
12. Furuya, D., Yagihashi, A., Yajima, T., Kobayashi, D., Orita, K., Kurimoto, M., et al. (2000) An immuno-polymerase chain reaction assay for human interleukin-18. *J. Immunol. Methods.* **238,** 173–180.
13. Joerger, R. D., Truby, T. M., Hendrickson, E. R., Young, R. M., and Ebersole, R. C. (1995) Analyte detection with DNA-labeled antibodies and polymerase chain reaction. *Clin. Chem.* **41,** 1371–1377.
14. Kakizaki, E., Yoshida, T., Kawakami, H., et al. (1996) Detection of bacterial antigens using immuno-PCR. *Lett. Appl. Microbiol.* **23,** 101–103.
15. Komatsu, M., Kobayashi, D., Saito, K., et al. (2001) Tumor necrosis factor-alpha in serum of patients with inflammatory bowl disease as measured by a highly sensitive immuno-PCR. *Clin. Chem.* **47,** 1297–1301.
16. Liang, H., Cordova, S. E., Kieft, T. L., et al. (2003) A highly sensitive immuno-PCR assay for detecting Group A Streptococcus. *J. Immnol. Methods* **279,** 101–110.
17. Mahbubani, M. H., Schaefer, F. W., 3rd, Jones, D. D., et al. (1998) Detection of Giardia in environmental waters by immuno-PCR amplification methods. *Curr. Microbiol.* **36,** 107–113.
18. Maia, M., Takahashi, H., Adler, K., Garlick, R. K., and Wands, J. R. (1995) Development of a two-site immuno-PCR assay for hepatitis B surface antigen. *J. Virol. Methods* **52,** 273–286.
19. McElhinny, A. S., Exley, G. E., and Warner, C. M. (2000) Painting Qa-2 onto Ped slow preimplantation embryos increases the rate of cleavage. *Am. J. Reprod. Immunol.* **44,** 52–58.
20. McElhinny, A. S., Kadow, N., and Warner, C. M. (1998) The expression pattern of the Qa-2 antigen in mouse preimplantation embryos and its correlation with the Ped gene phenotype. *Mol. Hum. Reprod.* **4,** 966–971.

21. McElhinny, A. S. and Warner, C. M. (1997) Detection of major histocompatibility complex class I antigens on the surface of a single murine blastocyst by immuno-PCR. *Biotechniques* **23**, 660–662.
22. McKie, A., Samuel, D., Cohen, B., et al. (2002) Development of a quantitative immuno-PCR assay and its use to detect mumps-specific IgG in serum. *J. Immunol. Methods* **261**, 167–175.
23. McKie, A., Samuel, D., Cohen, B., et al. (2002) A quantitative immuno-PCR assay for the detection of mumps-specific IgG. *J. Immunol. Methods* **270**, 135–141.
24. McKie, A., Vyse, A., and Maple, C. (2002) Novel methods for the detection of microbial antibodies in oral fluid. *Lancet Infect. Dis.* **2**, 18–24.
25. Mweene, A. S., Ito, T., Okazaki, K., et al. (1996) Development of immuno-PCR for diagnosis of bovine herpesvirus 1 infection. *J. Clin. Microbiol.* **34**, 748–750.
26. Mweene, A. S., Okazaki, K., and Kida, H. (1996) Detection of viral genome in non-neural tissues of cattle experimentally infected with bovine herpesvirus 1. *Jpn. J. Vet. Res.* **44**, 165–174.
27. Niemeyer, C. M., Adler, M., and Blohm, D. (1997) Fluorometric polymerase chain reaction (PCR) enzyme-linked immunosorbent assay for quantification of immuno-PCR products in microplates. *Anal. Biochem.* **246**, 140–145.
28. Niemeyer, C. M., Adler, M., and Blohm, D. (1999) High Sensitivity Detection of Antigens Using Immuno-PCR. NUNC Tech. Note 5.
29. Niemeyer, C. M., Adler, M., Gao, S., et al. (2001) Nanostructured DNA-protein aggregates consisting of covalent oligonucleotide-streptavidin conjugates. *Bioconjug. Chem.* **12**, 364–371.
30. Niemeyer, C. M., Adler, M., Pignataro, B., Lenhert, S., Gao, S., Chi, L., et al. (1999) Self-assembly of DNA-streptavidin nanostructures and their use as reagents in immuno-PCR. *Nucleic Acids Res.* **27**, 4553–4561.
31. Niemeyer, C. M., Wacker, R., and Adler, M. (2001) Hapen-functionalized DNA-steptavidin nanocircles as supramolecular reagents in a novel competitive immuno-PCR. *Angew Chem. Int. Ed.* **40**, 3169–3172.
32. Niemeyer, C. M., Wacker, R., and Adler, M. (2003) Combination of DNA-directed immobilization and immuno-PCR: very sensitive antigen detection by means of self-assembled DNA-protein conjugates. *Nucleic Acids Res.* **31**, e90.
33. Ozaki, H., Sugita, S., and Kida, H. (2001) A rapid and highly sensitive method for diagnosis of equine influenza by antigen detection usine immuno-PCR. *Jpn. J. Vet. Res.* **48**, 187–195.
34. Ren, J., Chen, Z., Juan, S. J., et al. (2000) Detection of circulating gastric carcinoma-associated antigen MG7-Ag in human sera using an established single determinant immuno-polymerase chain reaction technique. *Cancer* **88**, 280–285.
35. Ren, J., Fan, D. M., and Zhou, S. J. (1994) Establishment of immuno-PCR technique for the detection of tumor associated antigen MG7-Ag on the gastric cancer cell line. *Chung Hau Chung Liu Tsa Chih.* **16**, 247–250.
36. Ren, J., Ge, L., Li, Y., et al. (2001) Detection of circulating CEA molecules in human sera and leukopheresis of peripheral blood stem cells with E. coli expressed bispecific CEAScFv-streptavidin fusion protein-based immuno-PCR technique. *Ann. NY Acad. Sci.* **945**, 116–118.

37. Ruzicka, V., Marz, W., Russ, A., and Gross, W. (1993) Immuno-PCR with a commercially available avidin system [letter]. *Science* **260,** 698–699.
38. Saito, K., Kobayashi, M., Sasaki, M., et al. (1999) Detection of human serum tumor necrosis factor-alpha in healthy donors, using a highly sensitive immuno-PCR assay. *Clin. Chem.* **45,** 665–669.
39. Sanna, P. P., Weiss, F., Samson, M. E., et al. (1995) Rapid induction of tumor necrosis factor alpha in the cerebrospinal fluid after intracerebroventricular injection of lipopolysaccharide revealed by a sensitive capture immuno-PCR assay. *Proc. Natl. Acad. Sci. USA.* **92,** 272–275.
40. Sperl, J., Paliwal, V., Ramabhadran, R., et al. (1995) Soluble T cell receptors: detection and quantitative assay in fluid phase via ELISA or immuno-PCR. *J. Immunol. Methods* **186,** 181–194.
41. Sugawara, K., Kobayashi, D., Saito, K., Furuya, D., Araake, H., Yagihashi, A., et al. (2000) A highly sensitive immuno-polymerase chain reaction assay for human angiotensinogen using the identical first and second polyclonal antibodies. *Clin. Chim. Acta* **299,** 45–54.
42. Suzuki, A., Hoh, F., Hinoda, Y., et al. (1995) Double determinant immuno-polymerase chain reaction: a sensitive method for detecting circulating antigens in human sera. *Jpn. J. Cancer Res.* **86,** 885–889.
43. Warner, C. M., McElhinny, A. S., Wu, L., et al. (1998) Role of the Ped gene and apoptosis genes in control of preimplantation development. *J. Assist. Reprod. Genet.* **15,** 331–337.
44. Wu, H. C., Huang, Y. L., Lai, S. C., et al. (2001) Detection of Clostridium botulinum neurotoxin type A using immuno-PCR. *Lett. Appl. Microbiol.* **32,** 321–325.
45. Zhang, Z., Irie, R. F., Chi, D. D., et al. (1998) Cellular immuno-PCR. Detection of a carbohydrate tumor marker. *Am. J. Pathol.* **152,** 1427–1432.
46. Zhou, H., Fisher, R. J., and Papas, T. S. (1993) Universal immuno-PCR for ultrasensitive target protein detection. *Nucleic Acids Res.* **21,** 6038–6039.
47. Laitinen, R., Malinen, E., and Palva, A. (2002) PCR-ELISA I: application to simultaneous analysis of mixed bacterial samples composed of intestinal species. *Syst. Appl. Microbiol.* **25,** 241–248.
48. Ge, B., Zhao, S., Hall, R., et al. (2002) A PCR-ELISA for detecting Shiga toxin-producing *Escherichia coli. Microbes Infect.* **4,** 285–290.
49. Daly, P., Collier, T., and Doyle, S. (2002) PCR-ELISA detection of *Escherichia coli* in milk. *Lett. Appl. Microbiol.* **34,** 222–226.
50. Chansiri, K., Khuchareontaworn, S., and Sarataphan, N. (2002) PCR-ELISA for diagnosis of Trypanosoma evansi in animals and vector. *Mol. Cell Probes* **16,** 173–177.
51. Bhaduri, S. (2002) Comparison of multiplex PCR, PCR-ELISA and fluorogenic 5' nuclease PCR assays for detection of plasmid-bearing virulent Yersinia enterocolitica in swine feces. *Mol. Cell Probes* **16,** 191–196.
52. Martin-Sanchez, J., Pineda, J. A., Andreu-Lopez, M., et al. (2002) The high sensitivity of a PCR-ELISA in the diagnosis of cutaneous and visceral leishmaniasis caused by Leishmania infantum. *Ann. Trop. Med. Parasitol.* **96,** 669–677.

53. Landgraf, A., Reckmann, B., and Pingoud, A. (1991) Direct analysis of polymerase chain reaction products using enzyme-linked immunosorbent assay techniques. *Anal. Biochem.* **198,** 86–91.

54. Gutierrez, R., Garcia, T., Gonzalez, I., et al. (1997) A quantitative PCR-ELISA for the rapid enumeration of bacteria in refrigerated raw milk. *J. Appl. Microbiol.* **83,** 518–523.

55. Gibbons, C. L., Ong, C. S., Miao, Y., et al. (2001) PCR-ELISA: a new simplified tool for tracing the source of cryptosporidiosis in HIV-positive patients. *Parasitol. Res.* **87,** 1031–1034.

12

Covalent Coupling of DNA Oligonucleotides and Streptavidin

Florian Kukolka, Marina Lovrinovic, Ron Wacker, and Christof M. Niemeyer

Summary

Semisynthetic DNA–protein conjugates are synthesized by covalent coupling of thiol-modified DNA oligonucleotides and streptavidin. The resulting conjugates have a binding capacity for four equivalents of biotin and one nucleic acid of complementary sequence. The conjugates are purified to homogeneity by ultrafiltration and chromatography and characterized by photometry and gel electrophoresis. Subsequently, the conjugates are applied as molecular linkers in the DNA-directed immobilization of a biotinylated enzyme on a microplate, containing complementary capture oligonucleotides.

Key Words: Biotin–streptavidin interaction; protein crosslinking; DNA crosslinking; bioconjugate purification; electrophoretic characterization of DNA–protein conjugates; microplate analyses; solid-phase DNA hybridization; DNA and protein microarray technology.

1. Introduction

The remarkable biomolecular recognition of the water-soluble molecule biotin (vitamin H) to the homotetrameric protein streptavidin (STV, molecular weight approx 56 kDa) is characterized by the extraordinary affinity constant of about 10^{14} dm^{3}mol^{-1}, representing the strongest ligand-receptor interaction currently known (*1*). Because biotinylated materials are often commercially available or can be prepared with a variety of mild biotinylation procedures, biotin–streptavidin conjugates form the basis of many diagnostic and analytical tests (*2*). Another great advantage of STV is its extreme chemical and thermal stability. STV is resistant to many proteases, including proteinase K, under physiological conditions and can be heated repeatedly at the temperatures needed for polymerase chain reaction cycling with no apparent damage. It

From: *Methods in Molecular Biology, vol. 283: Bioconjugation Protocols: Strategies and Methods*
Edited by: C. M. Niemeyer © Humana Press Inc., Totowa, NJ

survives extremes of pH and can still bind biotin, for example, even in the presence of 7 *M* urea.

Short deoxyribonucleic acid (DNA) oligonucleotides are also powerful tools in biomedical diagnostics because of their great specificity of stringent hybridization, which allows any unique DNA sequence—16 to 20 nt in a target with the complexity of a mammalian genome (approx 3×10^9 bp)—to be detected specifically and, in principle, isolated. The power of DNA as a molecular tool is enhanced by the ability to synthesize virtually any DNA sequence by automated methods and to amplify any DNA sequence from microscopic to macroscopic quantities by the polymerase chain reaction. Another very attractive feature of DNA is the great rigidity of short double helices (30–60 bp), so that they behave effectively like a rigid rod spacer between two tethered binding sites on both ends.

The covalent attachment of an oligonucleotide moiety provides the STV with a specific binding domain for complementary nucleic acids in addition to its four native binding sites for biotin. This bispecificity of the hybrid molecules allows them to serve as universal, efficient, and highly selective connectors in the oligonucleotide-directed assembly of proteins and other molecular and colloidal components to supramolecular aggregates *(3)*. In this chapter, the covalent attachment of a 5'-thiolated oligonucleotide to STV is described, using the heterobispecific crosslinker sulfosuccinimidyl-4-(*N*-maleimidomethyl)cyclohexane-1-carboxylate (sSMCC; **Fig. 1**). The ε-amino groups of Lys side chains of the STV are first derivatized with the sSMCC crosslinker to provide a maleimide functionality, which is subsequently reacted with the thiolated oligonucleotide. The crosslinked products are prepurified by ultrafiltration and then fractionated by anion-exchange chromatography. The latter step allows for the quantitative separation of DNA–STV conjugates differing in the number of DNA strands attached.

Subsequent to purification, the covalent DNA-STV conjugate is employed in the DNA-directed immobilization (DDI) of proteins. DDI enables the production of laterally microstructured protein arrays. Such devices are currently of tremendous interest because of the demands of high-throughput biomedical analysis and proteome research *(4)*. Although DNA microarrays can easily be fabricated by automated deposition techniques *(5)*, the stepwise, successive immobilization of proteins on chemically activated surfaces is often obstructed by the general instability of sensitive biomolecules, which generally reveal a high tendency for denaturation. The DDI method provides a chemically mild, site-selective process for the attachment of multiple delicate proteins to a solid support (**Fig. 2**; **ref. *6***). The DDI method uses DNA microarrays as a matrix for the simultaneous, site-selective immobilization of many different DNA-tagged proteins or other molecular compounds. Because the lateral surface patterning

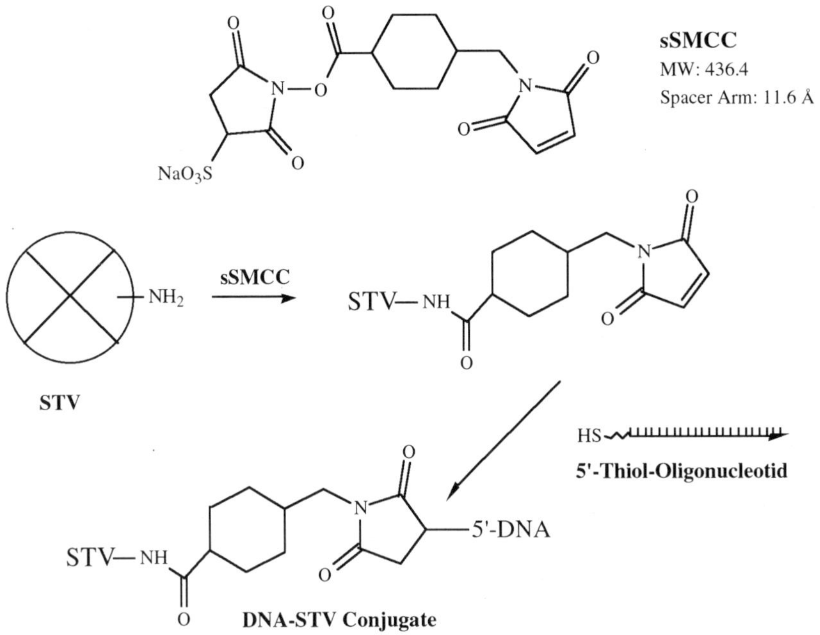

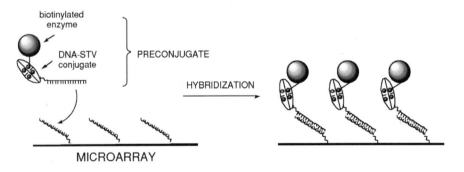

Fig. 1. Schematic drawing of the covalent crosslinking of STV and 5'-thiolated oligonucleotides.

Fig. 2. Schematic drawing of the DDI method.

can now be conducted at the level of the physicochemically stable nucleic acid oligomers, the DNA microarrays can be stored almost indefinitely, functionalized with proteins of interest via DDI immediately prior to use, and subsequent to hybridization, the DNA arrays can even be regenerated by alkaline denaturation of the double helical DNA linkers.

Here, a model DNA array on a microtiter plate is used as an immobilization matrix. To this end, biotinylated capture oligonucleotides are immobilized on a microplate coated with STV. Several capture probes are immobilized, which are all complementary to the covalent DNA–STV conjugate but differ in length. In addition, noncomplementary capture probes are bound to the plate as negative controls, allowing for estimation of the specificity of the DNA-directed immobilization. Biotinylated alkaline phosphatase (bAP) is used as the model protein to be immobilized, and various conjugates are prepared from different stoichiometric amounts of the bAP and the covalent DNA–STV conjugate. Subsequent to DDI, the immobilized proteins are detected by fluorescence measurements using a microplate reader.

2. Materials

1. Fast protein liquid chromatography (FPLC®) System (Amersham Pharmacia, Uppsala, Sweden).
2, Superdex® Peptide column (Amersham Pharmacia).
3. Gel filtration columns NAP5 and NAP10 (Amersham Pharmacia).
4. Molecular cut-off ultrafiltration unit (Centricon 30, Millipore, Bedford, MA).
5. Anion-exchange chromatography column MonoQ® HR5/5 (Amersham Pharmacia).

2.1. Buffers

1. TE: 10 mM Tris-HCl, 1 mM ethylenediamine tetra-acetic acid (EDTA), pH 7.4.
2. Phosphate-buffered saline (PBS): 16.7 mM K_2HPO_4, 83.3 mM K_2HPO_4, 150 mM NaCl, pH 7.3.
3. PBSE: 36.4 mM KH_2PO_4, 63.6 mM K_2HPO_4, 150 mM NaCl, 5 mM EDTA, pH 6.8.
4. Tris A: 20 mM Tris-HCl, pH 6.3.
5. Tris B: 20 mM Tris-HCl, 1 M NaCl, pH 6.3.
6. TBSE: 20 mM Tris-HCl, 150 mM NaCl, 1 mM EDTA, pH 7.3.
7. Sep. gel buffer: 1.5 M Tris-HCl, pH 8.8.
8. Stack. gel buffer: 1 M Tris-HCl, pH 6.8.
9. 6X Loading buffer: 50 mM Tris-HCl, 10% glycerol, 100 mM dithiothreitol (DTT), 0.1% bromophenol blue.
10. Running buffer: 25 mM Tris-HCl, 192 mM Glycin, pH 8.5.
11. TE: 10 mM Tris-HCl, 1 mM EDTA, pH 7.5.
12. Fixing solution: 40% ethanol, 10% acetic acid.
13. TBS: 20 mM Tris-HCl, 150 mM NaCl, pH 7.35.
14. TETBS: 20 mM Tris-HCl, 150 mM NaCl, 5 mM EDTA, 0.05% Tween®-20, pH 7.5.
15. MESTBS: 20 mM Tris-HCl, 150 mM NaCl, 4.5 % milk powder, 5 mM EDTA, 0.2% NaN_2, 1 mg/mL DNA, pH 7.35.
16. PBS: 5.552 g KH_2PO_4, 8.947 g K_2HPO_4, 400 mL ddH_2O, pH 7.5.
17. biotin-TETBS: 20 mM Tris-HCl, 150 mM NaCl, 800 μM dBiotin, 5 mM EDTA, 0.05% Tween®-20, pH 7.5.

18. Biotin–RDB: 20 mM Tris-HCl, 150 mM NaCl, 800 µM dBiotin, 0.45% milk powder, 5 mM EDTA, 0.02% NaN$_3$, 0.1 mg/mL DNA, pH 7.35.

2.2. Reagents

1. 123-bp ladder (Gibco BRL, Eggenstein, Germany).
2. Acrylamide:bis 37.5:1 solution (30%).
3. AttoPhos® (Roche, Mannheim, Germany).
4. Ammonium persulfate (APS, 10%).
5. Butanol.
6. *N,N*-dimethylformamide.
7. Dithiothreitol (DTT).
8. Mercaptoethanol.
9. Silver staining kit (Bio-Rad, Munich, Germany).
10. STV (Roche, Mannheim, Germany).
11. sSMCC (Pierce, Rockford, IL).
12. SybrGold® (Molecular Probes, Eugene, OR).
13. Tetramethylethylenediamine (TEMED).

3. Methods

The methods described below outline the conjugate synthesis, purification and quantitation (**Subheading 3.1.**), the characterization of the conjugates by native polyacrylamide gel electrophoresis (PAGE; **Subheading 3.2.**), and the solid-phase hybridization of different enzyme–DNA conjugates followed by fluorescence detection of the enzymatically formed fluorophor (**Subheading 3.3.**).

3.1. Conjugate Synthesis

The synthesis of DNA–STV conjugate, its purification, and quantitation are described in **Subheadings 3.1.1–3.1.7.** Prior to their conjugation, DNA and STV are activated and purified by gel filtration. The conjugates are then applied on an anion-exchange column and eluted with a salt gradient in the order of increasing DNA bound to the STV. Finally, the 1:1 (DNA:STV) conjugate is quantified photometrically.

3.1.1. Oligonucleotide Activation

1. To 100 µL (100 µM) of oligonucleotide A24 (sequence: TCC TGT GTG AAA TTG TTA TCC GCT) in TE buffer, pH 7.4 add 60 µL of a 1 M DTT solution.
2. Briefly mix and incubate 2 h at 37°C.

3.1.2. STV Activation

1. Dissolve about 2 mg of sSMCC in 60 µL of *N,N*-dimethylformamide (*see* **Note 1**).
2. Add the sSMCC solution to 200 µL of a 100 µM solution of STV in PBS buffer, pH 7.3.
3. Incubate in the dark at room temperature for 1 h.

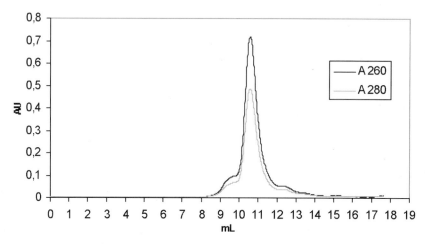

Fig. 3. Chromatograph of oligonucleotide purification by gel filtration.

3.1.3. Purification of Activated Oligonucleotide

Purify the activated oligonucleotide by gel filtration chromatography using a Superdex® Peptide column (Pharmacia) connected to a FPLC® system.

1. Inject the 160 μL of activated oligonucleotide and elute the sample with PBSE buffer, pH 6.8, using a flow-rate of 0.7 mL/min.
2. Detect the absorbance at 260 and 280 nm and collect peaks in 0.55-mL fractions.
3. Pool the collected fractions of the main peak (elution volume approx 10–12 mL (**Fig. 3**), which contain the activated oligonucleotide (*see* **Note 2**).

3.1.4. Purification of Activated STV

1. Remove the top and bottom caps of two disposable gel filtration columns (NAP5 and NAP10, Pharmacia) and pour off the conserving liquid.
2. Support the column over a suitable receptacle and equilibrate the columns by gravity flow-through of three complete fillings with PBSE, pH 6.8.
3. Apply the 260 μL of the activated STV just on top of the filter plate of the NAP5 column and allow the liquid to completely enter the gel bed.
4. Adjust the sample volume to 500 μL by applying 240 μL of PBSE, pH 6.8.
5. Elute the activated STV with 1 mL of PBSE, pH 6.8, and collect the 1 mL of filtrate.
6. Apply the collected filtrate sample onto the NAP10 column. Elute the activated STV with 1.5 mL of PBSE, pH 6.8 (*see* **Note 3**).

3.1.5. Crosslinking of STV and Oligonucleotide, Quenching, and Buffer Exchange

1. Mix the purified activated STV (**Subheading 3.1.4**) with the oligonucleotide fractions (**Subheading 3.1.3**) and incubate the solution for 1.5 h in the dark at room temperature.

Table 1
NaCl Gradient
of Anion-Exchange Chromatography

Volume (mL)	NaCl conc. (M)
0–4	0.3
4–8	0.3–0.36
8–41	0.45–0.8
41–61	1

2. Transfer the mixture into a molecular cut-off ultrafiltration unit (Centricon 30, Millipore) and reduce the volume to about 600 μL by alternating centrifugation at 2800g and shaking steps (*see* **Note 4**).
3. Add 1 μL of 1 M mercaptoethanol and further diminish the volume to approx 200 μL (*see* **Note 5**).
4. Add 1 mL of Tris A and repeat the concentration procedure until the sample volume is about 200 μL.

3.1.6. Purification of STV–Oligonucleotide Conjugate

1. Purify the STV–oligonucleotide conjugate by anion-exchange chromatography using an appropriate column (e.g., MonoQ® HR5/5, Pharmacia). Fractionate the sample by eluting with a NaCl gradient ranging from 0.3 M to 1 M, as indicated in **Table 1**.
2. Collect fractions of 0.55 mL in size. Record the absorbance at 260 and 280 nm.
3. Pool the peak fractions (**Fig. 4**), exchange buffer, and concentrate the conjugate by twofold ultrafiltration (*see* **Subheading 3.1.5.**) using each 500 μL of TBSE.
4. Store the fractions at 4°C until further use.

3.1.7. Conjugate Quantification

Quantitate the concentrated conjugate fraction by measuring the absorbance at 260 and 280 nm. Determine the ratio α and β of the absorbances at 260 and 280 nm of DNA (tA24) and STV. The concentration of the conjugate can be calculated by using a corrected absorbance value at 280 nm for STV (*see* **Note 6**).

$$A280_{STV} = (A_{280} - (A_{260} \times 1/\alpha))/(1 - \beta \times 1/\alpha) \tag{1}$$

Once the absorbance of STV at 280 nm is known, the concentration can be calculated using Lambert-Beers-Law:

$$c = A / (\varepsilon \times l) \tag{2}$$

Extinction coefficient (ε) of tetrameric STV at 280 nm: $\varepsilon = 142,400\ M^{-1}\ cm^{-1}$. The concentration of ssDNA can be calculated from the approximation that at 260 nm 1OD of ssDNA equals to about 33 μg/mL (*see* **Note 7**).

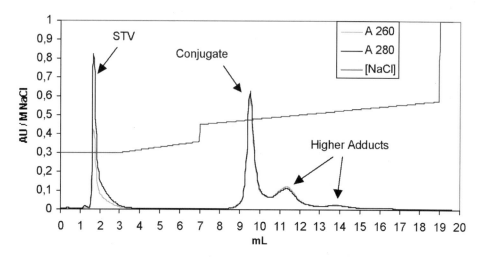

Fig. 4. Chromatograph of conjugate purification by anion-exchange chromatography. The conjugate has a stoichiometry of DNA:STV = 1:1, higher adducts are DNA:STV 2:1, 3:1, and so on.

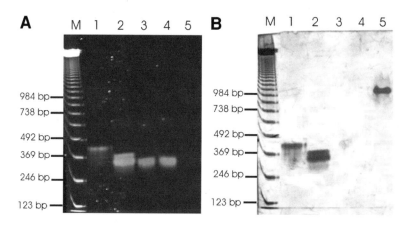

Fig. 5. Gel pictures of DNA–STV conjugates and controls. (**A**) Sybrgold stain, (**B**) Silver stain. M, 123-bp ladder; lane 1, DNA:STV 3:1; lane 2, DNA:STV 2:1 (containing 1:1); lane 3, DNA:STV 1:1; lane 4, DNA:STV 1:1 (control); lane 5, STV.

3.2. Characterization of the Conjugates by Nondenaturing PAGE

In this step, the conjugates are characterized following to their migrating properties in native PAGE. The conjugation of one DNA strand to STV will enhance its mobility from the addition of negative charge, whereas any further conjugated DNA reduces the mobility due to steric hindrance (**Fig. 5**). The

Table 2
Separation Gel

Separation gel, 8.5%	
Distilled H_2O	3.7 mL
1.5 M Tris, pH 8.8	2 mL
Acrylamide:bis 37.5:1 solution (30%)	2266 µL
Immediate before using	
10% APS	40 µL
TEMED	4 µL
Total	Approx 8 mL

preparation of the gel is described as well as the consecutive staining of DNA and protein containing bands (*see* **Note 8**).

3.2.1. Preparation of Separation Gel

1. Assemble the gel electrophoresis pouring device according to the manufacturer's instruction. Wear gloves when working with acrylamide.
2. Prepare the separation-gel in a glass beaker using the reagents specified in **Table 2** (*see* **Note 9**).
3. Pour the solution into the gap between the glass plates.
4. Carefully pipet 100 µL of butanol (aq.) solution on top of the gel.
5. Allow the gel to polymerize for at least 1 h.
6. Wash the top of the gel several times with ddH_2O.

3.2.2. Preparation of Stacking Gel

1. Fix the gel in the electrophoresis device according to the manufacturer's instruction. Insert a comb between the plates.
2. Prepare the stacking gel solution in a disposable plastic tube using the reagents specified in **Table 3**.
3. Pour the solution into the gap between the glass plates with a Pasteur pipet (*see* **Note 10**).

3.2.3. Sample Preparation

1. Dilute the peak fractions to a final STV concentration of 1.33 µM in a total volume of 17 µL TE and incubate 20 min at room temperature.
2. Add 3 µL of 6X loading buffer. Mix thoroughly and load 15 µL of each sample in the wells of the gel. In addition, a 123-bp DNA ladder is used as a marker at a final concentration of 100 µg/mL (*see* **Note 11**).

Table 3
Stacking Gel

Stacking gel, 8.5%	
Distilled H_2O	1600 mL
1.1 M Tris, pH 6.8	285 mL
Acrylamide:bis 37.5:1 solution (30%)	374 µL
Immediate before using	
10% APS	22.5 µL
TEMED	22.5 µL
Total	Approx 2.3 mL

3.2.4. Electrophoresis

Run the gel at constant 150 V for approx 90 min until the bromophenol blue dye front has moved to ca. 1.5 cm at the bottom of the gel.

3.2.5. SybrGold Staining

1. Dilute a 10,000X SybrGold stock solution 10,000-fold with TE to a final volume of 25 mL and fill this 1X staining solution in a container (*see* **Note 12**).
2. Carefully disassemble the electrophoresis device, remove the stacking gel, and transfer the separation gel in the staining solution. Gently agitate the gel at room temperature for approx 15 min.
3. Image the gel with a CCD camera, as described in **Subheading 3.2.8.**

3.2.6. Fixation

Subsequent to fluorescence imaging of the gel, fixate the gel for at least 20 min in fixing solution (40% ethanol, 10% acetic acid) by gently agitating on an orbital shaker.

3.2.7. Silver Staining

1. Decant the fixation solution (**Subheading 3.2.6.**). Prepare the oxidizing solution (according to manufacturer's instructions) in ddH_2O. Oxidize the gel for 5 min.
2. Rinse the gel several times with approx 400 mL ddH_2O. Prepare the staining solution in ddH_2O (manufacturer's instructions). Stain the gel for 20 min under orbital shaking. Rinse the gel briefly with ddH_2O.
3. Visualize the bands by covering the gel with a freshly prepared developing solution. Discard the developing solution and immediately stop the developing reaction by adding 5% acetic acid and incubating for at least 15 min (*see* **Note 13**).

3.2.8. Documentation

Image the stained acrylamide gel (*see* **Note 14**), using (a) for SybrGold stain, a transilluminator at 300 nm and a SybrGold camera filter and (b) for silver stain, use white light without a photographic filter.

3.3. DNA-Directed Immobilization (DDI)

In the DNA-Directed Immobilization (DDI), the 1:1 DNA–STV conjugate is used as a molecular adapters in the immobilization of biotinylated alkaline phosphatase, followed by an enzymatic assay. First, the coating of microplates with STV is described, followed by the functionalization of the STV-plates with complementary capture-oligonucleotides. Second, the synthesis of different DNA–enzyme conjugates and their immobilization followed by the enzymatic assay and the fluorescence detection is described.

3.3.1. Coating of Microplates With STV

1. Prepare 5 mL of a solution of STV (200 nM STV in PBS) by diluting a 19 µM stock solution in PBS.
2. With a multichannel pipet, add 50 µL of this solution to each well of a 96-well microtiter plate (Nunc) and incubate for at least 48 h at 4°C.
3. Wash the microplate three times for 1 min with 240 µL of TBS. Fill each well of the plate with 150 µL of MESTBS and incubate for at least 12 h at 4°C.

3.3.2. Immobilization of Capture Oligonucleotides

1. Wash the STV-coated microplate twice for 30 s and twice for 5 min with 240 µL of TETBS.
2. Prepare 2 mL of a 240 nM solution of each biotinylated capture oligonucleotide, as specified in **Table 4**, by adding 4.8 µL oligonucleotide to 1995 µL of TETBS. Add 50 µL of the oligonucleotide solution to different wells of the plate (**Table 4**).
3. Incubate the microtiter plate for 30 min at room temperature under orbital shaking.
4. Decant the oligonucleotide solution and wash the plate twice for 30 s and twice for 5 min, each with 240 µL of biotin-TETBS buffer (*see* **Note 15**).

3.3.3. Sample Preparation and Dilution Series

1. For preparation of the enzyme–DNA conjugates, mix various concentrations of the covalent STV–DNA conjugate and bAP as specified in **Table 5** (sample nos. 1–3).
2. Prepare a series of conjugates from native STV, bAP, and biotinylated oligonucleotide A24 (bA24) by simply mixing the three components (**Table 5**; sample nos. 4–6). The concentrations of the stock solutions are 19 µM for native STV, 50 µM for bAP, and 100 µM for bA24.

Table 4
Biotinylated Oligonucleotides

	bcA	bcAr	bcB
Length	24	14	26
Concentration (μM)	100	100	100
Sequence (5'–3')	AGC GGA TAA CAA TTT CAC ACA GGA	GCG GAT AAC AAT TT	ATG TGA CCT GTA TTG TTG GAT GTG AG
Complementary	Yes	Yes	No
Wells needed	32	32	16

Table 5
Ratio of Concentration

Sample no.	1	2	3	4	5	6
STV–DNA conjugate	10 pmol	10 pmol	10 pmol	–	–	–
bAP (50 μM)	10 pmol	2 pmol	50 pmol	10 pmol	1 pmol	10 pmol
STV (19 μM)	–	–	–	10 pmol	10 pmol	10 pmol
bA24 (100 μM)	–	–	–	10 pmol	10 pmol	1 pmol

3. Dilute the above samples to a final volume of 10 μL with buffer TE. Incubate the mixtures for 20 min at room temperature under orbital shaking.
4. Make up a serial dilution of the samples 1–3 with biotin–RDB buffer. The dilution series should include a constant gradient of four different STV concentrations in the range of 1 nM and 1 pM.
5. Adjust the samples 4–6 to a final STV concentration of 1 nM in biotin–RDB.
6. As a negative control, apply bAP to the wells at a concentration of 1 nM in biotin–RDB.

3.3.4. DDI

1. Ensure that each of the serial dilution samples of conjugate 1–6 is incubated on the complementary capture probes (bcA and bcAr), whereas only the highest concentrations of conjugates 1–6 are incubated in wells containing noncomplementary capture probe (bcB). Apply each conjugate in duplicate (*see* **Note 16**).
2. Remove the storage buffer, wash the microplate twice with 240 μL of TETBS for 1 min.
3. Add 50 μL of the serial dilutions (**Subheading 3.3.3.**) to the wells of the microplate. Incubate the microplate for 45 min at room temperature under orbital shaking.
4. Wash the plate twice for 30 s, twice for 5 min with 240 μL of TETBS and twice with 240 μL of TBS for 1 min.

3.3.5. Enzymatic Reaction

Add 50 μL of AttoPhos solution (Roche), prepared according to manufacturer's instructions, to each well of the microtiter plate. Incubate for 5, 10, and 20 min at room temperature under orbital shaking.

3.3.6. Fluorescence Detection

The fluorescent signals in the microplate are measured using a microplate multilabel reader (Perkin-Elmer) after 5, 10, and 20 min.

4. Notes

1. If the sSMCC doesn't dissolve readily, shake at 35°C and/or add more N,N-dimethylformamide.
2. Use the oligonucleotide as soon as possible in the crosslinking reaction because dimer formation through disulfide bridges might occur.
3. The twofold purification by gel-filtration chromatography is conducted to completely remove traces of the crosslinker, which otherwise will react with the thiolated oligonucleotide. Use the activated STV as soon as possible in the crosslinking reaction because the reactivity of the maleimide will be reduced over time.
4. Because the conjugates will lead to clogging of the ultrafiltration membrane during centrifugation, shaking the filtration unit in intervals will counteract and significantly reduce total filtration time.
5. The added mercaptoethanol reacts with free maleimide groups of the STV, preventing the addition of further oligonucleotide, because the reduction of the volume will result in an increase of the probability of higher conjugation. Addition of mercaptoethanol before concentrating reduced the yield of conjugate formation.
6. The formula is based on the assumption that the absorbance measured arises from the absorbance of both the STV and the oligonucleotide moiety present in the sample.

$$A280 = A280_{STV} + A280_{DNA} \tag{1}$$

$$A260 = A260_{STV} + A260_{DNA} \tag{2}$$

Assume that the absorbance ratio (A_{260}/A_{280}) for both the isolated STV and oligonucleotide is a constant value:

$$A260_{DNA}/A280_{DNA} = \alpha \tag{3}$$

$$A260_{STV}/A280_{STV} = \beta \tag{4}$$

Insertion of **Eqs. 3** and **4** into **1** and **2**, respectively, leads to **Eqs. 5** and **6**

$$A280 = A280_{STV} + (A260_{DNA} \times 1/\alpha) \tag{5}$$

$$A260 = A260_{DNA} + (A280_{STV} \times \beta) \tag{6}$$

Insertion of **Eq. 6** into **5** leads to **7a**

$$A280 = A280_{STV} + (A260 - A280_{STV} \times \beta) \times 1/\alpha \tag{7a}$$

Rearrangement of **Eq. 7a** gives the part of the absorption at 280 nm, which results from the STV:

$$A280_{STV} = (A280 - (A260 \times 1/\alpha))/(1 - \beta \times 1/\alpha) \tag{8}$$

7. Use a DNA concentration of about 1 μM for these measurements.
8. STV and STV conjugates containing bands often fail to get visualized during silver staining. They are, however, often visible as bands brighter then the background after prolonged development time.
9. Mix the components in the order shown. Mix gently but thoroughly, avoid air bubbles. The gel can be stored wrapped in a plastic film at 4°C for several weeks.
10. Mix the components in the order shown. Mix gently but thoroughly avoiding air bubbles. Allow the stacking gel to polymerize for 30 min before removing the comb. Flush the wells with running buffer to remove any unpolymerized acrylamide by pipetting up and down before loading the samples.
11. To further analyze the conjugates, samples of native STV with different molar equivalents of biotinylated oligonucleotide bA24 (STV:bA24 = 1:0, 1:0.5, 1:4) can be prepared.
12. In this step, all bands of the gel containing DNA are stained with the intercalating dye SybrGold (Molecular Probes). Protect the light-sensitive staining solution from light by covering it with aluminum foil.
13. In this step, all bands of the gel are stained with a silver staining kit (Bio-Rad). Use gloves for the silver staining. Cooling all reagents will reduce the background. Use water of the most pure quality for the reagents and the rinsing. Execute all steps on an orbital shaker avoiding folding of the gel because this will reduce the quality of the visualization. The oxidation will turn the gel yellow. When rinsing the gel with water, make sure the gel is completely discolored before adding the silver solution. Replacing the developer as soon as it begins to turn dark is also essential to reduce the background. After the complete silver staining procedure, the gel can be stored in ddH$_2$O.
14. Optimal photographic conditions need to be determined experimentally with the gel documentation device (AlphaImager, Biozym).
15. If necessary, the microtiter plate can be stored at 4°C after the addition of 240 μL of biotin–RDB to each well for several days.
16. Designing a pipetting scheme for applying the samples to the capture plate is strongly recommended.

References

1. Weber, P. C., Ohlendorf, D. H., Wendoloski, J. J., and Salemme, F. R. (1989) Structural origins of high-affinity biotin binding to streptavidin. *Science* **243**, 85–88.
2. Wilchek, M. and Bayer, E. A. (1990) Avidin-biotin technology. *Methods Enzymol.* **184**, 51–67.
3. Niemeyer, C. M. (2002) The developments of semisynthetic DNA-protein conjugates. *Trends Biotechnol.* **20**, 395–401.
4. Templin, M. F., Stoll, D., Schrenk, M., Traub, P. C., Vohringer, C. F., and Joos, T. O. (2002) Protein microarray technology. *Trends Biotechnol.* **20**, 160–166.
5. Pirrung, M. C. (2002) How to make a DNA chip. *Angew. Chem. Int. Ed.* **41**, 1276–1289; *Angew. Chem.* 2002, **114**, 1326–1341.

6. Niemeyer, C. M., Sano, T., Smith, C. L., and Cantor, C. R. (1994) Oligonucle-otide-directed self-assembly of proteins: semisynthetic DNA–STV hybrid molecules as connectors for the generation of macroscopic arrays and the construction of supramolecular bioconjugates. *Nucl. Acids Res.* **22,** 5530–5539.

13

Synthesis of Oligonucleotide–Peptide and Oligonucleotide–Protein Conjugates

David R. Corey

Summary

The conjugation of macromolecules offers a rapid and versatile route to improved function. Here, the methods for obtaining disulfide-linked oligonucleotide–peptide and oligonucleotide–protein conjugates are described. These hybrid molecules can be used deliver chemical functionality to specific sequences within deoxyribonucleic acid and ribonucleic acid. The peptides and proteins can also be used to enhance the hybridization properties of the attached oligonucleotides.

Key Words: Oligonucleotide; conjugation; disulfide; crosslinking; hybridization; semisynthetic enzyme.

1. Introduction

Synthetic oligonucleotides are convenient tools for the recognition of deoxyribonuclease (DNA) and ribonucleic acid (RNA; **ref. *1***). In this chapter, a simple approach to the chemical modification of oligonucleotides that can dramatically alter hybridization properties is described, as well as methods for the conjugation of oligonucleotides to peptides or proteins to afford hybrid molecules that combine properties from both constituents.

1.1. Goals for Improved Recognition

Any research program that aims to develop and apply chemically modified oligonucleotides must recognize that unmodified DNA oligonucleotides have already proven to be extraordinarily useful tools. Because existing techniques already work well, modified oligonucleotides will need to confer decisive advantages. Potential areas of improvement include hybridization rate, hybridization affinity, and discrimination against binding to mismatched targets. For

From: *Methods in Molecular Biology, vol. 283: Bioconjugation Protocols: Strategies and Methods*
Edited by: C. M. Niemeyer © Humana Press Inc., Totowa, NJ

recognition of cellular targets improved properties could be achieving by enhancing cellular uptake while in animals improved pharmacokinetic properties would be useful.

1.2. DNA–Peptide and DNA–Protein Conjugates

The synthesis of DNA conjugates is a powerful approach to improving DNA recognition. Conjugation allows the hybridization properties of the oligomer to be retained while incorporating properties from the peptide or protein. Earlier studies have demonstrated that oligonucleotides could be attached to staphylococcal nuclease (Snase; **ref. 2**). The oligonucleotide domain of the conjugate was able to deliver the nuclease to target sequences, whereupon the addition of calcium activated the nuclease for sequence-specific cleavage. The conjugates were able to bind and cleave single-stranded DNA and RNA targets *(2–6)*. In some cases, cleavage was catalytic (i.e., the nuclease was able to cut one target strand, dissociate, and then cut additional target strands; *4*). Turnover rates as high as 30 per minute could be achieved.

Oligonucleotide–SNase conjugates are also able to recognize and cleave sequences within either relaxed or supercoiled duplex DNA *(5,6)*. During these experiments, the surprising observation that attachment of SNase not only promotes DNA cleavage but also promotes invasion of duplex DNA by the attached oligonucleotide was made *(7)*. As a result, SNase should be considered a domain that both promotes DNA cleavage and DNA binding.

To separate the promotion of DNA binding from promotion of DNA cleavage, peptides were modeled after the surface of SNase and attached to oligonucleotides. These peptides were successful, enhancing the rate of hybridization to supercoiled DNA as much as 48,000-fold relative to the rate achieved by unmodified oligomers *(8–10)*. Enhanced hybridization may prove useful for recognition or chromosomal DNA and improved protocols for polymerase chain reaction or diagnostics.

2. Materials

1. Many different oligonucleotide supply houses sell oligonucleotides with either 3' or 5' thiol groups. Check current catalogs for updated availability and pricing (*see* **Note 1**).
2. Alternatively, derivatized controlled pore glass that can be used to introduce thiol groups at the 3' position can be commercially obtained and then used by the investigator or an institutional core laboratory. 3'-Thiol-modified solid support can be obtained from Glen Research (Sterling, VA).
3. 3'-Thiol-modified solid support can also be synthesized using underivatized controlled pore glass (Fairfield, NJ) and 1-*O*-(4,4'-dimethoxytrityl)-3,3'-thiopropanol as described *(6)*.

4. Similarly, reagents that can add thiol groups at the 5' position are also commercially available from Clonetech (Palo Alto, CA) as well as from other suppliers.

5. Peptides can be obtained from any commercial peptide supplier, from in-house peptide synthesis facilities, or synthesized by standard methods from commercially available protected amino acids.

6. 2,2'-Dithiodipyridine can be obtained from Aldrich (Milwaukee, WI).

7. Dithiothreitol (DTT) can be obtained from many commercial sources as a powder and stored desiccated at –20°C. Before use, it is dissolved in water at a concentration of 1 M and stored frozen at –20°C. We have observed that solid or dissolved DTT can lose its ability to reduce disulfide bonds after prolonged storage. We try to use solid DTT within 1 yr of first opening the bottle, and DTT solutions within one month of making them.

3. Methods

3.1. Synthesis of 3'-S-Thiopyridyl Oligonucleotides

1. The thiolated oligonucleotides on solid support were reduced by treatment with 20 mM DTT overnight at 37°C in 1 mM ethylenediamine tetraacetic acid, 10 mM Tris-HCl, pH 8.0. to yield free 3' thiol (**Fig. 1**).

2. Most of the DTT was removed by extraction with water-saturated n-butanol.

3. n-Butanol and residual DTT were removed by desalting on a Bio-Spin 6 column (Bio-Rad, Hercules, CA). **Steps 2** and **3** are redundant; once the investigator is comfortable with the protocol, one or the other can be eliminated.

4. The reduced oligonucleotide was added to an equal volume of 10 mM 2,2'-dithiodipyridine in acetonitrile and the mixture was incubated at room temperature for 30 min. Note that some DTT can be present in the solution prepared during **steps 2** and **3**. The 2,2'-dithiodipyridine is present in great excess, and its concentration is more than sufficient to both label the oligonucleotide and scavenge the free DTT (*see* **Note 2**).

5. The solution was extracted with diethyl ether (six times) to remove unreacted 2,2'-dithiodipyridine. In our experience, the ether extraction effectively removes virtually all 2',2'dithiodipyridine. The extraction is less able to remove the relatively hydrophilic thiopyridyl anion (which will yield a yellow organic layer), and if the oligonucleotide is assayed by UV spectroscopy, it is likely that a large absorbance will be observed at 343 nm.

6. 3'-S-thiopyridyl oligonucleotide was desalted on a Bio-Spin 6 column. This removes any remaining 2,2'-dithiodipyridine and thiopyridyl anion. Note, contamination with unreacted 2,2'-dithiodipyridine will interfere with subsequent chemical crosslinking steps, so **steps 5** and **6** must be adhered to. Contamination with thiopyridyl anion is a less severe problem, but its presence may prevent successful monitoring of the crosslinking reaction by UV spectroscopy. It will also be remove by this step.

7. At this point, the experimenter should have a desalted solution of 3'-S-thiopyridyl oligonucleotide. It is ready for crosslinking and can be stored indefinitely at –20°C.

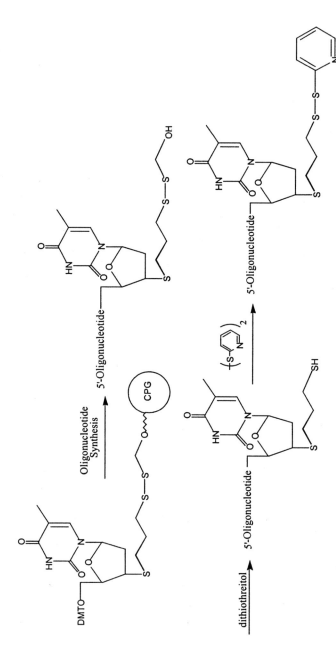

Fig. 1. Scheme for synthesis of 3'-labeled thiopyridyl oligonucleotides.

8. The concentration of *S*-thiopyridyl oligonucleotide can be calculated by treated an aliquot of oligomer with DTT (final concentration 100 μ*M*) to release thiopyridyl anion ($\varepsilon = 7060\ M^{-1}$ at 343 nm). This value should be approximately equal to the concentration of oligonucleotide calculated by measuring the absorbance at 260 nm (*see* **Note 3**).

3.2. Synthesis of 5'-SThiopyridyl Oligonucleotides

1. 5'-*S*-thiopyridyl oligonucleotides are prepared using C-6 thiolmodifier reagent from Clonetech.
2. The C-6 thiol modifier is introduced after completion of oligonucleotide synthesis at the 5'-terminal. The oligomer is cleaved from the resin by heating with concentrated ammonia for 6 h. The ammonia is then removed under reduced pressure and redissolved in 100 μL of 0.1 *M* triethylammonium acetate.
3. The modified oligomer contains a 5'-*S*-trityl group. This group is removed by adding 15 μL of 1 *M* aqueous silver nitrate with vortexing. The solution should be allowed to sit at room temperature for 30 min.
4. Aqueous DTT (1 *M*, 20 μL) is then added with vortexing. The solution is incubated at room temperature for 30 min. A precipitate will be immediately noticeable, and the solution will become gray. The precipitate is a complex of excess silver nitrate and DTT.
5. Centrifuge to remove precipitate. Wash pellet with 100 μL of 0.1 *M* triethylammonium acetate, centrifuge, and pool supernatants.
6. You now have a solution of reduced oligonucleotide. Add a solution of 2,2'-dithiodipyridine and proceed as described above (**Subheading 3.1., step 4**).
7. Note that it is possible that the oligonucleotide stock at the end of **Subheadings 3.1.** and **3.2.** may not be pure and may contain oligomer with an unreactive chemically blocked thiol. The possible presence of this contaminant is irrelevant, because it will not react with protein or peptide and can easily be removed after crosslinking.

3.3. Preparation of Thiol-Containing Protein

The following procedure describes methods for obtaining methods for obtaining staphylococcal nuclease containing cysteine for conjugation. Other proteins can be coupled by similar methods.

1. Plasmid pDC1 encoding staphylococcal nuclease containing the K116C mutation was provided by Dr. Peter G. Schultz (UC Berkeley) and was derived from plasmid pONF1 *(11)*. The introduction of cysteine into proteins that do not normally contain cysteine provides a uniquely reactive group that can readily form disulfide bonds to afford the required conjugates (**Fig. 2**).
2. K116C staphylococcal nuclease was expressed behind a *lac* promoter and an *omp*A signal. Expression was induced by addition of lactose and enzyme was purified as described *(11)*.

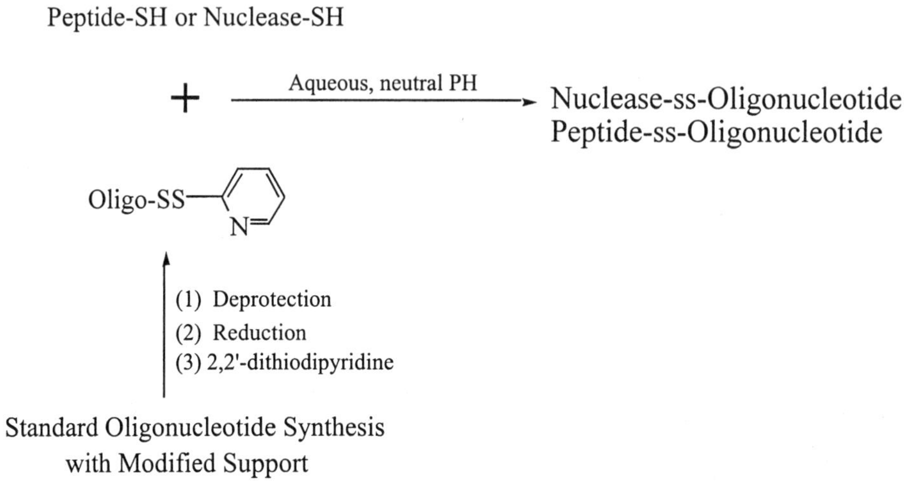

Peptide-SH or Nuclease-SH

$$+ \quad \xrightarrow{\text{Aqueous, neutral PH}} \quad \begin{array}{l}\text{Nuclease-ss-Oligonucleotide}\\\text{Peptide-ss-Oligonucleotide}\end{array}$$

Oligo-SS—⟨N=⟩

↑
(1) Deprotection
(2) Reduction
(3) 2,2'-dithiodipyridine
|

Standard Oligonucleotide Synthesis
with Modified Support

Fig. 2. Scheme for crosslinking peptides and proteins to oligonucleotides.

3. Staphylococcal nuclease was isolated as a mixture of monomer and disulfide-linked dimer. The enzyme was completely reduced to monomer by treatment with 50 mM DTT for 8 h at 37°C in 10 mM Tris-HCl, pH 8.0.

4. Monomeric enzyme was separated from DTT by Mono S cation exchange chromatography (Pharmacia) in 50 mM NaHEPES, pH 7.5, 1 mM ethylenebis(oxyethylenenitrilo)tetraacetic acid, and a gradient of 0.0 to 1.0 M NaCl.

5. Alternatively, the enzyme could be separated from DTT by desalting using two sequential G-25 size exclusion resin columns. It is essential that all DTT be removed from the protein prior to introducing the S-thiopyridyl oligonucleotide.

3.4. Preparation and Purification
of SNase–Oligonucleotide Conjugates

1. The reduced staphylococcal nuclease was mixed with 3'-S-thiopyridyl-oligonucleotide and the coupling was monitored at 343 nm.

2. The conjugate was purified by Mono Q anion exchange chromatography (Pharmacia) in 20 mM Tris-HCl, pH 8.0; 1 mM EGTA; and a gradient of 0.0 to 1.0 M NaCl.

3. The collected oligonucleotide–nuclease conjugate fractions were concentrated to 200 µL and desalted using BioSpin 6 spin columns (Bio-Rad).

3.5. Preparation of Thiol-Containing Peptides

1. Two to four milligrams of dry peptide are weighed out; dissolved in 10 mM Tris-HCl, pH 8.0 buffer; and incubated for 8 to 12 h at 37°C with 10 mM DTT to reduce all material to monomeric form.

2. The reduced peptide was purified by reverse-phase high-performance liquid chromatography (HPLC) using a C-18 Microsorb 5-μm 300 Å column (Rainin) and 0.1% trifluoroacetic acid in doubly distilled water (buffer A) and a gradient of 0 to 100% of buffer B (0.08% trifluoroacetic acid in 95:5 acetonitrile:doubly distilled water).
3. The solution of purified peptide was neutralized with 1/5 vol 100 m*M* Tris, pH 10.2, and immediately used for crosslinking reactions with *S*-thiopyridyl oligonucleotide. The reduced peptide is not stored for later use because oxidation and formation of peptide dimers will prevent successful conjugation (*see* **Note 4**).

3.6. Coupling of S-Thiopyridyl Oligonucleotides to Peptides

1. The peptide solution was added to a 1 mL quartz cuvet containing the 3'-*S*-thiopyridyl oligonucleotide (**Fig. 2**).
2. The reaction was monitored at 342 n*M* using a Hewlett Packard 8452 diode array spectrophotometer. Enough peptide and oligonucleotide were used to insure a distinct peak at 342 nm (>0.05 OD_{343}) upon completion of the reaction. When using cationic peptides, excess peptide can cause condensation of DNA and observation of a solid precipitate. Therefore, it is essential that no more than one equivalent of peptide be added (*see* **Notes 5** and **6**).

3.7. Purification of Oligonucleotide–Peptide Conjugates

1. After UV spectrometry indicates that crosslinking is complete, the conjugate is purified by anion-exchange chromatography with a Mono Q 5/5 column (Pharmacia) using 20 m*M* Tris-HCl, pH 8.0, and a gradient of 0.0 to 1.0 *M* NaCl. The attachment of the positively charged peptides caused the conjugates to migrate significantly faster than the parent oligonucleotide (*see* **Notes 7** and **8**).
2. The purified conjugate was concentrated to 200 μL by evaporation.
3. The concentrated solution of conjugate was desalted using BioSpin 6 columns (Bio-Rad). This removes the excess salt needed to elute the conjugate from the Mono Q column.
4. Conjugates are characterized by UV spectrophotometry. The absorbance maximum of the conjugates should be between 260 and 270 nm, as would be expected of a conjugate containing both DNA and protein.
5. The identity of conjugates can be further confirmed by treatment with DTT for 1 h at 37°C. This treatment should regenerate the free peptide or protein and free oligonucleotide. Reappearance of the free oligonucleotide can then be observed by fast protein liquid chromatography (FPLC; *see* **Notes 9–12**).

4. Notes

1. These protocols for derivitization of oligonucleotides with peptides and with the protein staphylococcal nuclease have proven to be readily reproducible. Equally good yields can be obtained with either 3' or 5' derivatized oligonucleotides. It should be possible to do 10 or more crosslinkings and purifications during a standard working day.

2. A major strength of this synthetic approach is the release of thiopyridyl anion during coupling. The increase at 343 nm, in combination with knowledge of the extinction coefficient for thiopyridyl anion, allows the amount of product to be calculated. A misleading increase at 343 will only be observed if the solution of peptide or protein that was treated with DTT is not thoroughly desalted to remove all DTT.

3. If a distinct peak does not appear at 343 nm, the coupling has not occurred. I strongly recommend that a full UV scan from 200 to 500 nm be performed so that the shape of the peaks do to DNA and thiopyridyl are clear. A full scan can also demonstrate that the baseline returns to zero beyond 400 nm. A non-zero baseline past 400 nm probably indicates precipitation. The usefulness of monitoring the 343 value cannot be stressed too highly.

4. Failure to obtain coupling is usually the result of a failure to generate a free thiol on the peptide, oligonucleotide, or protein. Corrective action includes use of fresh dithiothreitol in subsequent reductions. If a 5'-thiol is being introduced, care must be taken that the sequence of steps needed to introduce the *S*-thiopyridyl group are being performed properly. Occasionally, oligonucleotides resist reaction with S-thiopyridyl. Invariably, resynthesis of the oligomer produces better results. In these cases, we attribute failure to chemical modification of the thiol prior reaction with 2,2'-dithiodipyridine, probably through oxidation

5. Precipitation of the conjugate is a problem that is observed during coupling with cationic peptides. It is well known that oligonucleotides precipitated from solution in the presence of high concentrations of cations, and this is aggravated by the physical attachment of a cationic peptide to the oligonucleotide. We have observed that one to one complexes of peptide and oligonucleotide remain soluble. However, as the ratio of peptide added to solution relative to oligonucleotide passes 1:1, precipitation increases. This problem can be minimized adding peptide gradually and by ceasing addition once a 1:1 ratio has been reached.

6. We have less experience with neutral or anionic peptides or proteins. Because these lack electrostatic attraction for DNA or RNA, crosslinking is relatively slow. This may prove to be a complication for the syntheses of some conjugates, although one would expect that longer reaction times or higher peptide/protein concentrations would be adequate to drive the reaction to completion. It is also possible that attachment of neutral or anionic peptides may alter the purification properties of the conjugate less than the dramatic alteration that we observe for oligonucleotide-cationic peptide conjugates. However, given that the difference in the molecular characteristics of the oligonucleotide and the conjugate remain large, it is likely that adequate purification protocols can be identified.

7. Purification by anion or cation exchange chromatography is straightforward because the net charge of the conjugate is much different from the net charge of either the oligonucleotide or peptide/protein constituents.

8. If anion exchange is being used for purification, an important control to ensure the success of separation is to reduce the oligonucleotide and determine its retention time. This retention time will suggest when the conjugate will appear (i.e.,

several minutes earlier). We reduce the oligonucleotide prior to application to the column because we have found that the *S*-thiopyridyl protected oligomer can react with thiol-containing material on the HPLC or FPLC guard column and be retained.

9. The main limitation of our approach is that a disulfide bond is formed. This linkage is stable indefinitely upon storage but will degrade rapidly in the presence of reducing agents. As a result, use of these conjugates in solutions with a reducing environment is not recommended. Use of thioether linkages is a related strategy that can be used for these applications.

10. Conjugates can be assembled without resorting to use of expensive HPLC or FPLC instruments. Desalting columns and/or multiple extractions with organic solvents can be used to remove DTT and 2,2'-dithiodipyridine. The stoichiometry of the reaction can be adjusted so that almost complete conversion of starting material is achieved, making final purification unnecessary for some applications.

11. It is also possible to achieve successful conjugations without monitoring results by UV spectrophotometry. However, it cannot be stressed too strongly that failure to make use of monitoring of thiopyridyl anion greatly complicates the diagnosis of any problems that might be encountered. As mentioned above, investigators should invest the minimal time required to take a full scan from 200 to 500 nm.

12. The conjugates described here are linked by disulfide bonds that will be readily reduced under reducing conditions, such as those that prevail inside mammalian cells. This can be an advantage if one desires to have the peptide released after introduction into cells, but if one desires a stable linkage other coupling chemistries may need to be employed.

Acknowledgments

This work was supported by grants from the National Institutes of Health (GM60642) and by the Robert A. Welch Foundation (I-1244).

References

1. Braasch, D. A. and Corey, D. R. (2002) Novel antisense strategies for controlling gene expression. *Biochemistry* **41**, 4503–4510.
2. Corey, D. R. and Schultz, P. G., (1987) Generation of a hybrid sequence-specific single-stranded deoxyribonuclease. *Science* **238**, 1401–1403.
3. Zuckermann, R. N., Corey, D. R., and Schultz, P. G. (1988) Site-selective cleavage of RNA by a hybrid enzyme. *J. Am. Chem. Soc.* **110**, 1614–1615.
4. Corey, D. R., Pei, D., and Schultz, P. G. (1989) The generation of a catalytic oligonucleotide-directed nuclease. *Biochemistry* **28**, 8277–8286.
5. Corey, D. R., Pei, D., and Schultz, P. G. (1989) The sequence-selective hydrolysis of duplex DNA by an oligonucleotide-directed nuclease. *J. Am. Chem. Soc.* **111**, 8523–8525.
6. Pei, D., Corey, D. R., and Schultz, P. G. (1990) site-specific cleavage of duplex DNA by a semi-synthetic Nuclease via triple helix formation. *Proc. Natl. Acad. Sci.* **87**, 9858–9862.

7. Corey, D. R., Munoz-Medellin, D., and Huang, A. (1995) Strand invasion by oligonucleotide-nuclease conjugates. *Bioconjugate Chem.* **6,** 93–100.
8. Iyer, M., Norton, J. C., and Corey, D. R. (1995) Accelerated hybridization of oligonucleotides to duplex DNA. *J. Biol. Chem.* **270,** 14,712–14,717.
9. Corey, D. R. (1995) 48,000-fold acceleration of hybridization of chemically modified oligomers to duplex DNA. *J. Am. Chem. Soc.* **117,** 9373–9374.
10. Ishihara, T. and Corey, D. R. (1999) Rules for strand invasion by chemically modified oligonucleotides. *J. Am. Chem. Soc.* **121,** 2012–2020.
11. Takahara, M., Hibler, D. W., Barr, P. J., Gerlt, J. A., and Inouye, M. (1985) The ompA signal peptide directed secretion of Staphylococcal nuclease A by *Escherichia coli. J. Biol. Chem.* **260,** 2670–2674.

14

Synthesis of Peptide Nucleic Acid–Peptide Conjugates

Kunihiro Kaihatsu and David R. Corey

Summary

Synthetic oligonucleotides are versatile tools for recognizing ribonucleic acid and deoxyribonucleic acid. This chapter describes methods for enhancing recognition by derivatizing oligonucleotides with either proteins or peptides.

Key Words: Oligonucleotides; peptides; disulfide exchange; bioconjugate; enhanced hybridization.

1. Introduction

Peptide nucleic acids (PNAs) are deoxyribonucleic acid (DNA) analogs that bind with exceptionally high affinity to complementary sequences and have excellent potential for invasion of duplex DNA*(1)*. Here, we describe a simple approach to the development of PNAs that can dramatically alter hybridization properties and cellular uptake. We describe methods for (1) the conjugation of PNA to peptides to afford hybrid molecules that combine properties from both constituents and (2) the introduction of a fluorophore to PNA–peptides to follow their localization into the cells.

1.1. Goals for Improved Recognition

Synthetic molecules that recognize DNA and ribonucleic acid (RNA) sequences are promising tools for selectively controlling transcription and translation. PNAs can be synthesized readily, and it may be possible to further improve their properties through chemical modification. Potential areas of improvement include hybridization rate, hybridization affinity, and discrimination against binding to mismatched targets. For recognition of intracellular targets, improved properties could be achieved by enhancing cellular uptake, whereas in animals improved pharmacokinetic properties would be useful.

From: *Methods in Molecular Biology, vol. 283: Bioconjugation Protocols: Strategies and Methods*
Edited by: C. M. Niemeyer © Humana Press Inc., Totowa, NJ

1.2. PNA–Peptide Conjugates

The synthesis of PNA–peptide conjugates is a powerful approach to improving PNA recognition. PNAs are synthesized by protocols adapted from peptide synthesis, making attachment of peptides straightforward. Conjugation allows the hybridization properties of the PNA to be retained while incorporating properties from the peptide.

In our earliest studies, we demonstrated that PNA could be attached to various cationic peptides in high yield *(1–3)*. The PNA domain of the conjugate was able to recognize correct target sequences, whereas the cationic peptides enhance the binding efficiency to negatively charged phosphate backbone of DNA. PNA–peptides enhanced the rate of hybridization to DNA relative to the rate achieved by unmodified PNA *(4–6)*. We also demonstrated that peptide containing D-amino acids to PNAs also enhanced hybridization, providing a conjugate that should be able to promote DNA binding resist with resistance to degradation by nucleases and protease inside cells *(5)*.

PNA–peptide conjugates invade duplex DNA and inhibit RNA polymerase-mediated transcription at physiological ion strength and temperature *(7)*. Most recently, we have observed that PNA–peptides can enter cultured cancer cells and inhibit the protein expression (K. Kaihatsu, unpublished data). Our purpose here is to describe efficient methods for the synthesis of PNA–peptide conjugates to help investigators obtain molecules needed to fully explore the potential for using PNAs to control biological processes.

2. Materials

1. 9-Fluorenylmethoxycarbonyl-protected peptide nucleic acid monomers (A,T,C and G), base protected with benzhydryloxycarbonyl(Bhoc), Fmoc-A(Bhoc)-OH, Fmoc-T-OH, Fmoc-C(Bhoc)OH, and Fmoc-G(Bhoc)-OH were obtained from Applied Biosystems (Foster City, CA; *see* **Note 1**).
2. Linker 2-aminoethoxy-2-ethoxy acetic acid (Fmoc-AEEA-OH), activator 0.2 *M* *O*-(7-azabenzotriazol-1-yl)1,1,3,3,-tetramethyluronium. hexafluorophosphate in *N,N*-dimethylformamide (DMF), base solution 0.3 *M* 2,6-lutidine, and 0.2 *M* *N,N*-diisopropylethylamine in DMF, capping solution 5% (v/v) AcO$_2$, 6% (v/v) 2,6-lutidine in DMF, Deblock solution 20% (v/v) piperidine in DMF, and PNA diluent 1-methyl-2-pyrrolidinone (NMP) were obtained from Applied Biosystems.
3. Washing solution DMF, 1,1-dichloromethane, and trifluoro acetic acid (TFA), Isopropyl alcohol and diethyl ether were from Fisher Scientific (Pittsburgh, PA).
4. *m*-Cresol, allylalcohol, and hydrazine were from Aldrich (Milwaukee, WI).
5. Fmoc-XAL-PEG-PS synthesis columns (0.2 µmol prepacked) were from Applied Biosystems.
6. Amino acid monomers Fmoc-Ala-OH, Fmoc-Lys(Boc)-OH, Fmoc-D-Ala-OH, Fmoc-D-Lys(Boc)-OH, Fmoc-His-OH, Fmoc-Arg(Pbf)-OH, Fmoc-Pro-OH,

Fmoc-Val-OH, Fmoc-Lys(ivDde)-OH, Boc-Lys(Boc)-OH, and Boc-Pro-OH were from Novabiochem (San Diego, CA).
7. 5-(and-6)-Carboxytetramethylrhodamine was from Molecular Probes (Eugene, OR).
8. Polytetrafluoroethylene (1.5 mL; available in 0.22-μm size) or regenerated cellulose spin column was obtained from Fisher Scientific.

3. Methods
3.1. Preparation of Reagents for Fmoc Synthesis
1. Fmoc monomers and activator are stored at –20°C.
2. The bottles containing monomers and activator are warmed up to room temperature in a sealed container containing Dririte desiccant before solubilization.
3. Fmoc-PNA monomers were solubilized by adding 3.25 mL of NMP diluent directly to each amber bottle (final concentration of 216 mM). Allow the mixture to sit undisturbed for 30 min at room temperature to completely solubilize monomer. NMP and all other solvents are the driest grade available and should be stored in a desiccator.
4. Dissolve AEEA (2-aminoethyl-2-ethoxy acetic acid) in 2.4 mL of NMP diluent.
5. Dissolve amino acids by adding 4.6 mL of DMF to each clean and dry amber bottle (final concentration of 216 mM amino acid). Allow the mixture to sit undisturbed for 30 min at room temperature.

3.2. Synthesis of PNA–Peptides
1. Load reagents onto an Expedite 8909 synthesizer (*see* **Note 2**).
2. Attach clean and dry filters to all ports.
3. Install each amino acid solution to the PNA synthesizer.
4. We have observed that amino acids can crystallize in the synthesizer tubing. This can cause the synthesizer to become blocked and nonfunctional. Because amino acid monomers are relatively inexpensive, we dispose of solutions if we have no immediate plans to use them. By contrast, solutions of PNA monomers can be kept on the machine for as long as two weeks (*see* **Notes 3** and **4**).
5. Prime each amino acid solution and check the volume of flow through before starting the synthesis.
6. Before start the synthesis, prime Wash B solution (DMF, Applied Biosystems) to clean up the column line (*see* **Note 5**).
7. Attach Fmoc-XAL-PEG-PS columns on the synthesizer.
8. Perform the synthesis according to manufacturer's program and specification. We recommend double coupling the reactions for amino acids, sequences over 15 bases, poly-purine regions, GC-rich regions, and multiples of the same PNA base (more than three). This is done because these couplings tend to be less efficient, and double coupling helps to ensure that useful amounts of full-length product are obtained.
9. To prevent unexpected intermolecular reaction among PNA–peptides, leave N-terminal Fmoc group until a decision has been made to cleave the PNA from resin or add another group.

3.3. Cleaving of PNA–Peptide From Solid-Phase Resin

1. After finishing the synthesis, the Fmoc group on the N-terminus of the PNA is deprotected by piperidine (Deblock solution) on the PNA synthesizer. Invert the column and do this procedure again. We found that the deprotection efficiency of PNA–peptides that are more than 20 couplings were relatively low compare to shorter ones.

2. Remove the column from the synthesizer and wash it with 5 mL of DMF four times, reversing the direction of flow through the column each time. This washing step is done to remove remaining reactants on the resin.

3. Wash the column with 5 mL of isopropyl alcohol four times, reversing the direction of flow through the column each time. This washing step is performed to remove DMF from the resin. Removal of DMF simplifies to transfer of the resin from the column to spin column (*see* below) and give us good precipitation of PNA–peptides.

4. Dry the resin by blowing filtered house air across the column for 10 min, reversing the ends frequently.

5. Transfer the dried, resin-bound, deprotected PNA to a 1.5 mL, 0.2-μm PTFE or regenerated cellulose spin column.

6. Add 250 mL cleavage cocktail (TFA: *m*-Cresol = 1:0.25) to the resin for deprotecting Bhoc, Boc and Pbf group from PNA–peptides and cleave PNA–peptides from resin. Incubate the solution at room temperature for 90 min.

7. Centrifuge 2 min at 1300g for a PTFE filter or 2 min at 8400g for a regenerated cellulose filter.

8. Repeat **steps 6** and **7** but reduce cleavage time to 5 min. This removes any remaining PNA from the resin.

9. Collect the cleavage filtrate, remove the filter unit, and precipitate the PNA by adding 1 mL of cold (–20°C) diethyl ether. Invert the tube several times to ensure complete precipitation. Appearance of a precipitate is an indication that the PNA synthesis may be successful. Failure to observe a precipitate almost certainly indicates that the synthesis has failed.

10. Centrifuge precipitated PNA 2 min at 1300g. Discard the supernatant.

11. Wash the pellet three times with 1 mL diethyl ether, vortexing to suspend the pellet.

12. Centrifuge 2 min at 8400g to repack the pellet.

13. Remove as much of the supernatant as possible by aspiration and then air dry the pellet for 5 to 10 min in a chemical fume hood.

Add 200 μL of sterile water to the pellet (for a 2 μmol synthesis) and allow the tube to remain undisturbed for 10 to 15 min at 65°C. Because of the residual trifluoroacetic acid, the solution should be slightly acidic. This acidity causes the PNA to be protonated, increasing its solubility and allowing the experimenter to store the PNA as a concentrated stock solution.

3.4. Purification and Analysis of PNA–Peptide

PNA–peptide conjugates can be purified by reverse-phase high-performance liquid chromatography (RP-HPLC) followed by matrix-assisted laser desorption/ionization time-of-flight mass spectrometry (MALDI-TOF-MS).

3.4.1. Purification of PNA–Peptide by HPLC

1. Centrifuge a PNA solution 3 min at 12,000g, room temperature, to remove particulate.
2. Turn the water bath on and warm the C18 reverse-phase HPLC column (300-A Microsorb-MV column; Varian Analytical Instruments) up to 55°C by circulating the water. Heating prevents aggregation of PNAs on the column and gives us higher resolution of final products.
3. Observe the optical density at 260 nm.
4. Set up a gradient of 0% to 5% (v/v) RP-HPLC buffer B in buffer A for 6 min followed by 5 to 100% buffer B in buffer A for 24 min.
5. Collect each fractions corresponding to each major peak. A typical PNA synthesis should be quite pure, with only one major peak, but less pure syntheses can also provide useful quantities of material and should not be discarded.
6. Analyze all fractions by MALDI-TOF-MS (*see* **Subheading 3.4.2.**).
7. Combine fractions that contain substantial amounts of purified PNA–peptide into a 15 mL tube and freeze in the cold ethanol bath or –80°C freezer.
8. Lyophilize the samples overnight to remove TFA and acetonitrile.
9. Dissolve the pellet with 300 µL of milli-Q water.
10. Adjust the pH with buffer as needed.

3.4.2. MALDI-TOF-MS Analysis

MALDI-TOF mass spectrometer (Voyager-DE workstation; Applied Biosystems).

1. Overlay the sample with 1 µL of matrix consisting of 10 mg/mL a-cyano-4-hydroxy-cinnamic acid (Sigma) in 25/75 (v/v) RP-HPLC buffers A/B. If the molecular weight of PNA–peptides is bigger than 7000 or they include more than five charged amino acids, we recommend to use Sinapic acid as the matrix.
2. Leave the samples on the plate until samples are dried.
3. Activate the laser and collect data. PNA–peptide and fluorophore–PNA–peptide requires higher laser intensity (2300–2500) compared with PNA molecules. The laser energies are used 1500–2000 mV for PNA–peptides.

3.5. Preparation of Fluorophore–PNA–Peptide

This procedure is available for synthesizing PNA that has peptide on its N-term and fluorophore on its C-term. To retain the function of peptide, position of fluorophore on PNA–peptide has to be considered.

3.5.1. Introduction of a Reactive Amine on PNA–Peptide

1. Dissolve 1 mmol of Fmoc-Lys(ivDde)-OH [*N*-α-Fmoc-*N*-ε-1-(4,4-dimethyl-2,6-dioxocyclohex-1-ylidene)-3 methylbutyl-L-lysine] in 4.6 mL of dry DMF (final concentration of 216 m*M*).
2. Place a Fmoc-Lys(ivDde)-OH at the C terminus of PNA–peptide (*see* **Note 6**).
3. To minimize steric hindrance, provide three spacer (AEEA) molecules between fluorophore and PNA.
4. Place a Boc protected amino acid on the end of N-terminus of PNA–peptide.
5. Start PNA–peptide synthesis as usual.
6. After the synthesis has done, attach a 5 mL luer slip-tip syringe to one end of the column.
7. To deprotect ivDde group, prepare 2% hydrazine in dry DMF solution.
8. Draw 3 mL of hydrazine solution into a second 5-mL syringe and attach this to the other end of the column. When removing ivDde in the presence of allyl-based protection groups, allyl alcohol should be included in the deprotection to prevent reduction of the allyl group.
9. Slowly push the solution back and forth through the column for 10 min.
10. Transfer this cleavage solution into 15-mL tube.
11. To follow ivDde cleavage from PNA–peptides, the cleaving solutions were analyzed by UV spectrophotometry.
12. Continue this procedure until the absorbance of ivDde group (290 nm) disappeared from the cleaving solution.
13. Remove the column from the synthesizer and wash it with 5 mL of dry DMF four times, reversing the direction of flow through the column each time.
14. Dry the resin by blowing filtered house air across the column for 3 to 5 min, reversing the ends frequently.
15. Lyophylize the column overnight.

3.5.2. Addition of Rhodamine to the PNA–Peptide

1. Transfer the resin to an amber tube (1.5 mL). Fluorophore molecules are light sensitive, so they should be stored and used under dark conditions.
2. To activate the ε-amino group of C-term lysine, add 500 μL of base solution and vibrate the tube vigorously.
3. Centrifuge the tube and remove the supernatant.
4. Repeat **steps 2** and **3** three times.
5. Solubilize 4.3 mg (10 mmol) of 5-(and-6)-carboxytetramethylrhodamine mixed isomer in 200 μL of dry DMF (232 mmol).
6. Add 200 μL of base solution, 200 μL of activator, and 400 μL of diluent. Incubate the mixture at 60°C for 10 min. In this step, the rhodamine color changes from clear pink to dark red.
7. Mix the rhodamine solution and PNA–peptide conjugated resin in an amber tube.
8. Shake the mixture into the Eppendorf thermomixer at 60°C for 4 h.

9. Centrifuge the reaction tube and decant the supernatant. Add DMF into the tube and vibrate it. Do this procedure three times to remove free rhodamine. If the efficiency of the coupling is good, the color of resin is dark red. If it's not, the color of resin looks pink or orange.
10. Double couple and triple couple as needed.
11. Put the resin back to an empty column using a pipet. Wash the column with DMF and isopropanol, air-dry it, cleave the rhodamine–PNA–peptide conjugate from resin using 20% *m*-cresol in TFA solution, and precipitate with cold diethylether.
12. Dissolve the pellet with Milli-Q water.
13. Purify it by HPLC using C18 column as the same way of normal PNA. To avoid contamination, the column has to be washed cleanly each purification step.
14. The flow through is analyzed by UV spectrophotometry. The ratio of A_{260} and A_{560} has to correspond the molar extinction coefficiency of PNA-peptide and rhodamine, respectively.
15. Analyze the samples by MALDI-TOF-MS.

4. Notes

1. These protocols for preparation of PNA with peptides and fluorophore have proven to be readily reproducible. Equally good yields can be obtained with either C-term- or N-term-derivatized PNAs. The coupling efficiency is more than 90%.
2. The Expedite 8909 PNA synthesizer has three extra ports for the amino acids and linker molecule. Therefore, it is straightforward to synthesize a PNA–peptide that is composed of less than three different amino acids. If peptide portion includes more than four different amino acids, introduce the fourth amino acid solution after third amino acid is conjugated.
3. Unless the branched PNA–peptides are required, do not use amino acid that has a Fmoc group on its side chain during Fmoc synthesis (e.g., Fmoc-Lys(Fmoc)-OH). For the addition of fluorophore on a specific position of PNA–peptide, its N-terminus shouldn't be protected by -Fmoc or O-Dmab. These groups can be deprotected by hydrazine.
4. Keeping each reagent fresh and dry is essential. When the reagents are removed from –20°C freezer and placed in room temperature, moisture will collect around the reagent bottles. To avoid this moisture, we warm them up in a sealed container containing Dririte desiccant before solubilization. PNA monomer should be stored in dark to avoid exposing them under UV light for a long time.
5. Lower coupling yield can arise from inefficient delivery or low purity of each reagent. When amino acids reagents are replaced, attach DMF bottle to the machine and wash off remaining reagents by priming. Then, attach the new reagents to machine and prime each line twice to wash off DMF. This procedure needs to be done for preventing low coupling yield.
6. IvDde has proven to be a valuable tool for the preparation of branched peptides by Fmoc synthesis. After the synthesis has performed, ivDde group is easily cleaved with 10 mL of 2% hydrazine in DMF. Deprotection of ivDde can be followed by A_{290} of UV spectrophotometry.

Acknowledgments

This work was supported by grants from the National Institutes of Health (GM60642) and by the Robert A. Welch Foundation (I-1244).

References

1. Mayfield, L. D. and Corey, D.R (1998) Automated synthesis of peptide nucleic acids and peptide nucleic acid-peptide conjugate. *Anal. Chem.* **268,** 401–404.
2. Braasch, D. A. and Corey, D. R. (2001) Synthesis, analysis, purification, and intracellular delivery of peptide nucleic acids. *Methods* **23,** 97–107.
3. Braasch, D. A. Nulf, C. J., and Corey, D. R. (2002) Synthesis and purification of peptide nucleic acids, in *Current Protocols in Nucleic Acid Chemistry.* John Wiley & Sons, New York, NY, pp. 4.11.1–4.11.18.
4. Zhang, X. Ishihara, T., and Corey, D. R. (2000) Strand invasion by mixed base PNA and a PNA-peptide chimera. *Nucleic Acid Res.* **28,** 3332–3338.
5. Kaihatsu, K. Braasch, D. A., Cansizoglu, A., and Corey, D. R. (2002) Enhanced strand invasion by peptide nucleic acid-peptide conjugates. *Biochemistry* **41,** 11,118–11,125.
6. Kaihatsu K. Shah R.H, Zhao X, and Corey D. R. (2003) Extending duplex recognition by peptide nucleic acids (PNAs): strand invasion and inhibition of transcription by tail clamp PNAs and tail clamp PNA–peptide conjugates. *Biochemistry* **42,** 13,996–14,003.
7. Zhao, X. Kaihatsu, K., and Corey, D. R. (2003) Inhibition of transcription by bisPNA–peptide conjugates *Nucleosides, Nucleotides Nucleic Acids.* **22,** 535–546.

III

GLYCOSYL AND LIPID CONJUGATES

15

Protein Lipidation

Jürgen Kuhlmann

Summary

This chapter describes the hydrophobic modification of peripheral membrane-anchored proteins by isoprenylation and S-acylation. The coupling of bacterially expressed protein moieties with chemically synthesized lipopeptides is described as an in vitro alternative for the generation of lipoproteins

Key Words: Lipoprotein; posttranslational modification; isoprenoid; fatty acids; membrane binding; Ras proteins; chemical synthesis; coupling.

Eucaryotic cells are characterized by a complex network of internal membranes that form boundaries and define different compartments (e.g., nucleus), protect the cell against toxic components (e.g., lysosomes, proteasomes), or are involved in transport (Golgi). In the meantime, another aspect of endogenous membranes became relevant: the localization of proteins in specific membranes can optimize biochemical reactions by restriction to two-dimensional diffusion *(1)*, prevent concurrent processes by separating biological macromolecules, and increase the performance of signaling events.

Beside the incorporation of transmembrane proteins, covalent modifications with hydrophobic groups allow the peripheral insertion of macromolecules in endomembranes. Here, isoprenoids and fatty acids are the two most common lipid moieties that allow membrane anchorage. Incorporation of hydrophobic groups occurs by posttranslational modification and is catalyzed by specific enzymes. Isoprenoids such as farnesyl (C15) or geranylgeranyl (C20) groups are attached to C-terminal cysteines of the protein by farnesyl-(FTase) or geranylgeranyltransferases (GGTase I, GGTase II) with a stable thioether. Fatty acids can be introduced via *N*-acylation by *N*-myristoyltransferase, which adds myristate, laureate, or unsaturated fatty acids of similar chain length at the N-

From: *Methods in Molecular Biology, vol. 283: Bioconjugation Protocols: Strategies and Methods*
Edited by: C. M. Niemeyer © Humana Press Inc., Totowa, NJ

terminal glycine residue. Whereas *N*-acylation creates a stable chemical bond, *S*-acylation results in a labile thioester between a cysteine residue and a fatty acid (most commonly a palmitate; **ref. 2**). Further hydrophobic modifications are rarely described (e.g., attachment of a cholesterol moiety at a lysine residue of the Hedgehog protein; **ref. 3**).

Posttranslational modification by isoprenoid groups and fatty acids is in particular essential for the function of several main protein players in signal transduction and vesicular transport. G_α subunits of heterotrimeric G proteins are palmitoylated and/or myristoylated, as well as nonreceptor tyrosine kinases like p60 src and p56 lck. Most small GTP-binding proteins of the Ras superfamily require at least an isoprenoid modification for biological activity that has to be completed in some cases by an additional *S*-acylation with a palmitate group. These modifications generate a hydrophic C-terminus that anchors the proteins in the appropriate endogenous membrane.

Insertion of lipophilic side chains into membranes can be supported by additional electrostatic interactions, for example, between positively charged lysine residues of a protein and negatively charged lipids as supposed for the plasma membrane of eukaryotic cells *(4)*.

A first round of investigations on the interplay between membranes and membrane-associated proteins used chemically synthesized lipopeptides as powerful probes. The chemical origin of the probes allowed the introduction of fluorescent *(5)* or radioactive labels *(6)* that enabled analysis of association and dissociation with membranes and the adjustment of equilibria. In addition lipopeptide probes gave information on trafficking and localization of the corresponding lipoproteins in cell biological setups *(7,8)*.

Irrespective of these capabilities, lipopeptides show insurmountable limitations: they lack biological functionality and in the majority of cases their physical properties differ significantly from those of the complete protein. Lipopeptides often have reduced solubility in aqueous systems, whereas the underlying protein is still dissolved because of the dominance of its large polypeptide moiety. However, efficient access to posttranslationally modified proteins of the Ras superfamily is difficult, for example, in vitro prenylation of bacterially synthesized Ras or Rab proteins is well established *(9)*, but no analog biochemical setup exists for the subsequent proteolytic cleavage, carboxymethylation, or *S*-acylation. Hence, eucaryotic expression of small GTP-binding proteins for a long time was the only passable way to synthesize small amounts of completely modified products. Nevertheless high expenses, extraordinary costs, and low yields made this approach uninviting *(10)*.

These chapters describe two alternatives for the generation of Ras and Rab lipoproteins that are based on a bio-organic synthesis of C-terminal lipopeptides and bacterial expression of the N-terminal protein moiety. Both building blocks can be produced in high amounts and allow the isolation of lipoproteins in the milligram range. Beyond this the chemical origin of the lipopeptide part opens the well-directed introduction of artificial modifications as hydrolysis stable substitutions for acyl thioesters or the incorporation of fluorescence markers that are not feasible by any biological approach.

For Ras lipoproteins, coupling of a C-terminal-truncated protein moiety with an activated lipopeptide was established. Here, lipopeptides carry a N-terminal maleimidocaproyl group that reacts specifically with mercapto groups of proteins by conjugate addition of the thiol to the α,β-unsaturated carbonyl compounds *(11)*. The thiol is contributed by a carboxy terminal cysteine of the Ras protein. For Rab lipoproteins, chemoselective addition of lipopeptides to a recombinant-produced protein moiety by expressed protein ligation is introduced. By replacing a thioester in the C-terminus of the recombinant protein by the thiol of a N-terminal cysteine in the peptide, followed by a *S-N* acyl shift, the native peptide bond of the cellular protein is achieved with this technique *(12,13)*.

References

1. Kadereit, D., Kuhlmann, J., and Waldmann, H. (2000) Linking the fields - The interplay of organic synthesis, biophysical chemistry, and cell biology in the chemical biology of protein lipidation (Rev.). *ChemBioChem* **1**, 144–169.
2. Dunphy, J. T. and Linder, M. E. (1998) Signalling functions of protein palmitoylation (Rev.) *BBA Mol. Cell Biol. Lipids* **1436**, 245–261.
3. Chamoun, Z., Mann, R. K., Nellen, D., von Kessler, D. P., Bellotto, M., Beachy, P. A., and Basler, K. (2001) Skinny Hedgehog, an acyltransferase required for palmitoylation and activity of the Hedgehog signal. *Science* **293**, 2080–2084.
4. Murray, D., Ben-Tal, N., Honig, B., and McLaughlin, S. (1997) Electrostatic interaction of myristoylated proteins with membranes: simple physics, complicated biology. *Structure* **5**, 985–989.
5. Leventis, R. and Silvius, J. R. (1998) Lipid-binding characteristics of the polybasic carboxy-terminal sequence of K-Ras4b. *Biochemistry* **37**, 7640–7648.
6. Ghomashchi, F., Zhang, X. H., Liu, L., and Gelb, M. H. (1995) Binding of prenylated and polybasic peptides to membranes—affinities and intervesicle exchange. *Biochemistry* **34**, 11,910–11,918.
7. Schroeder, H., Leventis, R., Rex, S., Schelhaas, M., Nagele, E., Waldmann, H., and Silvius, J. R. (1997) S-acylation and plasma membrane targeting of the farnesylated carboxyl-terminal peptide of N-ras in mammalian fibroblasts. *Biochemistry* **36**, 13,102–13,109.
8. Waldmann, H., Schelhaas, M., Nagele, E., Kuhlmann, J., Wittinghofer, A., Schroeder, H., and Silvius, J. R. (1997) Chemoenzymatic synthesis of fluorescent N-ras lipopeptides and their use in membrane localization studies in vivo. *Angewandte Chemie Int. Ed.* **36**, 2238–2241.

9. Seabra, M. C. and James, G. L. (1998) Prenylation assays for small GTPases, in *Transmembrane Signaling Protocols* (Bar-Sagi, D., ed.), Humana Press, Totowa, NJ.
10. Page, M. J., Hall, A., Rhodes, S., Skinner, R. H., Murphy, V., Sydenham, M., and Lowe, P. N. (1989) Expression and characterization of the Ha-ras p21 protein produced at high levels in the insect/baculovirus system. *J. Biol. Chem.* **264,** 19,147–19,154.
11. Hermanson, G. T. (1996) Tags and probes, in *Bioconjugate Techniques*, Academic Press, Inc., London, UK, pp. 297–418.
12. Muir, T. W., Sondhi, D., and Cole, P. A. (1998) Expressed protein ligation—a general method for protein engineering. *Proc. Natl. Acad. Sci. USA* **95,** 6705–6710.
13. Severinov, K. and Muir, T. W. (1998) Expressed protein ligation, a novel method for studying protein-protein interactions in transcription. *J. Biol. Chem.* **273,** 16,205–16,209.

16

Synthesis of Lipidated Peptides

Ines Heinemann, Martin Völkert, and Herbert Waldmann

Summary
This chapter describes general methodologies for the synthesis of lipidated, that is, prenylated and/or palmitoylated peptides. Standard operating procedures are given for peptide synthesis both on the polymeric support and in solution.

Key Words: Prenylation; palmitoylation; solid-phase peptide synthesis; solution-phase peptide synthesis.

1. Introduction

Several strategies for the synthesis of lipidated peptides both in solution and on solid support have been developed and reviewed *(1,2)*. For peptides with longer amino acid chains, synthesis on solid support has nearly always been performed, whereas shorter peptides have been synthesized both on solid support and in solution. Particularly, for hexa- and heptapeptides corresponding to the Ras- and Rab-C-termini, respectively, the latter had been the case (**refs. *3,4*,** and *see* **Note 1**).

In general, different lipid groups and other modifications, such as linker or reporter groups, can be introduced into a peptide either by coupling modified building blocks or by modification of a selectively deprotected peptide. The strategies for peptide synthesis on solid support and in solution strongly depend on the lability of the products caused by the modifications. Particularly, the synthesis of peptides with different modifications has to go hand in hand with the development of a suitable protecting group strategy. Acid labile-protecting groups and blocking functions cleaved under hydrogenolytic conditions have to be avoided if prenyl moieties have been introduced into the peptide. Palmitoyl groups attached to cysteine side chains as thioesters are labile towards bases and nucleophiles, excluding synthesis strategies that require the

From: *Methods in Molecular Biology, vol. 283: Bioconjugation Protocols: Strategies and Methods*
Edited by: C. M. Niemeyer © Humana Press Inc., Totowa, NJ

deprotection of such peptides under basic conditions or in the presence of nucleophiles. An N-terminal amino-deprotected and *S*-palmitoylated cysteine is prone to an undesired *S,N*-acyl shift of the palmitoyl group. Consequently, if applied, palmitoylation has to be performed either on the entire peptide or at least on a larger building block.

In addition, acid labile prenylated dipeptide esters may be prone to diketopiperazine formation upon deprotection *(5)*. In solution, fragment coupling might be accompanied by racemization of the activated C-terminal amino acid. Therefore, fragment coupling should preferably be conducted at a C-terminal glycine or a proline.

A schematic overview over peptide synthesis on the solid support is given in **Fig. 1**. In addition to cleaving entirely deprotected peptides, the methods for solid-phase synthesis of lipidated peptides can be divided into two different approaches: One is the synthesis of unmodified peptide fragments, which are then condensed in solution with lipidated building blocks to yield the desired lipidated peptides. The other approach is to perform the entire synthesis of the lipidated peptide on the solid support. The introduction of palmitoyl thioesters on the polymeric support has been reported several times *(6–8)*, and the farnesylation on solid support was reported only recently *(9)*. Using this methodology, the entire synthesis of farnesylated and palmitoylated peptides could be performed on the polymeric support.

The proper choice of the linker group is of utmost importance for a successful synthesis. The fragment condensation strategy allows one to use a variety of different linker groups. In our own experience, the 2-chlorotrityl group is particularly advantageous. The target peptide is cleaved from this linker with dilute acid. The global solid-phase strategy has been performed only with the 4-hydrazinobenzoyl group, which is an oxidation-labile linker that can be cleaved either using Cu(II)-salts or *N*-bromosuccinimide *(10)*. Both linkers allow for application of the Fmoc group-based peptide synthesis strategy.

2. Materials

2.1. General Reagents

1. Diethylamine.
2. Triethylamine.
3. Diisopropylethylamine (DIPEA).
4. Piperidine.
5. Pyridine.
6. Trifluoroacetic acid.
7. Acetic acid.
8. Dimethylbarbituric acid.
9. Dithiothreitol.
10. Thioanisole.

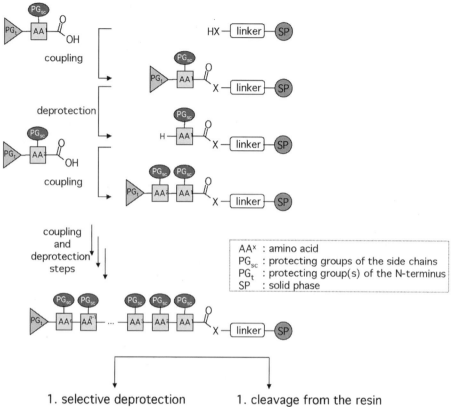

Fig. 1. Strategies for the assembly of lipidated peptides on the solid support.

11. Dimethylsulfide.
12. Triethylsilane.
13. [Pd(PPh$_3$)$_4$].
14. Copper(II)acetate.
15. *N*-Bromosuccinimide (NBS).
16. *N*-Chlorosuccinimide (NCS).

2.2. Building Blocks

1. Fmoc-protected amino acids were purchased from Novabiochem, Senn or synthesized according to published literature procedures.
2. Palmitoyl chloride (Sigma).
3. Farnesyl bromide (Fluka).

4. Maleimidocaproic acid (MIC-OH) (Fluka).
5. Geranylgeranyl alcohol (GerGerOH) also was prepared as described by literature procedures.

2.3. Coupling Reagents

1. Acetic anhydride, pivalic anhydride, 2-(1H-benzotriazole-1-yl)-1,1,3,3-tetramethyl-uronium hexafluorophosphate (HBTU), N-hydrosuccinimide (HOSu), 3-hydroxy-3,4-dihydro-4-oxo-1,2,3-benzotriazine (HODhbt), and 1-hydroxybenzotriazole hydrate (HOBt) were purchased from Fluka, Aldrich, and Novabiochem.
2. 1-Ethyl-3-(3-dimethylaminopropyl) carbodiimide hydrochloride (EDC) was a kind donation of Bayer AG.

2.4. Solvents

1. Dichloromethane was dried by distillation from CaH_2, tetrahydrofuran (THF) was dried by distillation from Na/K (*see* **Note 2**).
2. N,N-Dimethylformamide (DMF) was used as a peptide-grade reagent from Biosolve Ltd.
3. N-Methyl-pyrrolidone (NMP) also was used as peptide grade reagent from Biosolve Ltd.
4. Chloroform, methanol, trifluoroethanol, and 2 N NH_3 in MeOH were purchased from Aldrich and Fluka.

2.5. Stationary Phases for Purification

1. Silica (Merck).
2. Sephadex® LH 20 (Amersham Pharmacia).

2.6. Resins and Linkers

1. 2-Chlorotrityl–chloride resin was purchased from CBL Patras SA.
2. 4-Fmoc-hydrazinobenzoyl AM Novagel was purchased from Novabiochem.

3. Methods

3.1. Building Blocks (see Note 3)

3.1.1. Chlorination of Prenyl Alcohols

1. Dissolve N-chlorosuccinimide (recrystallized in water) (1.1 eq) in dry CH_2Cl_2 (2.5 mL/mmol) and cool the solution to –35 to –40°C with an acetonitrile/dry ice bath *(11)*.
2. Add dimethylsulfide (1.2 eq) and let the reaction stir for 5 min at 0°C.
3. Cool again to –35 to –40°C, add a solution of the prenyl alcohol in dry CH_2Cl_2, and let the reaction stir for 1.5 h at –15 to –20°C and 30 min at 0°C.
4. Pour the reaction mixture on ice cold brine, dilute with CH_2Cl_2, wash the organic phase five times with brine, and dry over Na_2SO_4.
5. Remove the solvent under reduced pressure.

3.1.2. Prenylation of Cysteine or Cysteine Methyl Ester

1. At 0°C, dissolve the cysteine or cysteine methyl ester in 2 N ammonia in methanol (*12*; *see* **Note 5**).
2. At −18°C, add a solution of prenyl chloride in dry THF and let the reaction stir for 30 min. Warm to 0°C and stir for an additional 1.5 h.
3. Remove the solvent by azeotropic distillation with toluene under reduced pressure.

3.1.3. Palmitoylation of Protected Dipeptides With a C-Terminal Cystine

1. Dissolve the dipeptide in CH_2Cl_2, add triethylamine (2.05 eq) and dithiothreitol (5 eq) and let the reaction stir for 60 min at room temperature.
2. Wash the reaction mixture three times with 1 N HCl and dry over Na_2SO_4.
3. Dissolve the crude product in CH_2Cl_2, add triethylamine (2.05 eq) and palmitoyl chloride (5 eq) and let the reaction stir for 2 h at room temperature.
4. Wash the reaction mixture with 1 N HCl, 1 N NaHCO$_3$ and brine.
5. After drying over Na_2SO_4 and filtering, concentrate the solution under reduced pressure.

3.2. Solution Phase Synthesis

3.2.1. Standard Peptide-Coupling Conditions

1. Dissolve both coupling partners 1:1 (molar) and HOBt or HODhbt (1.5 eq) in CH_2Cl_2.
2. At −15 to −20°C, add EDC (1.2 eq), let the reaction warm to room temperature and leave the reaction to stir for 18 h.
3. Add CH_2Cl_2, wash with water, dry over Na_2SO_4 and remove the solvents. Alternatively, remove the solvent by azeotropic distillation with toluene under reduced pressure.
4. After drying over Na_2SO_4 and filtering, concentrate the solution under reduced pressure.

3.2.2. Standard Procedure for the Removal of the Fmoc-Protecting Group

1. Dissolve the protected peptide in CH_2Cl_2/amine (Et$_2$NH or piperidine) 3–5:1 (v/v) at room temperature and leave the reaction to stir for 1–2 h. Monitor the reaction by thin-layer chromatography (TLC).
2. Remove the solvent by azeotropic distillation with toluene and chloroform.

3.2.3. Standard Procedure for the Removal
of Boc Groups or Tert-Butyl Esters (see **Note 6**)

1. Dissolve the protected peptide and thioanisole (2 eq) in CH_2Cl_2 and add trifluoroacetic acid to a concentration of 33–50% (v/v).
2. Leave the reaction to stir for 1–2 h at room temperature. Monitor the reaction by TLC.
3. Remove the solvent by azeotropic distillation with toluene and chloroform.

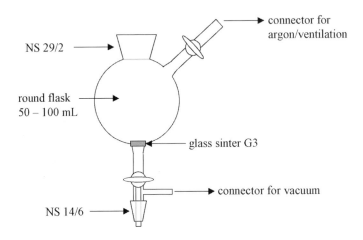

NS 29/2 ⟶

connector for
argon/ventilation

round flask
50 – 100 mL ⟶

glass sinter G3

connector for vacuum

NS 14/6 ⟶

Fig. 2. Shaking apparatus for solid-phase reactions.

3.2.4. Standard Procedure for the Removal of Aloc Groups or Allyl Esters

1. Dissolve the protected peptide and dimethylbarbituric acid (0.55 eq) in dry THF.
2. At room temperature, add a catalytic amount of tetrakis(triphenylphosphine)-palladium(0) and leave the reaction to stir for 2 h.
3. Monitor the reaction by TLC and finally remove the solvent under reduced pressure.

3.3. Solid-Phase Synthesis

All solid-phase peptide couplings and modifications are performed in a shaking apparatus modified after a description by Lewalter (**ref. *13*; *see* Fig. 2**). Agitation is achieved by placing the apparatus onto an orbital shaker. It is advantageous to preswell the resin 10 min in CH_2Cl_2 before use and after each vacuum drying. In general, polystyrene resins tend to swell better in CH_2Cl_2; therefore, the washing procedures should be performed with this solvent. For peptide couplings, however, the solubility of the reagents in CH_2Cl_2 is not high enough and DMF or NMP are applied. All reactions are performed at room temperature.

3.3.1. Solutions

3.3.1.1. Fmoc Cleavage (Solution A)

A 20% solution of piperidine in DMF (v/v), should be prepared freshly every day.

3.3.1.2. CAPPING (SOLUTION B)

A 10% solution of acetic anhydride in pyridine (v/v) should be prepared freshly every day. With the hydrazide resin, use pivalic anhydride instead of acetic anhydride to avoid blocking of the linker nitrogens *(14)*.

3.3.2. Determination of the Fmoc Loading

1. m = 3–4 mg of the dried resin are shaken with 5.0 mL of Solution A for 10 min.
2. Pipet 1 mL of this solution into a 3-mL UV-cuvet (d = 1 cm).
3. Add 2 mL of Solution A. Shake for mixing.
4. Use another UV cuvet with solution A as a reference and determine $\delta_{Abs301nm}$
5. Fmoc-loading is . $L = \dfrac{\Delta_{Abs301mn} \times 15 \times 1000}{7800 \times m}$ [mmol/g] Insert m reading in mg.
6. The theoretical loading after a reaction leading to the molecular mass difference δM [g/mmol] can be estimated using the formula $L = \dfrac{L_{old}}{1 + DM \times L_{old}}$ [mmol/g].

3.3.3. Loading of the Resins and Attachment of the First Amino Acid

3.3.3.1. 2-CHLOROTRITYL RESIN

1. Vacuum-dried 2-chlorotrityl chloride resin is shaken in an oven-dried reactor with 1.5 eq of the desired Fmoc-protected amino acid (*see* **Note 7**) and 4 eq DIPEA in 5–15 mL of dry CH_2Cl_2 per mmol functional groups on the resin for 60 min.
2. After removal of the solution, wash the resin five times with CH_2Cl_2.
3. The resin is capped by shaking 30 min with each 5 eq MeOH and DIPEA in CH_2Cl_2.
4. Remove liquid and wash five times with CH_2Cl_2. Dry resin *in vacuo* and determine loading (*see* **Subheading 3.3.2.**). Check mass balance.
5. Swell resin in CH_2Cl_2 and cleave Fmoc (*see* **Subheading 3.3.5.**).
6. Wash five times with CH_2Cl_2.

3.3.3.2. HYDRAZIDE RESIN

1. Cleave the Fmoc group (*see* **Subheading 3.3.5.**).
2. Proceed with chain elongation (*see* **Subheading 3.3.4.**).

3.3.4. General Chain Elongation Procedure

1. Dissolve 4 eq Fmoc-protected amino acid, 3.6 eq HBTU, and 4.8 eq. HOBt in DMF (*see* **Note 8**; 5–10 mL per mmol amino acid), add 8.0 eq. DIPEA. Shake mixture until all is dissolved (*see* **Note 9**).
2. Add solution to the resin.
3. Shake 45 min.
4. Remove liquid and wash five times with CH_2Cl_2.
5. Capping (optional): shake 5 min with solution B (5–15 mL per mmol reactive sites on the resin) to block unreacted amino functions. Wash five times with CH_2Cl_2 (*see* **Note 10**).

3.3.5. N-Terminal Fmoc Cleavage

1. Shake the resin with solution A (5–15 mL per mmol loading) for 5–10 min.
2. Remove liquid.
3. Repeat **steps 1** and **2**.
4. Wash five times with CH_2Cl_2.

3.3.6. Farnesylation on the Solid Support

1. Remove side chain protecting group from prenylation site cysteine(s) (*see* **Note 11**). Add a solution of 5 eq Far-Br and 12 eq DIPEA in DMF (5–10 mL/mmol).
2. Shake 4 h.
3. Wash six times with CH_2Cl_2.

3.3.7. Palmitoylation on Solid Phase

1. Remove side chain protecting group from prenylation site cysteine(s) (*see* **Note 11**).
2. Add a solution of 20 eq Pal-Cl, 20 eq HOBt, and 22 eq NEt_3 in a mixture of DMF and CH_2Cl_2 (1:3 v/v, 5–10 mL/mmol).
3. Shake 15 h.
4. Wash six times with CH_2Cl_2.

3.3.8. Attachment of MIC–OH

MIC attachment can be performed as described in the general chain elongation procedure (*see* **Subheading 3.3.4.** and **Note 12**). An alternative procedure is as follows:

1. Add a solution of 6.4 eq MIC–OH, 3.6 eq DIC, 5.5 eq HOBt, and 3.6 eq NEt_3 in DMF (5–10 mL/mmol).
2. Shake 3 h.
3. Wash five times with CH_2Cl_2.

3.3.9. Cleavage From the Polymeric Support

3.3.9.1. 2-Chlorotrityl Resin

1. Add a solution of 10% trifluoroacetic acid in CH_2Cl_2 (5–10 mL/mmol) and shake 10 min (*see* **Note 13**).
2. Collect solution and repeat **step 1**.
3. Wash resin three times with CH_2Cl_2. Combine all filtrates.
4. Remove solvent under reduced pressure (*see* **Note 14**).

3.3.9.2. Hydrazide Resin

1. For the NBS method (*see* **Note 15**), shake the resin with a solution of 2 eq each of NBS and pyridine in dry CH_2Cl_2 (5–10 mL/mmol functional groups) for 5 min.
2. Wash resin three times each with dry CH_2Cl_2 and dry THF.

3. Add 5 eq dry nucleophile (e.g., methanol) in dry CH_2Cl_2.
4. Shake 4 h.
5. Filter the resin and collect the filtrate.
6. Wash five times with CH_2Cl_2 and combine solution with the filtrate.
7. Remove solvent under reduced pressure.

1. For the Cu(II) method, place the resin in a flask.
2. Add a solution of 0.5 eq $Cu(OAc)_2$, 10 eq pyridine, and 5 eq nucleophile (e.g., methanol) in dry CH_2Cl_2 (5–10 mL/mmol).
3. Close the flask with a septum. Supply oxygen by a balloon on a long cannula reaching into the suspension. Add a short vent cannula (0.15 mm in diameter) so that O_2 is permanently bubbled through the solution.
4. Filter the resin and wash five times with CH_2Cl_2.
5. Remove solvent from combined filtrates under reduced pressure.
6. Remove copper as follows (optional; *see* **Note 16**): dissolve the crude mixture in CH_2Cl_2, apply the solution to an SPE cartridge (Supelco), elute the peptide with CH_2Cl_2. The copper salts remain on the cartridge.

3.4. Purification

The purification depends very largely on the properties of the target peptides and often the purification problems are more difficult to solve than the synthesis problems. Any given purification problem will require its own measures.

3.4.1. Column Chromatography

The standard method for product purification is flash column chromatography on silica gel using mixtures of cyclohexane and ethyl acetate, CH_2Cl_2 and methanol, $CHCl_3$ and methanol, or ethyl acetate and methanol.

3.4.2. Size-Exclusion Chromatography

1. Use a glass column (2 m × 4 cm) of Sephadex® LH20 in chloroform/methanol 1:1.
2. Place a circle of filtration paper directly on top of the Sephadex® LH20 and carefully put on a solution of the crude product in chloroform/methanol 1:1 (v:v).
3. At a speed of one drop every 6–10 s collect fractions of 3 mL.

3.4.3. High-Performance Liquid Chromatography

Use a RP C-4 column as stationary phase and water and acetonitrile, both with 0.1% TFA or formic acid as mobile phases. A linear gradient of 10–100% in 15 min is a good starting point for further optimization.

3.5. Analytics

The analytic characterization is performed using standard techniques of NMR spectroscopy, matrix-assisted laser desorption/ionization time-of-flight mass spectrometry, fast atom bombardment mass spectrometry (FAB-MS), and high-performance liquid chromatography–electrospray ionization–mass spectrometry (HPLC-EIS MS).

4. Notes

1. A more general but detailed overview about peptide synthesis is found in **ref. 5**.
2. THF is highly flammable, and Na/K reacts violently with water leading to evolution of hydrogen. Be careful with the distilling process.
3. All reactions were conducted under an argon atmosphere.
4. Originally published with 4 *N* ammonia in methanol, this reaction has been successfully performed using the commercially available solution of 2 *N* ammonia in methanol as well.
5. In most cases, purification of the free acids has been performed by washing the solid with diethylether repeatedly. Suspension of the solid was carried out with sonification.
6. If the loading proceeds with poor efficiency, the amino acid might not have been dry. An azeotropic drying can be performed as follows: suspend the amino acid in toluene or dioxane and remove solvent under reduced pressure. Repeat two to three times.
7. NMP is advantageous when low coupling yields caused by the formation of superstructures are to be expected.
8. Sequences containing methionine can be coupled under an atmosphere of argon to reduce the sulfoxide formation. The argon can either be applied by attaching a balloon, or it can be bubbled through the glass sinter, making further agitation unnecessary.
9. Capping is advisable in the synthesis of longer peptides (10 or more amino acids) or of peptides with difficult sequences. As such, consider sequences incorporating amino acids with steric hindrance, in particular amino acids that are *N*-alkylated or carry bulky side chain protecting groups.
10. Be aware of the fact that the free thiol at this step will be prone to oxidation, readily forming the disulfide. Proceed with the alkylation immediately after deprotection and avoid exposure to oxygen. A good cysteine side chaim-protecting group for this purpose is monomethoxytrityl. Monomethoxytrityl is cleaved using 1% TFA and 2% triethylsilane in CH_2Cl_2. Other options are the more stable Trt-group and the mixed disulfide with *t*-butylthiol. The latter is cleaved under reducing conditions, for example, with PBu_3.
11. Keep in mind that MIC is introduced to serve as a Michael acceptor. Consequently, between introduction of the MIC-group and the ligation strong nucleophiles, in particular free thiols have to be avoided.
12. When dealing with acid labile fragments, the TFA concentration can be reduced to 3–5%. Then, the cleavage might have to be repeated more often to obtain all material from the solid support.
13. For acid labile fragments, remove the TFA by azeotropic distillation with toluene (two times for one volume).

14. Because traces of water cleave the formed acyl diazine, causing the formation of undesired side products and a loss of material, all efforts should be undertaken to ensure that the solvents and reagents used in this procedure are really dry.
15. Depending on the intended purification method, flash chromatography will remove the copper salts. A further method is washing a solution of the crude product in CH₂Cl₂ with 1 *N* HCl.

References

1. Naider, F. R. and Becker, J. M. (1997) Synthesis of prenylated peptides and peptide esters. *Biopolymers* **43**, 3–14.
2. Kadereit, D. and Waldmann, H. (2000) Chemoenzymatic synthesis of lipidated peptides. *Monatshefte Chemie* **131**, 571–584.
3. Kuhn, K., Owen, D. J., Bader, B., Wittinghofer, A., Kuhlmann, J., and Waldmann, H. (2001) Synthesis of functional ras lipoproteins and fluorescent derivatives. *J. Am. Chem. Soc.* **123**, 1023–1035.
4. Kuhlmann, J., Tebbe, A., Völkert, M., Wagner, M., Uwai, K., and Waldmann, H. (2002) Photoactivatable synthetic Ras proteins: "baits" for the identification of plasma-membrane-bound binding partners of Ras. *Angewandte Chemie* **41**, 2546–2550.
5. Atherton, E. and Wellings, D. A. (2002) *Houben-Weyl Methods of Organic Chemistry, Vol. E 22a* (Goodman, M., Felix, A., Moroder, L., and Toniolo, C., eds.), Thieme, Stuttgart.
6. Mayerfligge, P., Volz, J., Kruger, U., Sturm, E., Gernandt, W., Schafer, K. P., et al. (1998) Synthesis and structural characterization of human-identical lung surfactant sp-c protein. *J. Peptide Sci.* **4**, 355–363.
7. Denis, B. and Trifilieff, E. (2000) Synthesis of palmitoyl-thioester T-cell epitopes of myelin proteolipid protein (PLP). Comparison of two thiol protecting groups (StBu and Mmt) for on-resin acylation. *J. Peptide Sci.* **6**, 372–377.
8. Creaser, S. P. and Peterson, B. R. (2002) Sensitive and rapid analysis of protein palmitoylation with a synthetic cell-permeable mimic of Src oncoproteins. *J. Am. Chem. Soc.* **124**, 2444–2445.
9. Ludolph, B., Eisele, F., and Waldmann, H. (2002) Solid-phase synthesis of lipidated peptides. *J. Am. Chem. Soc.* **124**, 5954–5955.
10. Wieland, T., Lewalter, J., and Birr, C. (1970) Nachträgliche Aktivierung von Carboxyl-Derivaten durch Oxydation und ihre Anwendung zur Peptidsynthese an fester Phase sowie zur Cyclisierung von Peptiden. *Liebigs Annalen* **740**, 31–47.
11. Corey, E. J., Takeda, M., and Kim, C. U. (1972) Method for selective conversion of allylic and benzylic alcohols to halides under neutral conditions. *Tetrahedron Lett.* **23**, 4339.
12. Brown, M. J., Milano, P. D., Lever, D. C., Epstein, W. W., and Poulter, C. D. (1991) Prenylated proteins. A convenient synthesis of farnesyl cysteinyl thioethers. *J. Am. Chem. Soc.* **113**, 3176–3177.
13. Lewalter, J. (1970) *Peptidsynthesen mit nachträglich aktivierbaren Carboxyl-Schutzgruppen in Lösung sowie an fester Phase.* PhD thesis, Heidelberg, p. 78.

14. Rosenbaum, C. and Waldmann, H. (2001) Solid phase synthesis of cyclic peptides by oxidative cyclative cleavage of an aryl hydrazide linker—synthesis of stylostatin 1. *Tetrahedron Lett.* **42,** 5677–5680.

17

In Vitro Semisynthesis and Applications of C-Terminally Modified Rab Proteins

Thomas Durek, Roger S. Goody, and Kirill Alexandrov

Summary

Expressed protein ligation is a powerful tool for the generation of natively folded proteins composed of recombinantly generated and chemically synthesized polypeptides. Using this approach, we developed protocols for the production of prenylated and/or otherwise-labeled Rab GTPase. The protocols are generally applicable to most small GTPases that can be supplied with a variety of new chemical functionalities. We used semisynthetic fluorescently labeled Rab7 GTPase as a molecular probe to study protein–protein interactions with components of the prenylation machinery

Key Words: Rab proteins; GTPase; prenylation; expressed protein ligation; peptides.

1. Introduction

Members of the Rab subfamily of Ras-related GTPases function as regulators of intracellular vesicle transport in all known eucaryotic cells *(1,2)*. Like other small GTPases, Rab proteins act as molecular switches cycling between guanosine triphosphate (GTP)-bound (active) and guanosine diphosphate (GDP)-bound (inactive) conformations. Although Rab proteins are synthesized in the cytosol as hydrophilic proteins, they gain the ability to reversibly associate with membranes as the result of a posttranslational lipid modification. This modification is essential for the biological activity of RabGTPases and involves the covalent addition of one or in most cases two isoprenoid (geranylgeranyl) moieties onto C-terminal cysteine residues via thioether linkages. This reaction is catalyzed by geranylgeranyltransferase type II (GGTase-II or RabGGTase). GGTase-II belongs to the family of protein prenyltransferases together with GGTase-I and farnesyltransferase (FTase), which prenylate members of the Rac/Rho and Ras GTPase family, respectively *(3)*.

From: *Methods in Molecular Biology, vol. 283: Bioconjugation Protocols: Strategies and Methods*
Edited by: C. M. Niemeyer © Humana Press Inc., Totowa, NJ

In contrast to the other protein prenyltransferases, GGTase-II can recognize newly synthesized Rab protein substrates only when they are associated with a protein factor termed Rab escort protein (REP). After prenylation, REP delivers the Rab proteins to their destination membrane. The central role of the GGTase-II/REP machinery is illustrated by the finding that all of the 60 Rab proteins identified so far in mammalian cells appear to be processed by this enzyme *(3,4)*.

The involvement of Rab proteins in a large number of intracellular trafficking steps and the critical importance of Rab prenylation for the intracellular vesicular transport have led to a significant demand for prenylated and often labeled Rab proteins for a variety of biological applications. Because prokaryotic organisms lack protein–prenyltransferases, prenylated Rab proteins were traditionally expressed and purified from animal tissues or from insect cell cultures infected with recombinant baculovirus *(5,6)*. These methods are laborious, costly, inflexible, and suffer from low yields. The situation was to some extent improved by reconstituting the prenylation reaction in vitro, but expression and purification of individual components remains challenging *(7)*.

We have developed an alternative approach for rapid generation of prenylated and/or labeled Rab proteins in amounts sufficient for most biological applications including crystallization. This approach is based on the recently developed expressed protein ligation methodology. Central to this method is the chemo- and regioselective native peptide-bond formation between two unprotected peptides or proteins, with one bearing a C-terminal α-thioester and the other carrying an N-terminal cysteine residue *(8–10)*.

Full-length Rab proteins contain ca. 210 amino acids, with the prenylateable cysteine residues situated in the last four C-terminal amino acids. C-terminally truncated Rab proteins can be expressed in *Escherichia coli* fused to an engineered intein followed by a chitin binding domain (CBD). The CBD allows easy affinity purification of the fusion protein on chitin beads. Several vectors containing a multiple cloning site upstream or downstream of an intein–CBD assembly are available commercially *(11)*.

The amide linkage between truncated Rab and the intein is in equilibrium with a thioester bond involving the intein's active site cysteine (*see* **Fig. 1A**). Therefore, Rab proteins can be cleaved off the intein as C-terminal α-thioesters via transthioesterification by addition of a thiol-containing compound. The α-thioester can then be reacted with a peptide containing an N-terminal cysteine mimicking the deleted Rab C-terminus (**Fig. 1B**). The latter reaction can be performed under mild conditions (i.e., in aqueous solutions, at moderate temperatures) and proceeds with good yields regardless of the sequence flanking the reactive groups. The C-terminal peptide can be chemically synthesized

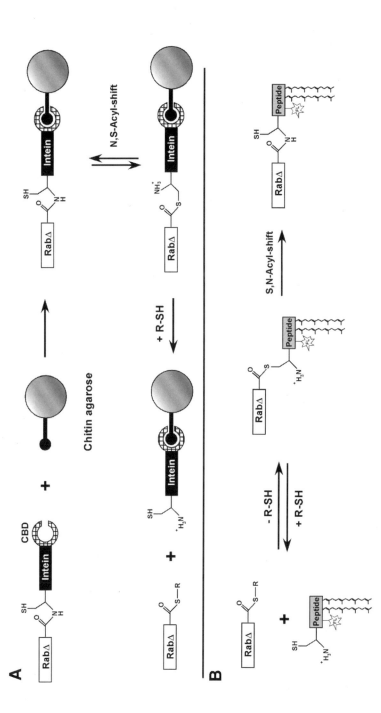

Fig. 1. Schematic representation of the expressed protein ligation strategy for generation of semisynthetic Rab proteins. (A) Purification of a recombinant truncated Rab–Intein–CBD fusion protein using Chitin–agarose beads and generation of the Rab–C-terminal thioester by thiol-induced cleavage. (B) Selective reaction of Rab-thioester with a peptide mimicking the C-terminus of Rab. The peptide is initially thiol captured by transthioesterification and forms a native amide linkage after rearrangement. The peptide may contain various modifications, such as fluorescent labels and prenyl groups.

and may be supplied with posttranslational modifications, fluorophores, affinity-tags, and so on, for a variety of experimental designs.

Here we describe procedures for the purification of Rab α-thioesters and their ligation to peptides complementing the C-terminus. Because prenylated peptides and Rab proteins are poorly soluble in aqueous solutions, we devised different protocols for production of prenylated and unprenylated proteins. These procedures can be considered generally applicable to essentially any Rab protein or other small GTPase. In addition, we provide examples and protocols for the use of the semisynthetic fluorescent Rab proteins in studies of prenylation reaction mechanisms.

2. Materials

2.1. Preparation of Recombinant Thioester-Tagged Rab7 Proteins

1. *E. coli* strain BL21(DE3) was from Novagen. pTYB or pTWIN bacterial-expression vectors were from New England Biolabs.
2. Standard microbiological media and reagents: ampicillin, isopropylthio-β-D-galactoside and phenylmethanesulfonylfluoride (PMSF) were from Gerbu (Gaiberg, Germany). Triton X-100 and 2-mercaptoethanesulfonic acid (MESNA) were from Sigma. Chitin binding beads were from New England Biolabs.
3. Wash buffer: 10 mM Na–Phosphate, pH 7.2; 0.1 M NaCl. Lysis buffer: 25 mM Na–phosphate, pH 7.5; 0.5 M NaCl, 0.5 mM PMSF, 2 mM MgCl$_2$; and 2 μM GDP. Ligation buffer: 10 mM Na–phosphate, pH 7.5; 0.1 mM MgCl$_2$; 2 μM GDP.

2.2. Expressed Protein Ligation of Rab7-Thioesters With Prenylated Peptides

1. 3-[(3-cholamidopropyl)dimethylammonio]-1-propanesulfonate (CHAPS) and Cetyltrimethylammonium bromide (CTAB) were from Roth (Karlsruhe, Germany).
2. Preparation of prenylated peptides containing a N-terminal cysteine is described in detail elsewhere *(12,13)*. Specifically, the synthesis and purification of the peptide Cys-Lys(dansyl)-Ser-Cys-Ser-Cys(GG)-OMe can be found in **ref. 14**. The peptide stock solution was prepared in a suitable solvent that ensures high solubility (at least 20 mM), such as methanol, acetonitrile, or aqueous detergent solution (40 mM CTAB, 2% CHAPS).
3. Rab7 protein thioester was prepared as described in **Subheading 3.1.** and concentrated to at least 10 mg/mL (approx 500 μM). REP-1 protein prepared as described in **ref. 15** (*see* **Note 1**).
4. Denaturation buffer: 100 mM Tris-HCl, pH 8.0; 6 M guanidinium-HCl; 100 mM DTE; 1% CHAPS; 1 mM ethylenediamine tetraacetic acid (EDTA). Renaturation buffer: 50 mM N-hydroxyethylpiperazine-N'-2-ethanesulfonate (HEPES), pH 7.5; 2.5 mM DTE; 2 mM MgCl$_2$, 10 μM GDP; 1% CHAPS; 400 mM arginine–HCl; 400 mM trehalose; 0.5 mM PMSF; 1 mM EDTA. Dialysis buffer: 25 mM HEPES,

pH 7.5; 2 mM MgCl$_2$; 2 µM GDP; 2.5 mM DTE; 100 mM (NH$_4$)$_2$SO$_4$; 10% glycerol; 0.5 mM PMSF; 1 mM EDTA. Gel filtration buffer: 25 mM HEPES, pH 7.5; 2 mM MgCl$_2$; 10 µM GDP; 2.5 mM DTE; 100 mM (NH$_4$)$_2$SO$_4$, 10% glycerol.

2.3. Expressed Protein Ligation of Rab7-Thioesters With Unprenylated Peptides

1. Thioester-tagged Rab protein was prepared as described in **Subheading 3.1.** and concentrated to at least 10 mg/mL.
2. Peptides with N-terminal cysteines are prepared by standard solution- or solid-phase peptide synthesis. The peptide Cys-Lys(dansyl)-Ser-Cys-Ser-Cys was custom synthesized and purified by Thermo Hybaid (Ulm, Germany) and peptide stock solution was prepared in 2% CHAPS.
3. Gel filtration buffer: 25 mM HEPES, pH 7.2; 40 mM NaCl; 2 mM MgCl$_2$; 10 µM GDP; 2.5 mM DTE.

2.4. Characterization of Semisynthetic Rab7 Proteins

1. REP-1 and GGTase-II prepared as described (*see* **Note 1**).
2. Buffer A: 50 mM HEPES, pH 7.2; 50 mM NaCl; 5 mM DTE. Buffer B: 25 mM HEPES, pH 7.2; 40 mM NaCl; 2 mM MgCl$_2$; 100 µM GDP; 2 mM DTE.

3. Methods

3.1. Preparation of Recombinant Thioester-Tagged Rab7 Proteins

C-terminally truncated Rab7 proteins were C-terminally fused to an assembly of intein–CBD domains and expressed in *E. coli* using the pTYB1 vector according to the instructions of the manufacturer. The coding region of the canine Rab7 gene truncated by six amino acids was amplified by polymerase chain reaction using pET3a Rab7 plasmid as a template (*16*). The polymerase chain reaction product was gel purified, digested with *Kpn*I and *Nde*I and ligated into the pTYB1 vector precut with the same enzymes. The resulting plasmid (pTYB1-Rab7δC6) was transformed into *E. coli* BL21(DE3) cells and transformants were selected on ampicillin (50 mg/L) agar plates. The fusion protein could be easily purified by a single affinity step using chitin beads. Cleavage of the fusion protein and generation of the C-terminal thioester is induced by addition of thiol-containing reagents, such as MESNA.

3.1.1. Expression of Rab7–Intein–CBD Fusion Proteins in E. coli

1. Inoculate a 5-mL culture of LB containing 125 mg/L ampicillin with BL21(DE3) bacteria containing the desired plasmid and incubate overnight in a 37°C shaker.
2. Use this preculture to seed 2 L of culture and incubate at 37°C in a shaker until the absorbance at 600 nm reaches 0.5–0.7. Add isopropylthio-β-D-galactoside to a final concentration of 0.5 mM and incubate cultures overnight at 20°C.

3. Harvest cells by centrifugation ($5000g$, 20 min, 4°C) and wash once in wash buffer. Cells can be stored frozen at this point.

3.1.2. Purification and Cleavage of Fusion Proteins on Chitin Beads

1. Resuspend the cells in 50–100 mL of lysis buffer and break the cells by passing them twice through a Fluidizer (Microfluidics) or by another standard method (e.g., sonication, lysozyme digestion). Add a new portion of 0.5 mM PMSF and Triton X-100 to a final concentration of 1%. Clarify the lysate by centrifugation ($30,000g$, 1 h, 4°C).

2. Transfer the supernatant into 50-mL Falcon tubes. Determine the total amount of expressed fusion protein by sodium dodecyl sulfate-polyacrylamide gel electrophoresis (SDS-PAGE) and add corresponding amounts of chitin beads to the supernatant (1 mL of beads can bind about 2 mg of fusion protein). Incubate the mixture on a rotating wheel at 4°C for 2 h.

3. Wash beads four times with lysis buffer containing 1% Triton X-100 and then four times with lysis buffer without the detergent.

4. Collect beads into one Falcon tube and induce cleavage of the fusion protein by adding powdered MESNA to a final concentration of 0.5 M and incubate overnight at room temperature on a rotating wheel. Some proteins may precipitate at high concentrations of MESNA. In such cases, lower the MESNA concentration to 100 mM.

   Collect the supernatant and exchange the buffer to ligation buffer by dialysis or gel filtration. Concentrate the protein to at least 10 mg/mL and store at –80°C in multiple aliquots. This procedure typically yields about 10 mg of Rab7-thioester per liter of bacterial culture.

3.2. Expressed Protein Ligation of Rab7-Thioesters With Prenylated Peptides

The expressed protein ligation reaction is usually initiated by simply mixing protein thioester and peptide containing a N-terminal cysteine in aqueous solutions. Because of the low solubility of the lipidated peptide in these solutions, detergents are essential for efficient ligation. Moreover, prenylated Rab proteins are not soluble in detergent free solutions, unless they are complexed to one of their natural "molecular chaperones" (REP or GDP dissociation inhibitor). The protocol therefore includes complex formation with REP-1.

3.2.1. Ligation and Separation of Reaction Product and Unligated Peptide

1. Add CTAB from a stock solution (200 mM) to the Rab7 thioester in ligation buffer to a final concentration of 50 mM. Start the reaction by adding at least 5 molar equivalents of peptide in a suitable solvent (*see* **Note 2**).

2. Incubate the reaction mixture overnight with agitation at 37°C.

3. Spin down the reaction mixture to remove protein and peptide precipitates. In order to remove excessive peptide wash the precipitate once with 1 mL metha-

nol, four times with 1 mL of methylenchloride, four times with 1 mL of methanol, and four times with 1 mL of distilled water at room temperature (*see* **Note 3**).

4. Dissolve the precipitate in denaturation buffer to a final protein concentration of 0.5 mg/mL and incubate overnight at 4°C. The solution can be stored at this point at –80°C.

3.2.2. Refolding of Ligation Products and Complex Formation With REP

1. Dilute the denatured protein at least 25-fold by adding it drop wise into refolding buffer at room temperature with gentle stirring. Incubate 30 min further at room temperature without stirring (*see* **Note 4**). Take a 100-µL aliquot and precipitate the proteins with 10% trichloracetic acid. Wash the precipitate once with ice-cold acetone and dissolve it in 20 µL of SDS sample buffer. Determine the amount of ligation product by SDS-PAGE.
2. Add an equimolar amount of REP protein to the refolding solution and incubate for 1 h on ice.
3. Dialyze overnight against two 5-L changes of dialysis buffer.
4. Concentrate dialyzed material to approx 2 mg/mL using a size-exclusion concentrator. Remove any insoluble material by centrifugation at 13,000g for 5 min at 4°C and load the supernatant on a Superdex-200 gel filtration column (Pharmacia) equilibrated with gel filtration buffer.
5. Pool peak fractions containing both REP and Rab7, concentrate to 10 mg/mL, and freeze in multiple aliquots at –80°C. The recovery yield is approx 10–50% with respect to starting Rab7-thioester.

3.3. Expressed Protein Ligation of Rab7-Thioesters With Unprenylated Peptides

Ligation of unprenylated peptides to the Rab-C-terminus can be performed under nondenaturing conditions. However, depending on the solubility of the peptide it might be necessary to add mild detergents in low concentration to the ligation mixture (e.g., CHAPS, *n*-octyl glucoside).

1. Add at least 5 equivalents of peptide to the Rab-thioester in ligation buffer. Incubate over night at room temperature with slight agitation.
2. Spin down the reaction mixture (13,000, 5 min, 4°C). Remove unreacted peptide and detergent by passing the supernatant over a desalting column (e.g., PD-10 from Pharmacia) equilibrated with gel filtration buffer.
3. Concentrate to 10 mg/mL and freeze in multiple aliquots at –80°C. Under the described conditions, the yields are usually higher than 75% with respect to Rab7-thioester.

3.4. Characterization of Semisynthetic Rab7 Proteins

The progression of the ligation reaction can be determined by SDS-PAGE and matrix-assisted laser desorption/ionization I mass spectrometry (*see* **Fig. 2**). When a fluorescently labeled peptide is used, the unstained, 10% acetic acid

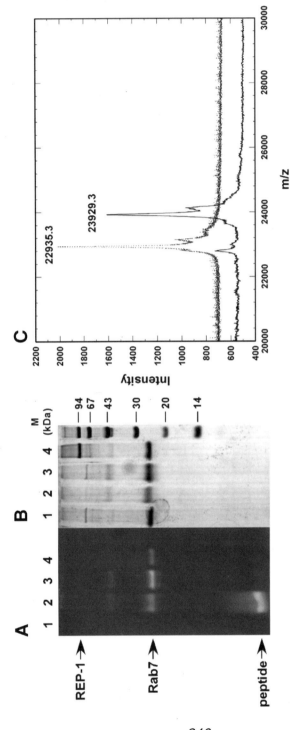

Fig 2. SDS-PAGE gel of Rab7δC6-MESNA thioester before (lane 1) and after (lane 2) ligation to a Cys-Lys(dansyl)-Ser-Cys-Ser-Cys(GG)-OMe peptide, after removal of unligated peptide (lane 3), and after complex formation with REP-1 and Superdex-200 gel filtration purification (lane 4). The gel was photographed either in UV light (A) or visible light after Coomassie blue staining (B). (C) Matrix-assisted laser desorption/ionization mass spectrometry spectrum of Rab7δC6-MESNA thioester (M_{calc} = 22932 Daltons, dotted line) and ligated and purified Rab7δC6-CK(dansyl)SCSC(GG)-OMe:REP-1 Complex (M_{calc} = 23,939, solid line).

fixed SDS-PAGE gel can be viewed in UV light. Typically, a band corresponding to the ligated Rab protein product is seen migrating at a position corresponding to ca. 24 kDa. Additionally, contaminating peptide can be detected
as a broad fluorescent band migrating close to the dye front. The stoichiometry
of the ligation reaction can be determined by comparing the concentration of a
chromophore moiety (e.g., a fluorescent label) to the protein concentration.
The molar concentration of the incorporated chromophore is obtained from the
Lambert-Beer-law $c = A/(\varepsilon \times 1)$, where A is the absorbance of the chromophore
determined on a absorption spectrophotometer, ε is the molar extinction coefficient of the chromophore (e.g., for dansyl $\varepsilon_{340\,nm} = 4000\ M^{-1}\ cm^{-1}$), and l is
the path length of the cuvette. The protein concentration is determined either
directly by using the absorbance at 280 nm and the same relationship as for the
chromophore or by standard colorimetric assays using bovine serum albumin
as a standard.

3.4.1. Characterization of the Interaction Between Semisynthetic Rab7 Proteins and the Components of the Prenylation Machinery Using Fluorescence Spectroscopy

1. Fluorescence measurements were performed either with an Aminco SLM 8100
   spectrofluorometer (Aminco, Silver Spring, MD) or a Spex Fluoromax-3 spectrofluorometer (Jobin Yvon, Edison, NJ). The sample was placed in a 1-mL quartz
   cuvet (Hellma), stirred continuously and thermostated at 25°C.
2. Fluorescent probes incorporated near the prenylation site at the C-terminus of the
   Rab proteins can be used to study interactions with GGTase-II or REP-1. On
   interaction of Rab7δC6-CK(dansyl)SCSC with REP-1, there is an increase of the
   direct dansyl fluorescence by a factor of almost 6 (**Fig. 3A**). Further addition of
   saturating amounts of GGTase-II resulted in a decrease of the fluorescence signal
   by ca. 20%. Using these signals, we were able to characterize important steps
   during the assembly of the ternary enzyme-substrate complex. Typically, we usually start with approx 200 n*M* Rab7δC6-CK(dansyl)SCSC in buffer B and the
   excitation and emission monochromators set to 338 nm and 490 nm, respectively.
   REP-1 is then titrated into the cuvette in 50 n*M* steps, until the increase in the
   fluorescence signal is saturated (**Fig. 3B**).
3. The change in fluorescence is plotted as a function of the total REP-1 concentration (**Fig. 3C**). Assuming a single binding site on the Rab7 protein for the interaction partner (REP-1), the K_d for the binding can be determined using the
   equation:

$$F = F_{min} + [K_d + P_0 + L - [(K_d + P_0 + L)^2 - 4P_0L]^{1/2}](F_{max} - F_{min})/2P_0$$

where F is the observed fluorescence after each step of titrator addition, F_{min} is
the initial value at $L = 0$, F_{max} is the final value at saturation, L is the total concentration of REP-1, and P_0 is the Rab7 protein concentration. The data was fitted
using GraFit 4.0 (Erithacus Software).

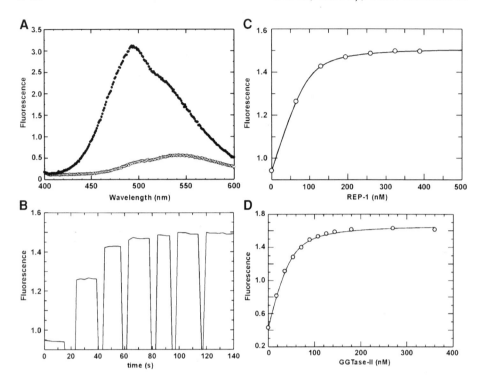

Fig. 3. (**A**) Fluorescence emission spectra of 380 n*M* Rab7δC6-CK(dansyl)SCSC alone (open circles) and after addition of 500 n*M* REP-1 (closed circles). The excitation was 338 nm. (**B**) Fluorescence titration of REP-1 to 250 n*M* Rab7δC6-CK(dansyl)SCSC. At each breakdown of the fluorescence signal, the concentration of REP-1 is increased by adding REP-1 from a stock solution to the cuvette. Excitation and emission wavelength were set to 338 nm and 490 nm, respectively. (**C**) Plot of the observed fluorescence signal as a function of REP-1 concentration. The solid line represents a fit to the binding equation with a value of 4 n*M* for the K_d. (**D**) Fluorescence titration of GGTase-II to 200 n*M* Rab7δC6-CK(dansyl)SCSC(GG)-OMe:REP-1 complex. The data were fitted using the binding equation to give a K_d value of 9 n*M*. The signal was based on fluorescence resonance energy transfer. Excitation and emission were set to 280 nm and 495 nm, respectively.

4. The stochiometric complex of Rab7δC6-CK(dansyl)SCSC and REP-1 can further be titrated in a similar way with GGTase-II to obtain binding parameters for ternary complex formation (data not shown; *see* **ref. 17**).
5. Semisynthetic single prenylated and fluorescently labeled Rab proteins are genuine intermediates of the Rab double prenylation reaction catalyzed by GGTase-II. Interaction of the Rab7δC6-CK(dansyl)SCSC(GG)-OMe:REP-1 complex with GGTase-II is accompanied by a threefold increase of the fluorescence resonance

energy transfer from tryptophan to the dansyl fluorophore when the sample is excited at 280 nm. We used this signal to characterize the interaction of a single prenylated Rab reaction intermediate in complex with REP-1 with GGTase-II *(14)*.

6. To do so, 50–200 n*M* Rab:REP-1 complex in buffer A was placed in a cuvet. Excitation was at 280 nm, whereas data collection was performed at 495 nm. Small aliquots of GGTase-II were then added to the cuvet, until the increase in fluorescence signal was saturated (**Fig. 3D**). Data evaluation was performed as described in **step 3** of this section.

4. Notes

1. REP-1 and GGTase-II can be produced in a baculoviral expression system *(18)*. However, low yields and the considerable cost factor put limitations on large-scale preparations. To circumvent these problems, we established a yeast expression system for REP-1 *(15)* and a bacterial expression system for GGTase-II *(7)*.

2. The choice of the detergent is crucial for a successful ligation reaction. By screening a selection of ca. 80 detergents, we found CTAB to be the most efficient in supporting the ligation reaction. The yield of the reaction can be improved by increasing the peptide to protein ratio to 10. The optimal ratio should be tested by serial dilutions followed by SDS-PAGE analysis.

3. By testing different Rab proteins, we came to the conclusion that under these reaction conditions, the product usually precipitates quantitatively. However, under certain conditions small quantities of the protein stay in solution. Washing the aqueous phase with methylenchloride efficiently removes contaminating peptide and leads to precipitation of the protein. Peptide recovered from the organic phases can be used in another round of ligation reaction after evaporation of the organic solvent. It is advisable to check the progress of purification by SDS-PAGE analysis (*see* **Subheading 3.4.**; **Fig. 2A,B**) before resolubilization of the ligated protein, so that in case of contaminations additional washing steps can be performed.

4. Depending on the Rab protein and peptide used it might be necessary to optimize the composition of the refolding buffer. The field of in vitro protein refolding is the subject of several excellent reviews *(19,20)*.

References

1. Martinez, O. and Goud, B. (1998) Rab proteins. *Biochim. Biophys. Acta* **1404,** 101–112.
2. Zerial, M. and McBride, H. (2001) Rab Proteins as membrane organizers. *Nat. Rev. Mol. Cell Biol.* **2,** 107–119.
3. Casey, P. J. and Seabra, M. C. (1996) Protein prenyltransferases. *J. Biol. Chem.* **271,** 5289–5292.
4. Pereira-Leal, J. P., Hume, A. N., and Seabra, M. C. (2001) Prenylation of Rab GTPases: molecular mechanisms and involvement in genetic disease. *FEBS Lett.* **498,** 197–200.

5. Horiuchi, H., Ullrich, O., Bucci, C., and Zerial, M. (1995) Purification of posttranslationally modified and unmodified Rab5 protein expressed in *Spodoptera frugiperda* cells. *Methods Enzymol.* **257,** 9–15.

6. Kikuchi, A., Yamashita, T., Kawata, M., Yamamoto, K., Ikeda, K., Tanimoto, T., et al. (1988) Purification and characterization of a novel GTP-binding protein with a molecular weight of 24,000 from bovine brain membranes. *J. Biol. Chem.* **263,** 5289–5292.

7. Kalinin, A., Thomä, N. H., Iakovenko, A., Heinemann, I., Rostkova, E., Constantinescu, A. T., et al. (2001) Expression of mammalian geranylgeranyl-transferase Type-II in *Escherichia coli* and its application for in vitro prenylation of Rab proteins. *Protein Expression Purification* **22,** 84–91.

8. Goody, R. S., Alexandrov, K., and Engelhard, M. (2002) Combining chemical and biological techniques to produce modified proteins. *Chem. Bio. Chem.* **3,** 399–403.

9. Hofmann, R. M. and Muir, T. W. (2002) Recent advances in the application of expressed protein ligation to protein engineering. *Curr. Opin. Biotechnol.* **13,** 297–303.

10. Blaschke, U. K., Silberstein, J., and Muir, T. W. (2000) Protein engineering by expressed protein ligation. *Methods Enzymol.* **328,** 478–496.

11. Xu, M.-Q. and Evans, T. C. (2001) Intein-mediated ligation and cyclization of expressed proteins. *Methods* **24,** 257–277.

12. Naider, F. R. and Becker, J. M. (1997) Synthesis of prenylated peptides and peptide esters. *Biopolymers* **43,** 3–14.

13. Kadereit, D., Kuhlmann, J., and Waldmann, H. (2000) Linking the fields—the interplay of organic synthesis, biophysical chemistry, and cell biology in the chemical biology of protein lipidation. *Chem. Bio. Chem.* **1,** 144–169.

14. Alexandrov, K., Heinemann, I., Durek, T., Sidorovitch, V., Goody, R. S., and Waldmann, H. (2002) Intein-mediated synthesis of geranylgeranylated Rab7 protein in vitro. *J. Am. Chem. Soc.* **124,** 5648–5649.

15. Sidorovitch, V., Niculae, A., Kan, N., Ceacareanu, A.-C., and Alexandrov, K. (2002) Expression of mammalian Rab Escort Protein-1 and -2 in yeast *Saccharomyces cerevisiae*. *Protein Expression Purification* **26,** 50–58.

16. Simon, I., Zerial, M., and Goody, R. S. (1996) Kinetics of interaction of Rab5 and Rab7 with nucleotides and magnesium ions. *J. Biol. Chem.* **271,** 20470–20478.

17. Iakovenko, A., Rostkova, E., Merzlyak, E., Hillebrand, A., Thomä, N. H., Goody, R. S., et al. (2000) Semi-synthetic Rab proteins as tools for studying intermolecular interactions. *FEBS Lett.* **468,** 155–158.

18. Armstrong, S. A., Brown, M. S., Goldstein, J. L., and Seabra, M. C. (1995) Preparation of recombinant Rab geranylgeranyltransferase and Rab escort proteins. *Methods Enzymol.* **257,** 30–41.

19. Lilie, H., Schwarz, E., and Rudolph, R. (1998) Advances in refolding of proteins produced in *E. coli*. *Curr. Opin. Biotechnol.* **9,** 497–501.

20. De Bernardez Clark, E., Schwarz, E., and Rudolph, R. (1999) Inhibition of aggregation side reactions during in vitro protein folding. *Methods Enzymol.* **309,** 217–223.

18

Generation and Characterization of Ras Lipoproteins Based on Chemical Coupling

Melanie Wagner and Jürgen Kuhlmann

Summary

Chemically synthesized truncated Ras proteins are coupled to C-terminal Ras peptides via a maleimidocaproyl linker. The resulting product is isolated by extraction with Triton X-114. The biological activity of these oncogenic Ras lipoproteins can be determined in a cell-based differentiation assay by microinjection into PC12 cells.

Key Words: Ras proteins; coupling; lipopeptide; oncogene; PC12; differentiation.

1. Introduction

Proteins of the Ras superfamily are involved in regulation of cell growth and differentiation and act as prototypes of guanosine triphosphate (GTP)-binding proteins (*1*). The first steps in Ras pathway involve binding of an extracellular ligand to an outer binding site of a receptor tyrosine kinase, autophosphorylation of tyrosines at the cytoplasmatic moiety of the kinase, and coupling of SH2-containing adaptor molecules with the phosphorylated receptor tyrosine kinase. Thereby, a nucleotide exchange factor is translocated to the plasma membrane and catalyzes the nucleotide exchange from Ras*guanosine diphosphate to Ras*GTP. All of these events and consecutive interactions as binding and activation of Ras effectors like Raf kinase occur at the plasma membrane.

For this reason, the biological activity of all Ras isoforms is strictly dependent on correct localization at the plasma membrane that is achieved by posttranslational modifications of the polypeptide chain after synthesis at the ribosome. In K-Ras4B, a hydrophobic farnesyl anchor is adjacent to a polycationic hexalysine stretch at the C-terminus and allows a combined hydrophobic and electrostatic interaction with unspecific anionic lipid binding sites in the plasma membrane (*2*).

From: *Methods in Molecular Biology, vol. 283: Bioconjugation Protocols: Strategies and Methods*
Edited by: C. M. Niemeyer © Humana Press Inc., Totowa, NJ

The N- and H-isoforms of Ras undergo a series of four modification steps. Introduction of hydrophobic groups starts (as for K-Ras4B) with the prenylation of the C-terminal cysteine in the CaaX box of the nascent polypeptide chain by a protein farnesyl transferase. Subsequent removal of the last three C-terminal amino acids by a prenyl protein specific endoprotease and donation of a methyl group to the *S*-farnesylated cysteine by a prenyl protein specific methyltransferase occur at the endoplasmatic reticulum *(3)*. Modification is completed by palmitoylation of one or two more cysteines located in the Ras C-terminus. This reaction is presumably catalyzed by a prenyl protein-specific palmitoyltransferase that appears to be endoplasmatic reticulum associated as well *(4,5)*.

Because of insufficient access on posttranslationally modified Ras, the overwhelming amount of structural, biochemical, and biophysical data on these GTP-binding proteins results from studies with bacterial expressed material *(6)*. Strategies for the synthesis of modified Ras proteins are based on eucaryotic expression systems like SF9 cells for the completely processed protein *(7–9)* or use in vitro prenylation by recombinant farnesyl transferase if only isoprenylation is required *(10,11)*. Nevertheless these approaches could not overcome the problems of low yields (eucaryotic expression) or incomplete modification (in vitro farnesylation).

Here, we present the generation of lipoproteins by coupling C-terminally truncated Ras protein from bacterial expression with a lipopeptide bearing the features desired. The protein moiety ends in a free cysteine and attacks the N-terminal maleimidocaproyl group of the lipopeptide in a nucleophilic reaction by the cysteine SH group. The exposed position of the cysteine in the highly flexible C-terminus of Ras makes this reaction almost specific. Further, cysteines in the protein corpus are less accessible so that the coupling reaction yields the correct product without significant side reactions *(12)*. The hydrophobic anchor of the neo-lipoprotein allows purification of product by extraction with a Triton X-114 saturated solution that separates from the aqueous phase at temperatures above 30°C *(13)*. Subsequent detergent is removed by ion-exchange chromatography of the Ras-chimera followed by sodium dodecyl sulfate-polyacrylamide gel electrophoresis (SDS-PAGE) and mass spectrometrical analysis.

Qualification of the Ras lipoprotein constructs can be proven in cell biological system (e.g., differentiation potential for PC12 cells *[14]* and biophysical setups *[12]*). An extensive library of natural and non-natural lipopeptides has been synthesized *(15)* and applied for the generation of the corresponding lipoproteins. The chemical origin of the peptides not only allows introduction of suitable fluorophores for localization experiments but also of photoactivatable side groups for labeling experiments *(16)*.

Table 1
Oligonucleotides for Polymerase Chain Reaction

Oligonucleotide	5'-Sequence-3'
N-Ras *Eco*RI 5'	GGAATTCTATGACTGAGTACAAACTGGTGG
N-Ras *Sma*I181 3' R	TCCCCCGGGTTACTAACAACCCTGAGTCCCATCATCAC
N-Ras G12V Mut	GTGGTTGGAGCTGTAGGTGTTGGG

2. Materials

2.1. Cloning of Truncated N-Ras

1. Truncation of the full-length N-Ras cDNA (accession no. X02751) and the introduction of the point mutation G12V were achieved using the high-fidelity *Pfu* DNA polymerase (Stratagene, La Jolla, CA).
2. For the truncated version of N-Ras two stop codons were introduced to position 182 and 183 of the N-Ras cDNA. The oligonucleotides N-Ras *Eco*RI 5', N-Ras *Sma*I 181 3' R, and N-Ras G12V Mut (**Table 1**) were used as primers for polymerase chain reaction.
3. The resulting fragment N-RasΔ181 was purified and digested with *Eco*RI and *Sma*I. It was subcloned into the ptac expression vector and was transformed into the *Escherichia coli* strain CK600K (Stratagene).

2.2. Preparation of Recombinant Protein

1. Standard bacteriological media, reagents, and equipment. A M-110S Microfluidizer Processor from Microfluidics Corporation, Newton, CA, was used to disrupt bacteria cells.
2. Buffer A: 20 mM Tris-HCl, pH 7.4, 5 mM MgCl$_2$, 2 mM dithioerythriol (DTE). If not otherwise noted: the buffer is filtered through a 0.45-μm filter.
3. Buffer B: buffer A containing 200 mM NaCl, 10 μM guanosine diphosphate (the buffer is filtered through a 0.22-μm filter and degased).
4. DEAE column: 600 mL of DEAE-Sepharose (Amersham Biosciences, Freiburg, Germany).
5. Gel filtration column: HiLoad 26/60 Superdex 200 (Amersham Biosciences).

2.3. Coupling Reaction

1. Triton X-114 (T X-114) was purchased from Fluka (Seelze, Germany). The T X-114 buffer was prepared as described before (*17*). Buffer: 30 mM Tris-HCl, pH 7.4, 100 mM NaCl, 30 g/L T X-114.
2. 5 mL of HiTrap® gel filtration column (Amersham Biosciences).
3. Buffer C: 20 mM Tris-HCl, pH 7.4, 5 mM MgCl$_2$.
4. Buffer D: buffer C + 2 mM DTE.
5. Sample mixer: MX 2 (Dynal, Oslo, Norway).
6. DEAE column: 10 mL of DEAE–Sepharose (Amersham Biosciences).

7. Matrix-assisted laser desorption/ionization–time of flight mass spectrometry (MALDI-TOF MS): Voyager-DE PRO BioSpectrometry Workstation (Applied Biosystems, Foster City, CA).
8. Standard for MALDI-MS: C3 from PE Biosystems, Foster City, CA. Part of the Sequozyme Peptide Mass Standard Kit (P2-3143-00). Contains the following proteins: thioredoxin, apomyoglobin (horse), and insulin (bovine).
9. Matrix for MALDI-MS: 10 mg/mL sinapinic acid (Sigma, Seelze, Germany) resolved in 0.05% (v/v) trifluoroacetic acid and 50% (v/v) acetonitrile.

2.4. Cleavage of Protection Group

1. Prepare 500 µM DTE in water.
2. 5 mL of HiTrap® gel filtration column (Amersham Biosciences).
3. Phosphate-buffered saline (PBS): 2.7 mM KCl, 10.1 mM Na$_2$HPO$_4$, 1.8 mM KH$_2$PO$_4$, 137 mM NaCl.

2.5. PC12 Assay

2.5.1. Cell Culture

1. PC12 cells are available from ATCC, Manassas, VA (cat. no. CRL-1721).
2. Dulbecco's modified Eagle's medium (DMEM; cat. no. 31885) and all other additives were purchased from Invitrogen (Karlsruhe, Germany).
3. Complete DMEM: DMEM containing 10% horse serum, 5% fetal bovine serum, antibiotics (100 U/mL penicillin; 100 µg/mL streptomycin), and 2 mM L-glutamine.
4. Nerve growth factor (NGF) is obtained from Roche Diagnostics (Mannheim, Germany).
5. Cell culture ware: Falcon (Becton-Dickson, Heidelberg, Germany).
6. Polystyrene-coated culture dishes (cat. no. 25000) are available from Corning Glass Works, New York.

2.5.2. Microinjection

1. Glass capillary with filament: borosilicat glass, outer diameter 1.0 mm, inner diameter 0.58 mm, filament 0.133 mm; cat. no. 1103207 (Hilgenberg, Malsfeld, Germany).
2. Capillary puller PD 5-H (Narishige, Tokyo, Japan).
3. Injection marker: fluorescein dextran (70 kDa) is available from Sigma.
4. Microinjection system: Micoinjector 5242 (Eppendorf, Hamburg, Germany), computer-automated microinjection system, AIS (Zeiss, Oberkochem, Germany).
5. Microscope: Axiovert 135 TV (Zeiss).
6. Video system for Axiovert: 135 AVT (AVT Horn, Aalen, Germany).

3. Methods

3.1. Preparation of Recombinant Protein

1. Inoculate 3 × 50-mL cultures of Luria broth (LB)/ampicillin/kanamycin with CK600K bacteria containing the expression plasmid for N-RasG12V1-181 and incubate overnight in a shaker at 37°C.

2. Inoculate 6×2.5-L cultures of LB/ampicillin/kanamycin with 25 mL each of saturated overnight bacterial culture and incubate in a shaker at 37°C. Induce the expression at an OD_{600} of approx 0.6 with 500 μM of isopropyl-β-D-thiogalacto-side and continue incubation at 30°C overnight.

3. All following steps are conducted at 0–4°C. Collect bacteria by centrifugation at 7000g for 15 min. Resolve the pellet in buffer A (three times the volume of the pellet; *see* **Notes 1** and **2**). Add phenylmethanesulfonylfluoride to a final concentration of 0.1 mM and 3 U/mL DNAse I. Stir the suspension for 30 min to resolve any cell lump. The cells are destroyed using a microfluidizer at maximum pressure. The lysate should then be centrifuged at 100,000g for 30 min.

4. Apply the supernatant to the DEAE–Sepharose column equilibrated in buffer A. Wash the column with 2 column volumes of buffer A, followed by a linear gradient of 0–1 M NaCl in buffer A (five column volumes). The flow-rate should be 4 mL/min and 20-mL fractions are collected.

5. Pool fractions containing N-Ras (detected by SDS-PAGE and Western blot; *see* **Note 3**) and precipitate with 3 M ammonium sulfate (final concentration). Add small portions of the ammonium sulfate within 30 to 60 min to the stirred protein solution. Centrifuge the sample.

6. Resuspend the precipitate in 10 mL of buffer B and apply it to the gel filtration column, which was equilibrated with buffer B. Fractions (2 mL) are collected at a flow rate of 2 mL/min and analyzed by SDS-PAGE.

7. Precipitate protein in the pooled fractions with 3 M ammonium sulfate. Recover the protein by centrifugation and resuspend it in 3 mL of buffer B.

3.2. Coupling Reaction

1. Use stoichiometric amounts of peptide and protein for the coupling reaction (**Note 4**). An easy-to-manage protein amount is about 10–20 mg (**Note 5**). Pass the purified N-RasG12V1-181 through a HiTrap® gel filtration column to remove any excess of salts and DTE required for storage of the protein. Pre-equilibrate the HiTrap® column with buffer C before applying the protein. At all stages of the coupling reaction, all samples containing protein, unless otherwise stated, were kept at or below 4°C.

2. Dissolve the peptide (synthesis, *see* Chapter 16) in methanol (concentration: 20 mg/mL) and add 1 mL of buffer-saturated T X-114 *(17)*. If the peptide is not completely solved, sonicate the solution for approx 15 min in a water bath until a homogeneous solution is obtained.

3. Add Ras protein (volume about 1 mL) to the detergent solution containing the lipid. Cover the mixture with argon (**Note 6**) and incubate rotating on a sample mixer at 4°C for 16 h.

4. Centrifuge the sample and transfer the supernatant into a 15-mL centrifuge tube (**Note 7**). 3 mL buffer D is added. For phase separation of the detergent-rich phase and aqueous phase, warm the sample to 37°C and centrifuge at room temperature (**Note 8**). Collect the upper aqueous phase and extract it two more times with T X-114 (2×1 mL). Combine the detergent phases and wash two times with fresh buffer (3×7 mL).

5. Dilute T X-114 phase ten times with buffer D and apply it on a DEAE column (10 mL). After washing with buffer D elute the bound protein with a sodium chloride gradient (0 M to 1 M NaCl). Analyze the fractions using SDS-PAGE. Concentrate the protein in an ultracentrifuge unit. The product is analyzed by MALDI-MS.

6. For MALDI-MS, dilute the sample 1:1, 1:10, 1:100, and 1:1000 in sinapinic acid matrix. Apply 2 µL to the Maldi plate. Let it air-dry.

3.3. Cleavage of the Protection Group

Some of the peptides used possess a *S*-tertiary-butyl protection group at the free cysteine. This has to be cleaved of before the probes can be used for microinjection into PC12 cells.

1. Add 500 µM of DTE to a final concentration of 50 µM to the protein sample. Repeat this step after 1, 2, and 3 h. Incubate the probe at 37°C.

2. After 4 h, pass the purified N-RasG12V1-181 through a HiTrap® gel filtration column to remove any excess of DTE. Use PBS as exchange buffer.

3.4. PC12 Assay

PC12 cells are rat pheochromocytoma cells *(18)*. Under normal growth conditions they have a chromaffin cell-like morphology. The cells respond to NGF stimulation by differentiating into nonreplicating sympathetic neuron-like cells. Microinjection of oncogenic Ras protein leads to the same morphological changes (**Fig. 1**).

3.4.1. Cell Culture of PC12 Cells

1. PC12 cells were incubated in 10% CO_2 at 37°C.

2. Remove the old medium and wash the cells temporary with 1X trypsin/ethylene-diamine tetraacetic acid (EDTA) (1 mL for a T-75 flask). Remove the trypsin/EDTA and add the same volume of fresh trypsin/ EDTA. Incubate the cells for 2 min in the incubator.

3. Remove cells from the bottom by slightly tapping against the flask. Resuspend the cells in 5 mL of complete medium. Transfer the cell suspension to a sterile 15-mL conical centrifuge tube. Spin the cells down at 300g for 3 min. Discard the supernatant (**Note 9**). Add 1 mL of medium to the cell pellet and resuspend it with a 1000-µL pipet. At this point, a cell count may be performed using a hemocytometer.

4. Resolve cell lumps by trituration (breaking up of cell lumps; **Note 10**). Seed PC12 cells with a density of 10×10^4/mL in a 35 mm $\varnothing$ cell culture dish (**Note 11**). Add NGF (final concentration in medium 100 ng/mL) and incubate the cells for 3 d (**Note 12**; **ref. *19***). Then wash the cells two times by incubating them for 10 min in NGF-free medium in the incubator and keep them in NGF-free medium for another 2 d.

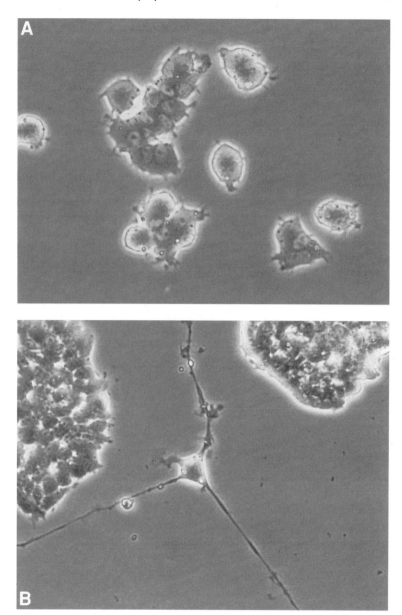

Fig. 1. PC12 cells before (**A**) and 2 d after (**B**) injection of 150 µ*M* N-RasG12Vfl.

5. One day before microinjection subculture the cells by trypsinization. Centrifuge
   and resuspend the cells. Triturate them and seed them onto a 35 ∅ mm polystyrene-coated cell culture dishes.

3.4.2. Microinjection

1. Dilute proteins to the desired concentration (**Note 13**) in PBS with 10 µ*M* fluorescein dextran (70 kDa) and centrifuge for 2 min at 4°C before injection.
2. Apply 0.5 µL of the solution to the top of a capillary prepared with the puller (**Notes 14 and 15**). Inject only single cells or cells at the edge of small lumps in order to make the evaluation more clear.
3. For significant statistics inject about 100 cells per dish and repeat the experiment at least three times (**Note 16**).
4. Forty hours after injection, the transformed cells show a typical differentiated phenotype with neurite-like outgrowth. Calculate the neurite outgrowth percentage by counting injected (fluorescent) and morphology changed cells. Cells are accepted as differentiated if they have developed at least one neurite like outgrowth with a minimum length of two times the maximum diameter of the cell.

4. Notes

1. A 15-L culture results in approximately a 80-g cell pellet, which has to be resuspended in about 240 mL of buffer A.
2. You can freeze the pellet in liquid nitrogen and store it at –80°C for several months.
3. Quite a lot of proteins elute from the DEAE column; therefore, for the first trial, a Western blot is necessary to detect the right protein band. Later on, you will find Ras easily as the heart-shaped protein band on the SDS-PAGE.
4. The maleimide group of the peptide can react with all free thiol functions. For C-terminal truncated (1-181) H- and N-Ras, it could be shown that the remaining three more C-terminal cysteines are significant less accessible than Cys181. At stochiometries of more then 3:1 for peptide over protein a significant increase of side products (two peptide adducts) appears in the mass spectra.
5. Increasing the concentration of protein and lipopeptide above the value recommended causes problems in phase separation during extraction of product with TX-114 solution.
6. Covering the reaction solution with inert gas prevents oxidation of protein SH-groups.
7. Use a vessel that allows recognition of TX-114 and aqueous phase after separation (clear plastic).
8. For proper phase separation, use minimum 4000*g* and minimum or no brake.
9. The trypsin/EDTA solution has to be replaced always by washing and centrifugation because otherwise the PC12 cells lose their ability to change morphology upon NGF and Ras treatment.
10. PC12 cells tend to form lumps, which makes the evaluation difficult. Trituration will break most of the lumps. Triturate the cells forceful by using a 1000-µL pipet. Put the pipet tip on the bottom of the cell culture dish and pipet the cell suspension up and down three to four times.
11. For maintenance of the cell culture and pretreatment of the PC12 with NGF you can use uncoated cell culture ware from Falcon. Just for microinjection we rec-

ommend polystyrene-coated dishes in order to increase the adhesion of the cells to the surface.

12. The cells are prestimulated with NGF to synchronize them. This treatment leads to faster and more reproducible results.

13. Full-length oncogenic N-Ras causes maximum differentiation of PC12 cells above 20 μM concentration (in the injection needle).

14. There is no need to use microloaders because the solution migrates along the filament to the tip by itself.

15. Check the appropriate tip size by application of 100 μM of FITC-Dextran. Take a cell culture dish with water and focus the capillary tip with the microscope. Using an UV-lamp you can see if the tip is open and whether more or less solution outpours.

16. Microinjections can be performed on air within 15 min without changing the CO_2-buffered medium. For injections that last longer the medium has to be exchanged by a pH-stable buffer (e.g., HEPES). Severe changes in the pH because of inadequate CO_2 concentration will lead to death especially of the microinjected cells. Therefore we also recommend using one dish for each injection.

References

1. Takai, Y., Sasaki, T., and Matozaki, T. (2001) Small GTP-binding proteins. *Physiol. Rev.* **81,** 153–208.
2. Roy, M. O., Leventis, R., and Silvius, J. R. (2000) Mutational and biochemical analysis of plasma membrane targeting mediated by the farnesylated, polybasic carboxy terminus of K-Ras4B. *Biochemistry* **39,** 8298–8307.
3. Gelb, M. H. (1997) Protein biochemistry—protein prenylation, et cetera—signal transduction in two dimensions. *Science* **275,** 1750–1751.
4. Bartels, D. J., Mitchell, D. A., Dong, X. W., and Deschenes, R. J. (1999) Erf2, a novel gene product that affects the localization and palmitoylation of Ras2 in *Saccharomyces cerevisiae. Mol. Cell. Biol.* **19,** 6775–6787.
5. Lobo, S., Greentree, W. K., Linder, M. E., and Deschenes, R. J. (2002) Identification of a Ras palmitoyltransferase in *Saccharomyces cerevisiae. J Biol. Chem.* **277,** 41,268–41,273.
6. Kuhlmann, J. and Herrmann, C. (2001) Biophysical characterization of the Ras protein. *Topics Curr. Chem.* **211,** 61–116.
7. Page, M. J., Hall, A., Rhodes, S., Skinner, R. H., Murphy, V., Sydenham, M., and Lowe, P. N. (1989) Expression and characterization of the Ha-Ras p21 protein produced at high levels in the insect/baculovirus system. *J Biol. Chem.* **264,** 19,147–19,154.
8. Rubio, I., Wittig, U., Meyer, C., Heinze, R., Kadereit, D., Waldmann, H., Downward, L., and Wetzker, R. (1999) Farnesylation of Ras is important for the interaction with phosphoinositide 3-kinase gamma. *Eur. J. Biochem.* **266,** 70–82.
9. Inouye, K., Mizutani, S., Koide, H., and Kaziro, Y. (2000) Formation of the Ras dimer is essential for Raf-1 activation. *J. Biol. Chem.* **275,** 3737–3740.

10. Zimmerman, K. K., Scholten, J. D., Huang, C.-C., Fierke, C. A., and Hupe, D. J. (1998) High-level expression of rat farnesyl: Protein transferase in *Escherichia coli* as a translationally coupled heterodimer. *Protein Expression Purification* **14**, 395–402.

11. Seabra, M. C. and James, G. L. (1998) Prenylation assays for small GTPases, in *Transmembrane Signaling Protocols* (Bar-Sagi, D., ed.), Humana Press, Totowa, NJ, pp. 251–260.

12. Bader, B., Kuhn, K., Owen, D. J., Waldmann, H., Wittinghofer, A., and Kuhlmann, J. (2000) Bioorganic synthesis of lipid-modified proteins for the study of signal transduction. *Nature* **403**, 223–226.

13. Bordier, C. (1981) Phase separation of integral membrane proteins in Triton X-114 solution. *J. Biol. Chem.* **256**, 1604–1607.

14. Bar-Sagi, D. and Feramisco, J. R. (1985) Microinjection of the Ras oncogene protein into PC12 cells induces morphological differentiation. *Cell* **42**, 841–848.

15. Kuhn, K., Owen, D. J., Bader, B., Wittinghofer, A., Kuhlmann, J., and Waldmann, H. (2001) Synthesis of functional Ras lipoproteins and fluorescent derivatives. *J. Am. Chem. Soc.* **123**, 1023–1035.

16. Kuhlmann, J., Tebbe, A., Volkert, M., Wagner, M., Uwai, K., and Waldmann, H., (2002) Photoactivatable synthetic Ras proteins: "Baits" for the identification of plasma-membrane-bound binding partners of Ras. *Ang. Chem. Int. Ed.* **41**, 2546–2550.

17. Masterson, W. J. and Magee, A. I. (1992) Lipid modifications involved in protein targeting, in *Protein Targeting: A Practical Approach* (Magee, A. I. and Wileman, T., eds.), IRL Press at Oxford University Press, New York, NY, p. 242.

18. Greene, L. A. and Tischler, A. S. (1976) Establishment of a noradrenergic clonal line of rat adrenal pheochromocytoma cells which respond to nerve growth factor. *Proc. Natl. Acad. Sci. USA* **73**, 2424–2428.

19. Schmidt, G. and Wittinghofer, A. (2000) Priming of PC12 cells for semiquantitative microinjection studies involving Ras. *FEBS Lett.* **474**, 184–188.

19

Conjugation of Glycopeptide Thioesters to Expressed Protein Fragments

Semisynthesis of Glycosylated Interleukin-2

Thomas J. Tolbert and Chi-Huey Wong

Summary

This method describes the conjugation of a synthetic glycopeptide to the N-terminus of a recombinant human interleukin-2 (IL-2) protein fragment. The IL-2 protein fragment is produced as an affinity-tagged fusion protein in *Escherichia coli* and then cleaved with the highly selective TEV protease to remove the affinity tag and uncover an N-terminal cysteine. The N-terminal cysteine is then used in native chemical ligation reaction to join the IL-2 protein fragment to a glycosylated tripeptide thioester that had been previously synthesized to produce a glycosylated form of IL-2.

Key Words: Glycoprotein; chemoselective; native chemical ligation; expressed protein ligation; semisynthesis; glycopeptide; thioester; conjugation; interleukin-2.

1. Introduction

Glycoproteins are often difficult biomolecules to study because natural systems produce them as heterogeneous mixtures, with several types of oligosaccharides attached to a single protein sequence. Because of this, there is great interest in developing semisynthetic methods that allow the conjugation of well-defined synthetic glycopeptides to protein fragments to produce glycoproteins with homogeneous glycosylation for glycoprotein studies *(1,2)*. Chemoselective ligations, such as native chemical ligation (NCL) and expressed protein ligation (EPL), are cysteine–thioester ligations that have been used to incorporate glycosylation and synthetic modifications into proteins *(3–9)*, The semisynthesis of glycosylated, human interleukin-2 (IL-2) is presented here as an example of the use of cysteine–thioester ligations to

From: *Methods in Molecular Biology, vol. 283: Bioconjugation Protocols: Strategies and Methods*
Edited by: C. M. Niemeyer © Humana Press Inc., Totowa, NJ

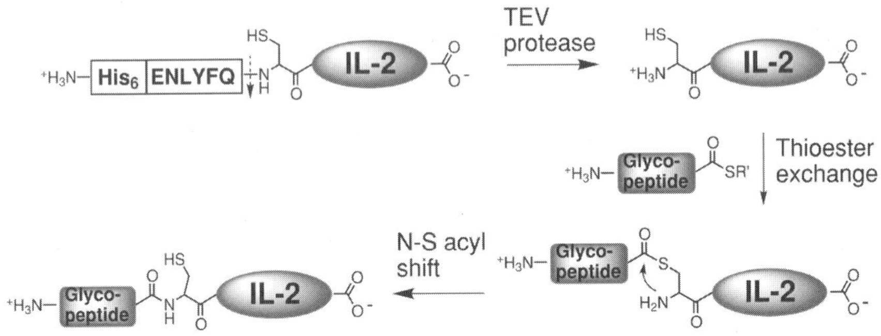

Fig. 1. Semisynthesis of IL-2.

conjugate a well-defined synthetic glycopeptide to a protein fragment that was produced in *Escherichia coli* (**Fig. 1**).

In this method, an IL-2 fragment with an N-terminal cysteine is produced from an affinity-tagged fusion protein with a modified TEV protease cleavage site inserted between the affinity tag and the desired protein fragment *(10)*. This allows the IL-2 protein fragment to be purified by affinity chromatography and then cleaved by the highly selective TEV protease to remove the affinity tag and unmask an N-terminal cysteine that can be used in a cysteine–thioester ligation *(3)*. Next, a synthetic glycopeptide thioester is mixed with the IL-2 protein fragment to initiate cysteine thioester ligation. First, a thioester exchange occurs, joining the IL-2 protein fragment and the glycopeptide, and then an *N-S* acyl shift results in the formation of a native peptide bond. In this way, well-defined glycopeptides can be efficiently conjugated to expressed protein fragments, such as the IL-2 fragment.

2. Materials

1. pTrcHisB (Invitrogen, Carlsbad, CA).
2. Restriction enzymes: *Kpn*I, *Bgl*II, *Sal*I, *Eco*RI (New England Biolabs Inc., Beverly, MA).
3. Oligonucleotide primers.
4. T4 polynucleotide kinase and T4 DNA ligase (New England Biolabs Inc.).
5. *E. coli* XL1Blue (Stratagene, La Jolla, CA).
6. pLW46 (ATCC 39452; **ref. *11***).
7. *E. coli* DH5αF' (Invitrogen).
8. Luria-Bertani (LB) medium (Sigma, St. Louis, MO).
9. Ampicillin, sodium salt (Sigma).
10. 0.5 *M* isopropyl-β-D-thio-galactopyranoside (IPTG) solution, filter-sterilized through a 0.2-micron syringe filter, isopropyl-β-D-thio-galactopyranoside (Sigma).

11. Lysis buffer: 50 mM sodium phosphate, pH 8; 20 mM β-mercaptoethanol.
12. Denaturing resuspension buffer: 50 mM sodium phosphate, pH 8.0; 10 mM β-mercaptoethanol; 7 M guanidine hydrochloride.
13. 0.2-Micron filter (Fisher Scientific, Pittsburgh, PA).
14. Superflow Ni-NTA affinity resin (Qiagen, Valencia, CA).
15. Denaturing wash buffer: 50 mM sodium phosphate, pH 8.0; 10 mM β-mercaptoethanol; 300 mM NaCl; 5% glycerol; 10 mM imidazole; 6 M urea.
16. Denaturing elution buffer: 50 mM sodium phosphate, pH 8.0; 10 mM β-mercaptoethanol; 200 mM imidazole; 6 M urea.
17. Dithiothreitol (Sigma).
18. Dialysis tubing 1000 MWCO (Spectrum Laboratories Products Inc., New Brunswick, NJ).
19. Oxidation buffer: 50 mM sodium phosphate, pH 8; 7 M guanidine hydrochloride.
20. Phosphate buffer: 50 mM sodium phosphate, pH 7.8.
21. TEV protease (Invitrogen).
22. β-Mercaptoethanol (Sigma).
23. Boc-Asn(GlcNAc)-OH *(12)*.
24. H-Gly-Gly-OBn *p*-tosylate salt (Bachem, King of Prussia, PA).
25. DMF, Dimethylformamide (Fluka, Milwaukee, WI, 99.8%).
26. Diisopropylcarbodiimide (DIC; Aldrich, Milwaukee, WI, 99%).
27. 1-Hydroxybenzotriazole hydrate (HOBt; Aldrich).
28. *N*-methyl morpholine (Aldrich, 99%).
29. Methanol (Fisher Scientific).
30. Diethyl ether (Fisher Scientific).
31. Palladium 10 wt.% on activated carbon (Aldrich).
32. Hydrogen (Airgas, Radnor, PA).
33. Celite 545, diatomaceous earth (Fisher Scientific).
34. Silica gel 60, Geduran (EMD Chemicals Inc., Gibbstown, NJ).
35. Ethyl acetate (Fisher Scientific).
36. Ethyl 3-mercaptopropionate (Aldrich, 99%).
37. Trifluoroacetic acid (Aldrich, 99%).
38. Ammonium sulfate (Sigma).
39. Guanidine hydrochloride (Sigma).
40. 2-Mercaptoethanesulfonic acid, sodium salt (Aldrich).
41. Acetonitrile, high-performance liquid chromatography (HPLC) grade (Fisher Scientific).

3. Methods

The following method describes the semisynthesis of a glycosylated form of human IL-2 using native chemical ligation. The individual procedures consist of (1) construction of a His-tagged, TEV protease cleavable, truncated IL-2 expression plasmid, (2) expression and purification of the His-tagged IL-2 fusion protein, (3) oxidation and refolding of the His-tagged IL-2 fusion protein, (4) TEV protease cleavage of the fusion protein to remove the affinity tag and unmask an N-terminal cysteine to be used in ligation, (5) chemical synthesis of a model glycopeptide thioester to be used in the protein ligation, and (6) ligation of the glycopeptide thioester to the IL-2 fragment using native chemical ligation and characterization.

3.1. Affinity-Tagged IL-2 Expression Plasmid

Construction of an affinity-tagged IL-2 expression plasmid was conducted by first altering the commercially available pTrcHisB vector (Invitrogen) to insert a linker sequence and a *Sal*I restriction site into the multiple cloning site to create a parent vector and then inserting DNA encoding the TEV protease cleavage site fused to a truncated form of human IL-2 into that parent vector.

1. pTrcHisB (Invitrogen) was double digested with *Kpn*I and *Bgl*II restriction enzymes and the linearized product was gel purified.
2. A synthetic double-stranded fragment of DNA was formed by annealing the synthetic oligos 5'-GAT CTG ATT ACG ATA TCC CAA CGA CCG TCG ACG CTG GTA C and 5'-CAG CGT CGA CGG TCG TTG GGA TAT CGT AAT CA. The resulting double-stranded fragment was phosphorylated with T4 polynucleotide kinase to prepare it for ligation with T4-DNA ligase.
3. The parent vector pTrcHisB-linker was formed by ligation of the linearized pTrcHisB from **step 1** and the synthetic double-stranded DNA fragment from **step 2** using T4 DNA ligase. The ligation product was transformed into *E. coli* XL1Blue by electroporation, and the plasmids recovered from transformants were confirmed by restriction digestion with *Sal*I and DNA sequencing.
4. A DNA fragment encoding a TEV protease cleavage site with a cysteine mutation at the P1' position *(10)* fused to a fragment of human IL-2 (DNA encoding amino acids 7-133) was produced by polymerase chain reaction (PCR) using the primers 5'-CCG CGC GTC GAC GAA AAC CTG TAT TTT CAG TGC ACA AAG AAA ACA CAG CTA and 5'-CCG GCG GAA TTC TCA AGT CAG TGT TGA GAT GAT GCT and IL-2 encoding plasmid pLW46 (ATCC 39452; **ref. *11***) as the template (*see* **Note 1**).
5. The PCR fragment from **step 4** was inserted into the pTrcHisB-linker parent vector from **step 3** using *Sal*I and *Eco*RI restriction sites and T4 DNA ligase. Correct insertion of the PCR fragment DNA was verified by DNA sequencing, and the resulting plasmid was named pCys6IL2.

3.2. Expression and Purification of the Affinity-Tagged IL-2 Fusion Protein

The affinity-tagged IL-2 fusion protein encoded by pCys6IL2 was expressed as inclusion bodies in *E. coli*. The protein was solubilized from the inclusion bodies under denaturing conditions, and nickel-NTA affinity chromatography was also conducted under denaturing conditions to yield purified affinity-tagged IL-2 fusion protein (**Fig. 2**; **ref. *10***).

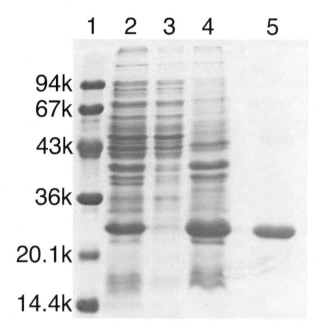

Fig. 2. Purification gel of His-tagged IL-2 fragment. Lane 1, molecular weight markers;lane 2, whole cell extract; lane 3, cell lysate; lane 4, inclusion bodies; lane 5, purified protein after Ni–NTA column.

1. Plasmid pCys6IL2 was transformed into *E. coli* DH5αF' cells. A colony of pCys6IL2/DH5αF' was inoculated into a 10-mL LB culture supplemented with 0.4% glucose and ampicillin was grown overnight at 37°C.
2. A 1-L flask containing LB media and ampicillin was inoculated with the 10 mL overnight culture from **step 1** and incubated at 37°C with shaking until the OD_{600nm} of the culture was 0.4.
3. Once OD_{600nm} = 0.4, the culture was induced with 0.2 m*M* IPTG for 6 h at 37°C.
4. Cells were harvested after 6 h of IPTG induction by centrifugation at 8000*g* for 15 min, and cell pellets were stored at –20°C before purification.
5. Cell pellets were resuspended in approx 25 mL of lysis buffer.
6. Cells were lysed by French press, passing cells at least three times through the press at 1500 Psi to insure complete lysis.
7. Cell debris and inclusion bodies were collected by centrifugation at 12,000*g* for 30 min.
8. The cell debris and inclusion body pellet was resuspended in approx 40 mL of denaturing resuspension buffer by vigorous stirring for 1 h.
9. To remove particulate matter that had not dissolved, the resuspended inclusion body pellet was centrifuged for 1 h at 12,000*g* and then filtered through a 0.2-micron syringe filter before loading onto the Ni-NTA affinity resin.

10. The clarified, resuspended inclusion body solution was loaded onto a Ni-NTA affinity column, and then the column was washed with approx 10 column volumes of denaturing wash buffer.
11. Purified protein was then eluted from the column with denaturing elution buffer.
12. Dithiothreitol was added to the purified protein solution to a final concentration of 10 mM, and the protein solution was dialyzed extensively against 50 mM sodium phosphate buffer pH 7.0 to remove denaturant and thiol. Precipitated, reduced IL-2 fusion protein was lyophilized and stored at –20°C.

3.3. Oxidation and Refolding of the His-Tagged IL-2 Fusion Protein

Purified, reduced IL-2 fusion protein was air-oxidized by dissolving the lyophilized protein in a denaturing buffer and exposing the protein solution to air to form IL-2's structurally important disulfide bond. After the disulfide bond has formed, the denaturant is dialyzed away to yield water-soluble oxidized IL-2 fusion protein.

1. Lyophilized, reduced IL-2 fusion protein from **Subheading 3.2.** was dissolved in oxidation buffer to a concentration of approx 1 mg protein per mL of buffer.
2. The dissolved protein solution was gently stirred under an air atmosphere for 3 d to facilitate air oxidation of the IL-2 protein.
3. The oxidized protein solution was then dialyzed against extensively against several changes of phosphate buffer with 1000 MWCO dialysis tubing to remove guanidine hydrochloride. A small amount of precipitated protein sometimes forms upon removal of the guanidine hydrochloride denaturant, and can be removed by centrifugation. The resulting protein solution contains oxidized, refolded His-tagged IL-2 suitable for TEV cleavage reactions.

3.4. TEV Protease Cleavage of the His-Tagged IL-2 to Generate an N-Terminal Cysteine

TEV protease cleavage of the oxidized, refolded His-tagged IL-2 fusion protein is next used to remove the histidine tag from the fusion protein and unmask an N-terminal cysteine that is utilized to join the IL-2 fragment to synthetic glycopeptide thioesters using a native chemical ligation reaction (**Fig. 3**; ref. *10*). The TEV protease cleavage of His-tagged IL-2 is conducted under dialysis conditions to minimize thiazolidine formation (*see* **Note 2**). In addition, TEV protease is a cysteine protease and requires a reducing agent to be added to the reaction buffer to maintain protease activity (*see* **Note 3**).

1. The oxidized, refolded His-tagged-IL-2 from **Subheading 3.3.** was placed in a 1000 MWCO dialysis bag and TEV protease (100 U of TEV protease per mg of His-tagged IL-2) was added to this solution (*see* **Note 4**). The dialysis bag was placed in phosphate buffer (50 mM sodium phosphate buffer, pH 7.8, with a volume at least 100 times larger than the cleavage reaction volume).

A   **B**   t/h 0   24

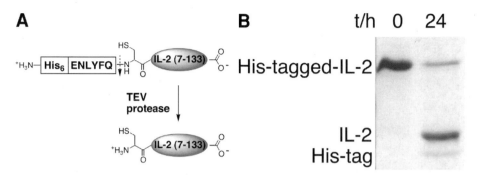

Fig 3. TEV protease cleavage of His-tagged IL-2. (**A**) Schematic of TEV protease cleavage. (**B**) Gel showing cleavage of His-tagged IL-2.

2. The cleavage reaction was initiated by adding β-mercaptoethanol to the dialysis buffer to a final concentration of 1 mM.
3. This reaction was gently stirred and incubated at 25°C for approx 24 h. The cleavage reaction was then analyzed by sodium dodecyl sulfate polyacrylamide gel electrophoresis and matrix-assisted laser desorption-mass spectometry (MALDI-MS) to confirm formation of the correctly cleaved IL-2 fragment.

3.5. Chemical Synthesis of a Glycopeptide Thioester for Protein Ligation

The glycosylated tripeptide thioester used in this method to produce a glycosylated form of IL-2 was produced using solution phase chemical synthesis (**Fig. 4**). The glycosylated amino acid Boc-Asn(GlcNAc)-OH (*12*) was coupled to H-Gly-Gly-OBn using diisopropylcarbodiimide and HOBt. The benzyl protecting group was then removed by hydrogenation over a palladium catalyst. The free carboxylic acid of the resulting tripeptide was coupled to ethyl 3-mercaptopropionate to form a thioester, and then the N-terminal boc protecting group was removed using a mixture of trifluoroacetic acid and water, yielding the glycosylated tripeptide thioester (*see* **Note 5**).

1. Boc-Asn(GlcNAc)-OH (0.91 g, 2.0 mmol) was activated with 2 eq of diisopropylcarbodiimide (0.63 mL, 4.0 mmol) and 2 eq of HOBt (0.61 g, 4.0 mmol) in DMF (35 mL) with stirring for 15 min. Then 1.1 eq of H-Gly-Gly-OBn *p*-tosylate salt (0.87 g, 2.2 mmol) and 4 eq of *N*-methyl morpholine (0.88 mL, 8.0 mmol) were added to the reaction. The reaction was stirred at room temperature for 4.5 h and then concentrated by rotovap. The residue was dissolved in a minimum amount of methanol and precipitated with diethyl ether. The precipitate was collected by filtration and washed with diethyl ether. The resulting precipitate was dissolved in methanol (75 mL) and 10% palladium on carbon (approx 50 mg) was added to the solution. This mixture was placed under a hydrogen atmosphere (approx 1 atmosphere pressure) and stirred for 5 h. The

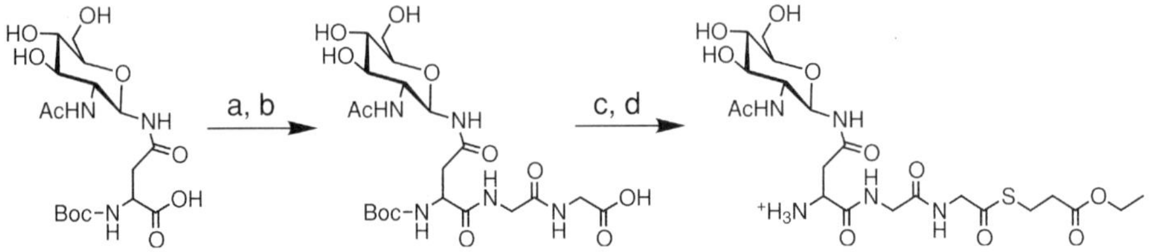

Fig. 4. Synthesis of H-Asn(GlcNAc)-Gly-Gly-thioester tripeptide. a, H-Gly-Gly-OBn, DIC, HOBt, *N*-methyl morpholine; b, MeOH, 10% Pd on C, H_2; c, Ethyl 3-mercaptopropionate, DIC, HOBt; d, trifluoroacetic acid/water (95/5).

palladium on carbon was removed by filtration through celite, and the resulting filtrate was concentrated by rotovap. The resulting residue was purified by flash chromatography using a mixture of ethyl acetate: methanol: water (70:20:10) as eluent to give 0.75 g of a white solid. ^{1}H NMR (400 MHz, CD3OD): δ 4.99 (d, 1H, J = 9.7), 4.45 (t, 1H, J = 5.7), 4.01–3.84 (m, 5H), 3.76 (t, 1H, J = 10.1), 3.67 (dd, 1H, J = 11.6, 4.6), 3.53 (t, 1H, J = 9.1), 3.39–3.30 (m, 2H), 2.72 (d, 2H, J = 5.9), 1.99 (s, 3H), 1.45 (s, 9H); MALDI-Fourier Transform Mass Spectrometry (FTMS) 572.2175, (calcd. 572.2180 Mna$^+$).

2. Boc-Asn(GlcNAc)-Gly-Gly-OH (0.35 g, 0.65 mmol) was placed into a flask and dissolved in *N,N*-dimethylformamide (25 mL). Diisopropylcarbodiimide (4 eq, 0.41 mL, 2.6 mmol), HOBt (2 eq, 0.20 g, 1.3 mmol), and ethyl 3-mercapto-propionate (5 eq, 0.42 mL, 3.2 mmol) were then added to this solution. This mixture was stirred at room temperature for 24 h, and then the solvent was removed under reduced pressure. The resulting residue was dissolved in a mini-mum amount of methanol and precipitated with diethyl ether. The precipitate was collected by filtration and washed with diethyl ether. The precipitate was then dissolved in a 10 mL solution of trifluoroacetic acid:water (95:5) and stirred for 15 min. The solvent was removed by rotovap, and the residue was azeotroped three times with toluene. The resulting residue was purified by flash chromatog-raphy using a mixture of ethyl acetate: methanol: water (70:20:10) as eluent to yield 0.21 g of the glycosylated tripeptide thioester as a white solid. ^{1}H NMR (400 MHz, D2O): δ_5.08 (d, 1H, J = 9.7), 4.21–4.15 (m, 4H), 4.06 (d, 1H, J = 17), 4.01 (d, 1H, J = 17), 3.91–3.74 (m, 4H), 3.63 (t, 1H, J = 9.1), 3.55-3.46 (m, 2H), 3.18 (t, 2H, J = 6.8), 2.79–2.62 (m, 4H), 2.03 (s, 3H), 1.27 (t, 3H, J = 7.2); ^{13}C NMR: δ 202.00, 178.01, 176.32, 175.89, 175.02, 173.88, 79.86, 79.20, 75.84, 71.08, 63.61, 62.11, 55.82, 53.04, 50.54, 44.03, 41.52, 35.41, 25.03, 23.65, 14.92; MALDI-FTMS 588.1965, (calcd. 588.1946 Mna$^+$).

3.6. Ligation of the Glycopeptide Thioester to the IL-2 Fragment

Native chemical ligation was used to link the TEV protease-cleaved IL-2 fragment from **Subheading 3.4.** to the glycosylated tripeptide thioester from **Subheading 3.5. (Fig. 5)**. In this procedure the IL-2 fragment is ammonium

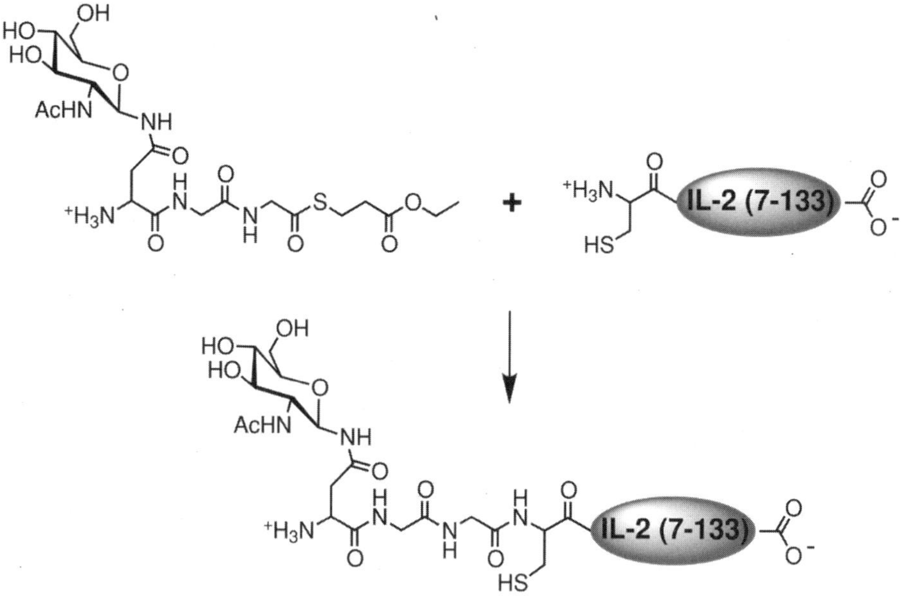

Fig. 5. Ligation of IL-2 fragment from **Subheading 3.4.** to the glycosylated tripeptide thioester from **Subheading 3.5.**

sulfate precipitated to concentrate the protein and allow the ligation reaction to be conducted in a smaller volume (*see* **Note 6**). In addition, the ligation reaction is conducted under denaturing conditions to keep the extremely hydrophobic IL-2 protein soluble under the strongly reducing conditions of native chemical ligation (*see* **Note 7**). Formation of the correctly ligated product, glycosylated IL-2, is observed by mass spectrometry.

1. Approximately 3 mL of the TEV protease-cleaved IL-2 fragment from **Subheading 3.4.** (protein concentration approx 1 mg/mL) was ammonium sulfate precipitated by adding solid ammonium sulfate to the protein solution and gentle mixing until the solution had a saturation of ammonium sulfate of 90%. The precipitated ammonium sulfate pellet was collected by centrifugation.
2. The ammonium sulfate precipitated IL-2 fragment from **step 1** was dissolved in 0.44 mL of a solution consisting of 50 m*M* sodium phosphate, pH 7.6; 6.8 *M* guanidine hydrochloride; 30 m*M* 2-mercaptoethanesulfonic acid; and 5 m*M* of the glycosylated tripeptide thioester from **Subheading 3.5.** This mixture was incubated at room temperature for 24 h. The ligation solution should be prepared immediately before use because thioesters can slowly hydrolyze in aqueous solution.
3. The resulting protein solution was desalted and purified by reverse-phase HPLC and the resulting ligated glycosylated IL-2 was analyzed by MALDI-time of flight MS (**Fig. 6**). Mass spectral results were in good agreement with the calculated

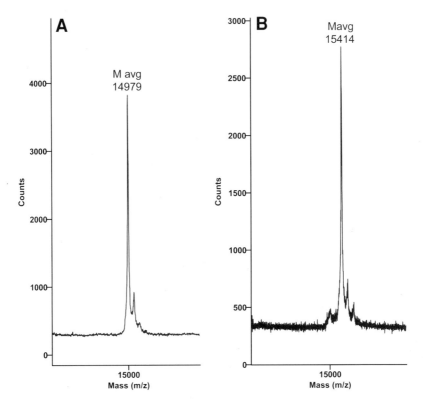

Fig. 6. MALDI-time of flight MS of (**A**) TEV protease-cleaved IL-2 fragment from **Subheading 3.4.** (expected mass 14,975; observed mass 14,979); (**B**) glycosylated IL-2 product from **Subheading 3.6.** (expected mass 15,406; observed mass 15,414).

mass for addition of glycosylated tripeptide to the IL-2 fragment (expected mass 15406, observed mass 15414).

4. Notes

1. Plasmid pLW46 encodes the human IL-2 protein with a single mutation, Cys125 to Ser. This mutation does not affect IL-2 activity and improves the recovery of active IL-2 during in vitro oxidation *(11)*.
2. N-terminal cysteines are very reactive chemical moieties that can react with aldehydes to form thiazolidines under mild conditions *(13–15)*. Thiazolidine formation will prevent a protein or peptide from reacting in cysteine–thioester ligations, such as native chemical ligation or expressed protein ligation *(16)*. Because many carbon based buffers, such as Tris, have small amounts of aldehyde contaminants *(17)*, a noncarbon-based buffer is used, phosphate buffer, and the TEV protease cleavage reaction is dialyzed against a large excess of phosphate buffer to dilute any aldehydes that are initially present in the reaction.

3. TEV protease cleavage activity requires added reducing agent to keep the active site cysteine reduced. This can cause difficulties in some proteins, such as IL-2, that have structurally important disulfide bonds. Cleavage of proteins with disulfide bonds with TEV protease can give varying amounts of cleavage before the disulfide bond becomes reduced and the proteins begin to aggregate, but uncleaved affinity-tagged protein can always be separated from cleaved protein by utilizing affinity chromatography or reverse-phase HPLC.

4. For small-scale reactions, TEV protease can be obtained commercially from Invitrogen and also Recombinant Technologies LLC. For larger-scale protein cleavage reactions, TEV protease can be economically produced in *E. coli* (**refs. 10** and *18*; unit definition of TEV protease: one unit will cleave ≥95% of a 3-µg test substrate in 1 h at 30°C).

5. Thioesters can be very sensitive to basic conditions and can even be hydrolyzed slowly in nearly neutral aqueous conditions. Because of this, when synthesizing thioesters they should always be stored in a freezer, and exposure to aqueous, basic conditions should be avoided.

6. Concentration of the IL-2 protein solution allows the ligation reaction to be conducted in a smaller volume, which results in a much smaller amount of glycopeptide thioester being used in the ligation. This is advantageous because the glycopeptide thioester is expensive and takes several synthetic steps to produce.

7. IL-2 is an extremely hydrophobic protein and has a tendency to aggregate in aqueous solution under reducing conditions. Because of this, care must be taken to completely dissolve the IL-2 protein fragment in a strongly denaturing solution (6.8 *M* guanidine hydrochloride) to insure that the IL-2 protein fragment is in solution so it can react with the glycopeptide thioester during the ligation reaction.

References

1. Sears, P., Tolbert, T., and Wong, C.-H. (2001) Enzymatic approaches to glycoprotein synthesis. *Genet. Engin.* **23,** 45–68.
2. Davis, B. G. (2002) Synthesis of glycoproteins. *Chem. Rev.* **102,** 579–601.
3. Dawson, P. E. and Kent, S. B.H. (2000) Synthesis of native proteins by chemical ligation. *Annu. Rev. Biochem.* **69,** 923–960.
4. Erlanson, D. A., Chytil, M., and Verdine, G. L. (1996) The leucine zipper domain controls the orientation of AP-1 in the NFAT.cntdot.AP-1.cntdot.DNA complex. *Chem. Biol.* **3,** 981–991.
5. Marcaurelle, L. A., Mizoue, L. S., Wilken, J., Oldham, L., Kent, S. B., Handel, T. M., et al. (2001) Chemical synthesis of lymphotactin: a glycosylated chemokine with a C-terminal mucin-like domain. *Chemistry* **7,** 1129–1132.
6. Xu, M.-Q., and Evans, T. C., Jr. (2001) Intein-mediated ligation and cyclization of expressed proteins. *Methods* **24,** 257–277.
7. Kochendoerfer, G. G., Chen, S. Y., Mao, F., Cressman, S., Traviglia, S., Shao, H., et al. (2003) Design and chemical synthesis of a homogeneous polymer-modified erythropoiesis protein. *Science* **299,** 884–887.

8. Tolbert, T. J. and Wong, C.-H. (2000) Intein-mediated synthesis of proteins containing carbohydrates and other molecular probes. *J. Am. Chem. Soc.* **122,** 5421–5428.

9. Hofmann, R. M. and Muir, T. W. (2002) Recent advances in the application of expressed protein ligation to protein engineering. *Curr. Opin. Biotechnol.* **13,** 297–303.

10. Tolbert, T. J. and Wong, C.-H. (2002) New methods for proteomic research: preparation of proteins with N-terminal cysteines for labeling and conjugation. *Angew. Chem., Int. Ed.* **41,** 2171–2174.

11. Wang, A., Lu, S.-D., and Mark, D. F. (1984) Site-specific mutagenesis of the human interleukin-2 gene: structure-function analysis of the cysteine residues. *Science* **224,** 1431–1433.

12. Mizuno, M., Haneda, K., and Iguchi, R. (1999) Synthesis of a glycopeptide containing oligosaccharides: chemoenzymic synthesis of eel calcitonin analogs having natural N-linked oligosaccharides. *J. Am. Chem. Soc.* **121,** 284–290.

13. Zhang, L. and Tam, J. P. (1996) Thiazolidine formation as a general and site-specific conjugation method for synthetic peptides and proteins. *Anal. Biochem.* **233,** 87–93.

14. Zhao, Z. G., Im, J. S., Lam, K. S., and Lake, D. F. (1999) Site-specific modification of a single-chain antibody using a novel glyoxylyl-based labeling reagent. *Bioconjugate Chem.* **10,** 424–430.

15. Villain, M., Vizzavona, J., and Rose, K. (2001) Covalent capture: a new tool for the purification of synthetic and recombinant polypeptides. *Chem. Biol.* **8,** 673–679.

16. Erlanson, D. A. and Verdine, G. L. (1997) Chemical ligation of EDTA to recombinant proteins: a new strategy for affinity cleavage. *Protein Eng.* **10,** 47.

17. Shiraishi, H., Kataoka, M., Morita, Y., and Umemoto, J. (1993) Interactions of hydroxyl radicals with tris (hydroxymethyl) aminomethane and Good's buffers containing hydroxymethyl or hydroxyethyl residues produce formaldehyde. *Free Radical Res. Commun.* **19,** 315–321.

18. Lucast, L. J., Batey, R. T., and Doudna, J. A. (2001) Large-scale purification of a stable form of recombinant tobacco etch virus protease. *BioTechniques* **30,** 544, 546, 548, 550, 554.

20

Subtilisin-Catalyzed Glycopeptide Condensation

Thomas J. Tolbert and Chi-Huey Wong

Summary

This method describes the use of subtilisin-catalyzed peptide condensation to form a 15-residue glycopeptide from two smaller synthetic peptides. A 12-residue peptide ester is synthesized by solid-phase peptide synthesis using a PAM-modified Rink amide resin that allows the formation of a peptide ester suitable for subtilisin ligation. The 12-residue acyl donor peptide ester is then ligated to a 3-residue acyl acceptor glycopeptide amide using subtilisin (EC 3.4.21.62) in a buffered mixture of water and DMF (1:9).

Key Words: Protease; subtilisin; glycopeptide; solid-phase peptide synthesis; peptide; glycoprotein; protein; condensation; ligation.

1. Introduction

Subtilisin-catalyzed peptide condensation is a technique that has been used to ligate synthetic peptide fragments together to form large peptides, glycopeptides, and full-length proteins (*1–7*). Reverse proteolysis catalyzed by subtilisin is free of racemization, unlike some chemical condensation techniques and also does not require a cysteine at the junction between two peptides to be joined, as native chemical ligation calls for (*8*). Application of subtilisin-catalyzed peptide condensation to glycopeptide and glycoprotein synthesis offers a useful route to these difficult to synthesize biomolecules.

This method describes the use of a kinetically controlled subtilisin ligation reaction for the formation of a glycopeptide (**Fig. 1**). In kinetically controlled subtilisin condensations, one peptide, the acyl donor, is activated as an ester and the condensation reaction is conducted in a mixture of water and an organic cosolvent, such as dimethylformamide (DMF), dimethyl sulfoxide (DMSO), or MeOH (*9*). The reaction is a competition between hydrolysis of the acyl donor with water and aminolysis of the acyl donor with the amine of an acyl acceptor peptide to form the desired product. The amount of hydrolysis

From: *Methods in Molecular Biology, vol. 283: Bioconjugation Protocols: Strategies and Methods*
Edited by: C. M. Niemeyer © Humana Press Inc., Totowa, NJ

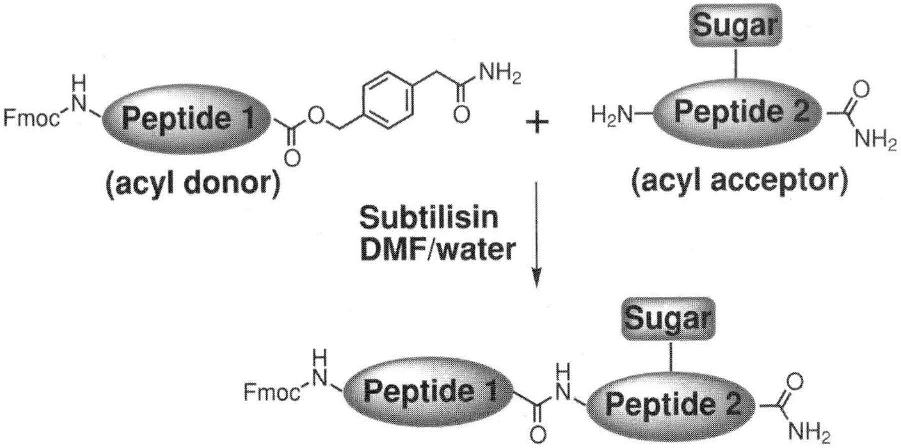

Fig. 1. Subtilisin-catalyzed glycopeptide condensation. Aminolysis of the activated peptide PAM ester (Peptide 1) with a glycopeptide amide (Peptide 2) catalyzed by subtilisin in an organic cosolvent/water mixture yields a ligated glycopeptide.

can be minimized by decreasing the concentration of water in the reaction by increasing the amount of cosolvent, increasing the aminolysis rate by increasing the concentration of the acyl acceptor peptide, and use of mutated forms of subtilisin that have been optimized for peptide bond formation or stability in organic solvents *(10–13)*. Self-condensation reactions of peptides can be prevented by blocking the N-terminus of the acyl donor peptide with an amine-protecting group, such as Fmoc, and blocking the C-terminus of the acyl acceptor peptide as an amide.

The scope of subtilisin-catalyzed glycopeptide coupling reactions has been extensively studied, and examples of subtilisin condensation of glycopeptides containing *O*-linked and *N*-linked glycans, as well as acylated and unprotected sugars have been reported *(3,4)*. The active site of subtilisin has been carefully mapped out to determine which residues of substrate peptides can be glycosylated (**Fig. 2**). It was found that many areas of the subtilisin active site, including the S4, S3, S2', S3', and S4' sites, would accept glycosylated amino acid residues, whereas some areas very near to the cleavage/ligation junction, the S2, S1, and S1' sites, would not accept glycosylation *(4)*.

In this method a dodecapeptide ester is synthesized by solid-phase peptide synthesis using a PAM (2-[1'-(hydroxymethyl)phen-4'-yl]-acetamide) modified Rink amide resin. The PAM-modified Rink amide resin allows Fmoc-based synthesis to be used to construct the dodecapeptide on the resin and then combines deprotection of the peptide side chains, cleavage from the resin, and

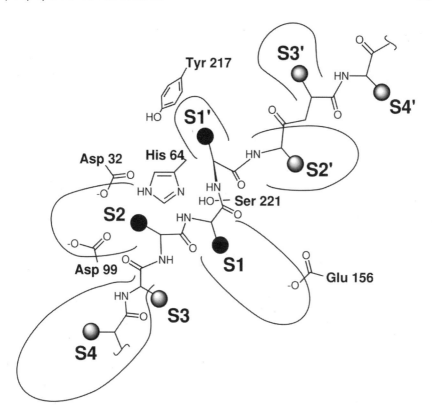

Fig. 2. Diagram of the subtilisin (EC 3.4.21.62)-active site. Amino acid side chains represented as black dots (S2, S1, and S1') represent sites where subtilisin will not accept glycosylation. All of the other sites shown (S4, S3, S2', S3', and S4') will accept glycosylated amino acids (*4*).

formation of the peptide ester suitable for subtilisin ligation into a single step. The dodecapeptide ester is then ligated to a glycosylated tripeptide amide derived from solution phase synthesis by subtilisin-catalyzed peptide condensation to form a 15-residue glycopeptide.

2. Materials

1. Amino acids used in solid-phase peptide synthesis: Fmoc-Ala-OH, Fmoc-Ile-OH, Fmoc-Lys(Boc)-OH, Fmoc-Val-OH, Fmoc-His(Trt)-OH, Fmoc-Asn(Trt)-OH, Fmoc-Gln(Trt)-OH, Fmoc-Thr(OtBu)-OH (Novabiochem, San Diego, CA).
2. 4-(Bromomethyl)phenylacetic acid phenacyl ester (Aldrich, Milwaukee, WI).
3. Bu$_4$NHSO$_4$, Tetrabutylammonium hydrogen sulfate (Aldrich).
4. CH$_2$Cl$_2$, methylene chloride (Fisher Scientific, Pittsburgh, PA).

 5. Saturated, aqueous $NaHCO_3$ solution.
 6. $MgSO_4$, magnesium sulfate.
 7. Ethyl acetate (Fisher Scientific).
 8. Hexanes (Fisher Scientific).
 9. Silica gel 60, Geduran (EMD Chemicals Inc., Gibbstown, NJ).
10. Zinc dust (Aldrich).
11. Acetic acid (Aldrich).
12. Celite 545, diatomaceous earth (Fisher Scientific).
13. Et_2O, diethylether (Fisher Scientific).
14. 1 *M* HCl solution.
15. Fmoc-Rink amide AM resin (Novabiochem).
16. DMF (Fluka, Milwaukee, WI, 99.8%).
17. Morpholine (Aldrich, 99%).
18. 1-Hydroxybenzotriazole hydrate (HOBt; Aldrich).
19. *N*-methyl morpholine (NMM; Aldrich, 99%).
20. HBTU, 2-(1 *H*-benzotriazole-1-yl)-1,1,3,3-tetramethyluronium hexafluoro-phosphate (Novabiochem).
21. Acetic anhydride (Ac_2O; Aldrich).
22. Pyridine (Aldrich, 99%).
23. Piperidine (Aldrich, 99.5%).
24. Trifluoroacetic acid (Aldrich, 99%).
25. Triethylsilane (Aldrich, 99%).
26. MeOH, methanol (Fisher Scientific).
27. Cbz-Ser($Ac_3GlcNAc\beta$)-OH *(14)*.
28. EDC,1-[3-(dimethylamino)propyl]-3-ethylcarbodiimide hydrochloride (Aldrich, 98%).
29. Tetrahydrofuran (THF; Fisher Scientific).
30. Aqueous ammonia, ammonium hydroxide 28–30% solution (Aldrich).
31. Palladium hydroxide ($Pd(OH)_2$), 20 wt.% Pd, wet, Degussa type (Aldrich).
32. Hydrogen (H_2; Airgas, Radnor, PA).
33. Cbz-Gly-Gly-OH (Bachem, King of Prussia, PA).
34. Ethanol (Aldrich).
35. Triethanolamine buffer (Sigma, St. Louis, MO).
36. Subtilisin BPN', EC 3.4.21.62 (Sigma, Protease Type XXVII).
37. Phenylmethanesulfonyl fluoride (PMSF; Sigma, 99%).

3. Methods

The following method describes the synthesis of a 15-residue glycopeptide using subtilisin-catalyzed glycopeptide condensation to join two smaller peptides together (*see* **Note 1**). Procedures for (1) synthesis of a solid-phase resin that will allow formation of esters that can be used in subtilisin ligation, (2) solid phase synthesis of a peptide ester using this resin, (3) synthesis of a glycopeptide amide for subtilisin ligation, and (4) subtilisin-catalyzed peptide condensation are described.

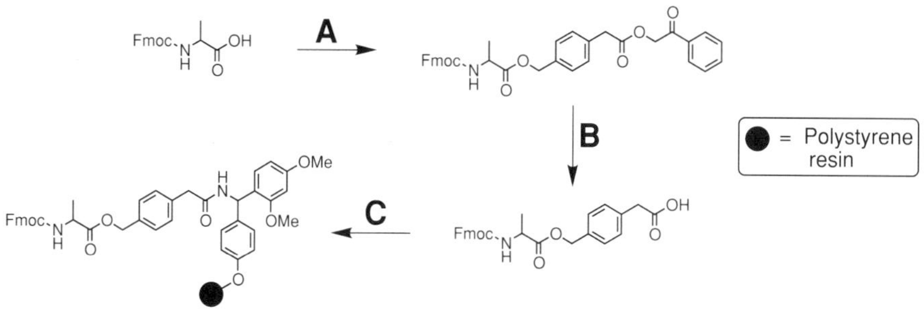

Fig. 3. (**A**) 4-(bromomethyl)phenylacetic acid phenacyl ester, Bu_4NHSO_4, aqueous $NaHCO_3/CH_2Cl_2$ (1:1); (**B**) zinc dust, 85% acetic acid; (**C**) Rink amide AM resin, HBTU, HOBt, NMM, DMF.

3.1. Immobilization of Amino Acids Onto Rink Amide Resin Through PAM Linkers

An amino acid-loaded Rink Amide resin that has a PAM linker (*see* **Note 2**) between the amino acid and the resin is synthesized in this procedure (**Fig. 3**). The N-terminally protected amino acid is first coupled to a Pac (*see* **Note 3**) protected form of the PAM linker [4-(bromomethyl)phenylacetic acid phenacyl ester], and then the Pac protecting group of the PAM linker is removed using reductive acidolysis to yield the free carboxylic acid. The free carboxylic acid is then coupled to Rink amide AM resin using 2-(1H-benzotriazole-1-yl)-1,1,3,3-letramethyluronium hexafluorophosphate (HBTU) to yield an Fmoc-Ala-PAM loaded Rink amide resin. When cleaved by 95% trifluoroacetic acid, this resin will yield peptide PAM amide ester that can be used in subtilisin ligation reactions.

1. Fmoc-Ala-OH (0.295 g, 0.95 mmol), 4-(bromomethyl)phenylacetic acid phenacyl ester (0.091 g, 0.19 mmol), and Bu_4NHSO_4 (0.065 g, 0.19 mmol) were placed into a flask containing 9.5 mL of saturated, aqueous $NaHCO_3$ solution and 9.5 mL of CH_2Cl_2. This mixture was stirred vigorously for 5 h, and then the organic layer was separated, washed, with saturated $NaHCO_3$ solution, washed with water, and dried over $MgSO_4$. The organic layer was then filtered and concentrated under reduced pressure. The resulting residue was purified by silica gel chromatography (eluent 67% ethyl acetate in hexane) to yield 0.095 g of the desired product. R_f: 0.40 (ethyl acetate:hexane, 2:1). ^{1}H-NMR (400 MHz, $CDCl_3$): δ 7.88 (d, 2H, $J = 7.4$), 7.76 (d, 2H, $J = 7.5$), 7.62–7.58 (m, 4H), 7.47, (t, 2H, $J = 7.7$), 7.39 (t, 2H, 7.4), 7.36–7.29 (m, 6H), 5.33 (s, 2H), 5.17 (s, 2H), 4.39 (m, 3H), 4.20 (t, 1H, $J = 7.1$), 3.82 (s, 2H), 1.44 (d, 3H, $J = 7.1$). HRMS (FAB, pos): 710.1155 ($M+Cs^+$, calcd. 710.1182).
2. The product from **step 1** (0.200 g, 0.35 mmol) was treated with 0.595 g of activated zinc dust (*see* **Note 4**) in 10 mL of 85% acetic acid. After all of the starting

material had reacted, as monitored by thin-layer chromatography, the zinc dust was removed by filtration through Celite. Et_2O (22 mL) and H_2O (18 mL) were then added to the filtrate and the aqueous and organic layers were separated. 1 *M* HCl was added to the aqueous phase until the pH was 1.0–1.5, and then the aqueous phase was extracted with Et_2O twice. The organic layers were combined and washed with H_2O six times. The solvents were removed under reduced pressure, and residual acetic acid was azeotroped with benzene. After drying under high vacuum, 0.120 g of product was obtained. R_f: 0.30 (*n*-Hex/EtOAc/AcOH, 1:2:0.05). ^{1}H-NMR (400 MHz, $CDCl_3$): δ 7.76 (d, 2H, *J* = 8.0), 7.59 (d, 2H, *J* = 6.0), 7.40 (t, 2H, *J* = 7.3), 7.33–7.26 (m, 6H), 5.35 (d, 1H, *J* = 6.4), 5.16 (m, 2H), 4.45–4.34 (m, 3H), 4.21 (t, 1H, *J* = 7.1), 3.64 (s, 2H), 1.44 (d, 3H, *J* = 7.0). ^{13}C-NMR: δ 178.5, 176.6, 173.9, 156.74, 142.3, 134.5, 129.7, 128.5, 127.7, 127.0, 125.0, 120.0, 67.1, 66.8, 50.8, 47.1, 40.3. HRMS (FAB, pos): 592.0719 ($M+Cs^+$, calcd. 592.0736).

3. Fmoc-Rink amide AM resin (0.25 mmol/g loading, 0.40 g, 0.1 mmol) was deprotected by a mixture of DMF (5 mL) and morpholine (5 mL) by mixing the resin in this solution for 1 h. The DMF/morpholine mixture was then removed and the resin was washed with DMF. The resin was then treated with Fmoc-Ala-Pam ester from **step 2** (0.060 g, 0.131 mmol), HOBt (0.030 g, 0.196 mmol), NMM (0.029 mL, 0.262 mmol), and HBTU (0.50 g, 0.131 mmol) in 11 mL of DMF for 16 h. The reactants were then removed by filtration, and the resin was washed with DMF. The resin was then treated with a mixture of Ac_2O (3 mL) and pyridine (9 mL) to cap any unreacted amines. After 10 min of shaking, the Ac_2O/pyridine mixture was removed by filtration and the resin was washed with DMF to yield Fmoc-Ala-PAM-linker-Rink amide resin.

3.2. Solid-Phase Peptide Synthesis and Cleavage From Resin to Form Peptides With PAM Esters

In this procedure the Fmoc-Ala-PAM-linker-Rink amide resin produced in **Subheading 3.1.** is used in solid phase peptide synthesis to produce a 12-mer peptide with a PAM amide on the C-terminus (**Fig. 4**). Standard Fmoc-synthesis conditions are used to produce this peptide, and the final Fmoc is left protecting the peptide's N-terminus to prevent polymerization of the peptide under subtilisin catalyzed condensation conditions. The final step is a treatment with trifluoroacetic acid and scavengers that serves to both cleave the peptide from the resin and remove the internal peptide protecting groups to yield the 12-mer Fmoc-peptide-PAM amide.

1. Fmoc deprotection: The resin was placed in a reaction vessel containing a 10-mL solution of DMF:piperidine (8:2) and shaken for 15 min. The DMF:piperidine solution was removed by filtration, and the resin was washed with DMF to remove any residual piperidine.

2. Coupling of Fmoc amino acids: A three- to fivefold excess of the Fmoc amino acid was added to the resin as a 0.12 *M* solution, which also contained

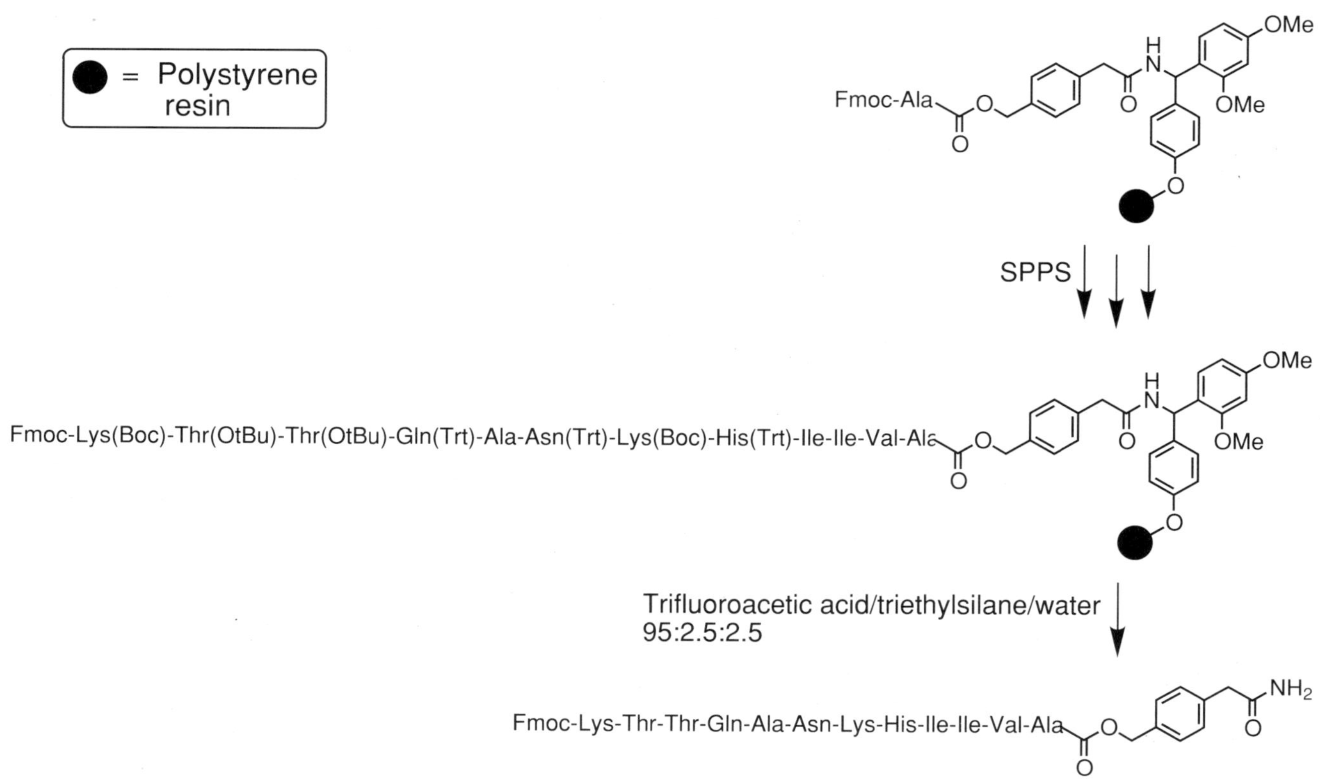

Fig. 4. Solid-phase synthesis of the 12-amino acid peptide PAM ester.

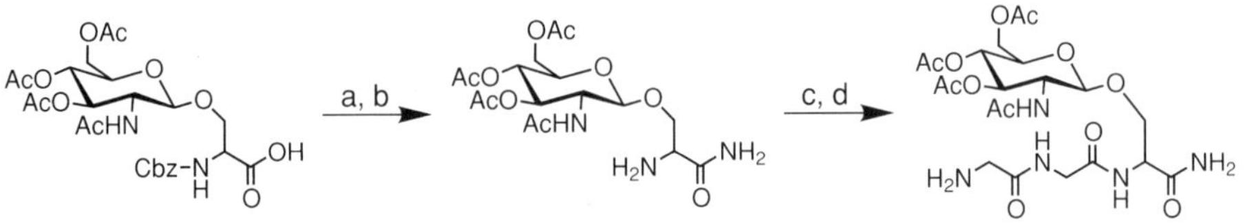

Fig. 5. a, EDC, HOBt, NH4OH, THF; b, Pd(OH)$_2$ (Degussa type), H$_2$, EtOH; c, Cbz-Gly-Gly-OH, HBTU, HOBt, NMM, DMF; d, Pd(OH)$_2$ (Degussa type), H$_2$, EtOH.

    1.5 equivalents of HOBt, 2 equivalents of NMM, and 1.0 equivalent of HBTU per equivalent of Fmoc amino acid. The reaction was shaken for 3–5 h, and then the solution was removed by filtration and washed extensively with DMF.

3. Capping of free amines after amino acid coupling: Immediately after amino acid coupling, the uncoupled free amines were capped using a mixture of pyridine and acetic anhydride (3:1). The resin was treated with 8 mL of this capping solution for 6 min and then rinsed extensively with DMF.

4. **Steps 1–3** were conducted for each amino acid to be coupled to the resin. Starting with the Fmoc-Ala-PAM-linker-Rink amide resin produced in **Subheading 3.1.**, the following amino acids were coupled sequentially: Fmoc-Val-OH, Fmoc-Ile-OH, Fmoc-Ile-OH, Fmoc-His(Trt)-OH, Fmoc-Lys(Boc)-OH, Fmoc-Asn(Trt)-OH, Fmoc-Ala-OH, Fmoc-Gln(Trt)-OH, Fmoc-Thr(OtBu)-OH, Fmoc-Thr (OtBu)-OH, and Fmoc-Lys(Boc)-OH.

5. Once the final amino acid had been coupled to the resin, the peptide was treated with 10 mL of a solution containing 95% trifluoroacetic acid, 2.5% triethylsilane, and 2.5% H$_2$O for 20 min. The solution was collected, and the remaining resin was washed three times with trifluoroacetic acid. The combined filtrates were evaporated to dryness and the resulting residue was dissolved in 5 mL of methanol, filtered to remove insoluble material, and precipitated with 100 mL of Et$_2$O. The precipitate was isolated by centrifugation, taken up into H$_2$O, and lyophilized. The resulting residue was purified by reverse-phase high-performance liquid chromatography to yield 0.128 g of the 12-mer Fmoc-peptide-PAM amide product shown in **Fig. 5**. Matrix-assisted laser desorption mass spec. 1714 (M+Na$^+$ calcd. 1714).

3.3. Synthesis of Glycopeptide Amides for Subtilisin Condensation

    The synthesis of a glycosylated tripeptide amide is described in this procedure starting from the glycosylated amino acid Cbz-Ser(Ac3GlcNAcβ)-OH (**Fig. 6**; **ref. *14***). First, an amide is formed on the C-terminus of the glycosylated serine, and then the Z protecting group is removed, and a Cbz-Gly-Gly-OH dipeptide is coupled to the N-terminus. Removal of the Cbz group from the N-terminal glycine yields a glycosylated tripeptide amide, H-Gly-Gly-Ser(Ac$_3$GlcNAcβ)-NH$_2$, which is suitable for subtilisin-catalyzed ligation.

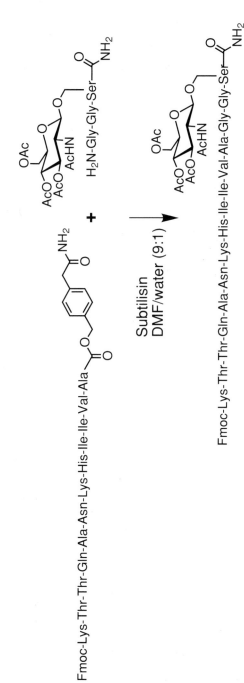

Fmoc-Lys-Thr-Thr-Gln-Ala-Asn-Lys-His-Ile-Ile-Val-Ala

Subtilisin
DMF/water (9:1)

Fmoc-Lys-Thr-Thr-Gln-Ala-Asn-Lys-His-Ile-Ile-Val-Ala-Gly-Gly-Ser

Fig. 6. Subtilisin-catalyzed glycopeptide condensation of peptide PAM ester from **Subheading 3.2.** and glycopeptide amide from **Subheading 3.3.**

275

1. Cbz-Ser(Ac3GlcNAcβ)-OH (0.150 g, 0.27 mmol), HOBt (0.062 g, 0.40 mmol), and EDC (0.052 g, 0.27 mmol) were placed in a flask and dissolved in THF (50 mL). This solution was stirred at room temperature for 20 min. Aqueous ammonia (180 μL, 2.7 mmol) was then added slowly by syringe. The reaction was then stirred for 1 h, and then solvent was removed under vacuum. The resulting residue was dissolved in CH_2Cl_2 and washed three times with a saturated $NaHCO_3$ solution, once with water, and the organic layer was dried over $MgSO_4$. Solvent was removed under reduced pressure to yield a white solid. The white solid was purified by silica gel chromatography (eluent 90% CH_2Cl_2/10% MeOH/ 0.1% NH_4OH) to yield 0.135 g of a white product (Cbz-Ser(Ac₃GlcNAcβ)-NH₂). R_f 0.6 (90% CH_2Cl_2: 10% MeOH:0.1% NH_4OH). ¹H-NMR (500 MHz, d₆- DMSO): δ 7.88 (d, 1H J = 9.0), 7.37–7.30 (m, 5H), 7.24 (s, 1H), 7.16 (s, 1H), 7.05 (d, 1H, J = 8.5), 5.06 (t, 1H J = 9.5), 5.01 (s, 1H), 4.82 (t, 1H, J = 9.5), 4.66 (d, 1H, H-1', J = 8.5), 4.18 (m, 1H), 4.12 (m, 1H), 3.99 (m, 1H), 3.85–3.81 (m, 2H), 3.70–3.67 (m, 2H), 1.99 (s, 3H), 1.96 (s, 3H), 1.89 (s, 3H), 1.72 (s, 3H). ¹³C-NMR: δ 171.38, 170.27, 169.53, 169.44, 169.02, 156.31, 136.92, 133.90, 133.18, 101.09, 73.56, 71.49, 67.02, 35.80, 62.30, 54.92, 53.54, 23.06, 21.08. HRMS (FAB, pos): 700.1142 (M+Cs⁺, calcd. 700.1119).

2. The product (Cbz-Ser(Ac₃GlcNAcβ)-NH₂) from **step 1** (0.117 g, 0.21 mmol) was mixed with 25 mL of ethanol to give a white suspension. Pd(OH)₂ (Degussa type, 10 mg) was added and the mixture was placed under a hydrogen atmosphere. The reaction was stirred for 2 h, during which the white suspension turned clear, and then filtered through Celite to remove the Pd(OH)₂. The solvent was removed under reduced pressure to give 0.89 g of a white solid (H-Ser(Ac₃GlcN Acβ)-NH₂), which was used directly in **step 3**.

3. 0.022 g of the product (H-Ser(Ac₃GlcNAcβ)-NH₂) from step 2 (0.15 mmol) was dissolved in DMF. Cbz-Gly-Gly-OH (0.40 g, 0.15 mmol), HOBt (0.022 g, 0.15 mmol), NMM (41 μL, 0.3 mmol), and HBTU (0.087 g, 0.23 mmol) were added to this solution. The reaction was stirred at room temperature for 12 h and then solvents were removed under reduced pressure. The residue was dissolved in CH_2Cl_2, washed three times with saturated $NaHCO_3$, once with water, and dried over $MgSO_4$. Solvent was removed under reduced pressure, and the residue was purified by chromatography (eluent 92% CH_2Cl_2, 8% MeOH, 0.1% NH_4OH) to yield 0.026 g of a white solid (Cbz-Gly-Gly-Ser(Ac₃GlcNAcβ)-NH₂). R_f 0.4 (92% CH_2Cl_2: 8% MeOH:0.1% NH_4OH). ¹H-NMR (500 MHz, d₆-DMSO):δ 8.14 (m, 1H), 7.85 (m, 2H), 7.53 (m, 1H), 7.35–7.32 (m, 5H), 7.22 (br d, 2H, J = 7), 5.07 (t, 1H, J = 9), 5.02 (s, 1H), 4.82 (t, 1H, J = 9), 4.68 (d, 1H, J = 8), 4.35 (m, 1H), 4.19 (m, 1H), 4.00 (m, 1H), 3.83 (m, 2H), 3.76–3.64 (m, 6H), 2.01 (s, 3H), 1.95 (s, 3H), 1.89 (s, 3H), 1.76 (s, 3H). ¹³C-NMR: δ 170.94, 170.19, 169.71, 169.67, 169.36, 168.83, 128.42, 127.89, 100.25, 70.81, 68.46, 65.62, 61.77, 50.08, 43.60, 22.75, 20.59, 20.48, 20.42. HRMS (FAB, pos): 704.2366 (M+Na⁺, calcd. 704.2391).

4. The product (Cbz-Gly-Gly-Ser(Ac3GlcNAcβ)-NH₂) from **step 3** (0.020 g, 0.029 mmol) was dissolved in 5 mL of ethanol. Pd(OH)₂ (Degussa type, 10 mg) was

added and the mixture was placed under a hydrogen atmosphere. The reaction was stirred for 4 h, and then filtered through Celite to remove the Pd(OH)$_2$. The solvent was removed under reduced pressure to give 0.014 g of a white solid (H-Gly-Gly-Ser(Ac$_3$GlcNAcβ)-NH$_2$), which was then used directly in **Subheading 3.4.**

3.4. Subtilisin Condensation of Glycopeptides to Peptides With PAM Esters

The use of subtilisin to catalyze the condensation of the dodecapeptide PAM ester from **Subheading 3.2.** with the glycosylated tripeptide amide from **Subheading 3.3.** is described in this procedure (**Fig. 6**). First, a DMF/buffer mixture of the peptides to be ligated is prepared, and then subtilisin is added as the last component to initiate the condensation reaction. Once the reaction is complete a protease inhibitor is added to inactivate the subtilisin to prevent proteolysis of the product peptide.

1. 1.4 mg (0.94 μmol) of the dodecapeptide PAM ester from **Subheading 3.2.** and 1.3 mg (2.8 μmol) of the glycosylated tripeptide from **Subheading 3.3.** were dissolved in 90 μL of DMF. To this solution was added 8 μL of 50 mM triethanolamine buffer, pH 7.8, and the solution was mixed thoroughly. 2 μL Subtilisin (EC 3.4.21.62) stock solution (10 mg/mL) was added to initiate the reaction, and the reaction was incubated at 37°C (*see* **Note 5**).
2. After incubating for 4 h, the reaction was quenched by adding 2 μL of a PMSF stock solution (1 mg/mL in acetonitrile) to inactivate the subtilisin (*see* **Note 6**). The products of the reaction were analyzed by matrix-assisted laser desorption mass spectrometry, and it was found that 79% of the dodecapeptide PAM ester from **Subheading 3.2.** had been ligated to form the 15-residue glycopeptide product (*see* **Note 7**).

4. Notes

1. This method is a modified version of a method reported by Witte et al. *(4)*.
2. The PAM linker is an acid and base stable linker that allows a variety of solid-phase chemistry to be conducted without cleaving the linkage between the synthesized peptide and the resin.
3. The Pac (phenacyl ester) protecting group can be orthogonally deprotected with reductive acidolysis in the presence many of the protecting groups normally used in Fmoc solid phase peptide synthesis. This allows a wide variety of Fmoc amino acids to be attached to Rink amide resin through PAM linkers in the manner described here.
4. Zinc dust was activated by washing it six times with 1M HCl, six times with H$_2$O, six times with ethanol, and six times with diethyl ether.
5. A high percentage of organic solvent (90% DMF), an excess of the glycosylated tripeptide acyl acceptor, and relatively high concentrations of both peptides are used in this reaction to promote aminolysis of the dodecapeptide PAM ester over

hydrolysis of the ester with water. Lower amounts of DMF, such as a DMF:buffer ratio of 7:3, gave significant amounts of hydrolysis of the dodecapeptide PAM ester.

6. Active subtilisin has the potential to cleave the ligated peptide product produced in this reaction if given enough time. To prevent this from occurring the subtilisin in the reaction is inactivated with PMSF, which is an irreversible inhibitor of serine proteases that sulfonylates active site serines.

7. This reaction using native subtilisin gave a 79% conversion to the ligated glycopeptide product with 20% hydrolysis of the dodecapeptide PAM ester or an aminolysis:hydrolysis ratio of 3.95. Improved yields of ligation can be achieved by using modified versions of subtilisin that have been selected for increased ligase activity or stability in elevated temperatures and high concentrations of organic solvents. Using a modified version of subtilisin, variant 8397 K256Y *(12)*, in this peptide condensation gave an 84% conversion to the 15-residue glycopeptide product with reduced hydrolysis of the peptide ester (aminolysis/hydrolysis ratio of 5.25).

References

1. Moree, W. J., Sears, P., Kawashiro, K., et al. (1997) Exploitation of subtilisin BPN' as catalyst for the synthesis of peptides containing noncoded amino acids, peptide mimetics and peptide conjugates. *J. Am. Chem. Soc.* **119,** 3942–3947.

2. Liu, C.-F. and Tam, J. P. (2001) Subtilisin-catalyzed synthesis of amino acid and peptide esters. Application in a two-step enzymatic ligation strategy. *Organic Lett.* **3,** 4157–4159.

3. Wong, C. H., Schuster, M., Wang, P., and Sears, P. (1993) Enzymic synthesis of N- and O-linked glycopeptides. *J. Am. Chem. Soc.* **115,** 5893–5901.

4. Witte, K., Seitz, O., and Wong, C.-H. (1998) Solution- and solid-phase synthesis of N-protected glycopeptide esters of the benzyl type as substrates for subtilisin-catalyzed glycopeptide couplings. *J. Am. Chem. Soc.* **120,** 1979–1989.

5. Vogel, K. and Chmielewski, J. (1994) Rapid and efficient resynthesis of proteolyzed triose phosphate isomerase. *J. Am. Chem. Soc.* **116,** 11,163–11,164.

6. Jackson, D. Y., Burnier, J., Quan, C., et al. (1994) A designed peptide ligase for total synthesis of ribonuclease A with unnatural catalytic residues. *Science* **266,** 243–247.

7. Witte, K., Sears, P., and Wong, C.-H. (1997) Enzymic glycoprotein synthesis: preparation of ribonuclease glycoforms via enzymic glycopeptide condensation and glycosylation. *J. Am. Chem. Soc.* **119,** 2114–2118.

8. Dawson, P. E., Muir, T. W., Clark-Lewis, I., and Kent, S. B. H. (1994) Synthesis of proteins by native chemical ligation. *Science* **266,** 776–779.

9. Barbas, C. F., III, Matos, J. R., West, J. B., and Wong, C.-H. (1988) A search for peptide ligase: cosolvent-mediated conversion of proteases to esterases for irreversible synthesis of peptides. *J. Am. Chem. Soc.* **110,** 5162–5166.

10. Nakatsuka, T., Sasaki, T., and Kaiser, E. T. (1987) Peptide segment synthesis catalyzed by the semisynthetic enzyme thiolsubtilisin. *J. Am. Chem. Soc.* **109,** 3808–3810.

11. Abrahmsen, L., Tom, J., Burnier, J., et al. (1991) Engineering subtilisin and its substrates for efficient ligation of peptide bonds in aqueous solution. *Biochemistry* **30,** 4151–4159.
12. Sears, P., Schuster, M., Wang, P., et al. (1994) Engineering subtilisin for peptide coupling: studies on the effects of counterions and site-specific modifications on the stability and specificity of the enzyme. *J. Am. Chem. Soc.* **116,** 6521–6530.
13. Zhong, Z. Z. and Wong, C. H. (1991) Development of new enzymic catalysts for peptide synthesis in aqueous and organic solvents. *Biomed. Biochim. Acta* **50,** S9–S14.
14. Seitz, O. and Wong, C.-H. (1997) Chemoenzymic solution- and solid-phase synthesis of O-glycopeptides of the Mucin domain of MAdCAM-1. A general route to O-LacNAc, O-Sialyl-LacNAc, and O-Sialyl-Lewis-X peptides. *J. Am. Chem. Soc.* **119,** 8766–8776.

IV

BIOFUNCTIONALIZATION OF SURFACES

21

Peptide Nucleic Acid Microarrays

Anette Jacob, Ole Brandt, Achim Stephan, and Jörg D. Hoheisel

Summary

A fast and economical procedure for the production of peptide nucleic acid (PNA) microarrays is presented. PNA oligomers are synthesized in a fully automatic manner in 96-well plates using standard Fmoc chemistry. Subsequently, the oligomers are released from the support and spotted onto glass or silicone slides, which were activated by succinimidyl ester. This process allows for a concomitant purification of the oligomers directly on the chip surface. Although the terminal primary amino groups of the full-length products bind selectively to this surface, none of the byproducts of synthesis, such as truncated sequences or cleaved side chain protection groups, will bind and are therefore washed away. In this chapter, protocols are presented for the whole production process as well as sample hybridization.

Key Words: Peptide nucleic acid; parallel PNA synthesis, surface derivativation, DNA hybridization.

1. Introduction

Deoxyribonucleic acid (DNA) microarrays have become an indispensable tool for the analysis of nucleic acids in a high-throughput format. Nowadays, hybridization analyses are routinely used for many purposes, particularly in combination with fluorescence detection. However, the performance of such analyses could be improved still, simplifying the processes involved or increasing sensitivity, for example. To such ends, we pursue the use of peptide nucleic acid (PNA) as the arrayed probe molecule (1). PNA is a synthetic substitute of DNA. Regular nucleobases are attached to a pseudopeptide backbone via a methylene-carbonyl spacer (2,3). PNA molecules exhibit excellent DNA and ribonucleic acid (RNA) binding capability, chemical stability, and resistance to enzymatic digestion (4). Because of the uncharged nature, PNA permits the hybridization of DNA samples in the absence of salt in the buffer because no interstrand repulsion as between two negatively charged DNA strands needs to

From: *Methods in Molecular Biology, vol. 283: Bioconjugation Protocols: Strategies and Methods*
Edited by: C. M. Niemeyer © Humana Press Inc., Totowa, NJ

be counteracted. As a consequence, the target DNA has fewer secondary structures and is more accessible to the probe molecules. Therefore, the fragmentation of sample molecules before hybridization can be avoided. Furthermore, the use of PNA permits the adoption of an alternative detection mode, for which no labeling is required *(5,6)*.

Reproducibility and reliability of PNA microarrays depend strongly on the quality of the oligomers on the chip surface. Also, relatively large numbers of oligomers are required for microarrays. However, only a rather small quantity of each oligomer is needed. For several years, we pursued the establishment of processes that permit the production of many PNA molecules of high quality at a reasonable cost. Here, we provide a detailed protocol based on our latest developments *(6,7)* for a fast and economical production of PNA microarrays. It combines parallel and automated PNA synthesis at a scale of 0.4 μmol with an on-chip purification technique. The whole process is summarized in **Fig. 1**. Because synthesis takes place on a resin that is placed in filter-bottom wells, the amount of PNA produced can be adapted to the actual needs by adjusting accordingly the amount of resin and the reagent volumes. Synthesis yields suffice for thousands of microarrays. PNA chips produced by this procedure have proven to be stable for a long time and can be reused in hybridization analyses many times over.

2. Materials

2.1. Parallel PNA Synthesis in Microwell Plates

2.1.1. Instrumentation

Synthesis is performed with an AutoSpot robot (INTAVIS Bioanalytical Instruments AG, Cologne, Germany) using 96-well plates with a frit in each well. Quality control is performed on a Reflex II MALDI-TOF mass spectrometer, (Bruker-Daltronik, Bremen, Germany).

2.1.2. Reagents

Rink resin LS (100–200 mesh, substitution of 0.2 mmol/g, *see* also **Note 1**) was obtained from Advanced ChemTech (Louisville, KY). Fmoc PNA monomers, Fmoc-AEEA–OH linker, and O-(7-azabenzotriazol-1-yl)-N,N,N',N'-tetramethyluronium hexafluoro-phosphate (HATU) were supplied by PE Biosystems (Framingham, MA). Fmoc-protected amino acids were obtained from Novabiochem–Calbiochem (Läufelfingen, Switzerland). Dimethylformamide (DMF; SDS, Peypin, France) and 1-methyl-2-pyrrolidone (NMP; Sigma Aldrich, Munich, Germany) were both in a purity grade used for peptide chemistry. N,N-diisopropylethylamine (DIEA), 2,6-lutidine, 1,2-dichloroethane, and

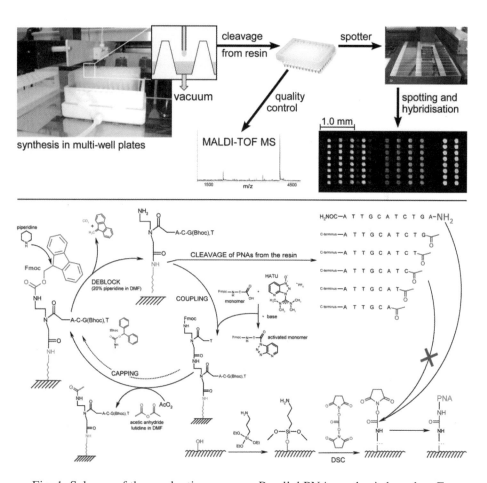

Fig. 1. Scheme of the production process. Parallel PNA synthesis based on Fmoc chemistry takes place in resin-filled microwell plates at a 0.4-µmol scale using an AutoSpot pipetting robot. Synthesis consists of iterative cycles of removing the Fmoc protection group (deprotection), activation of the monomers by adding HATU and base mix, followed by coupling to the growing chain and capping of unreacted, not elongated amino groups with acetic anhydride. Final products are released from the resin and transferred into another microwell plate. After quality control by MALDI-TOF MS, the crude oligomers are spotted onto glass or silicon slides by means of a pin spotter. Surfaces of the slides are silanized and activated with a succinimidyl ester that binds selectively the 5'-terminal primary amino group of the full-length product. The 3'-amino function represents an amide and is less reactive whereas acetylated amino groups of truncated sequences cannot bind at all. Thus purification of the crude products and spotting takes place simultaneously. The resulting microarrays are used in hybridization experiments.

trifluoroacetic acid (TFA) were supplied by Fluka (Steinheim, Germany); triisopropylsilane is from Sigma Aldrich (Munich, Germany).

For automated synthesis, the following solutions are prepared:

1. Fmoc monomer solutions of Fmoc-protected PNAs, amino acids, or AEEA–OH linker, respectively: each 0.3 M in NMP.
2. HATU solution: 0.6 M in DMF.
3. Base mix: 0.6 M DIEA and 0.9 M 2,6-lutidine in DMF.
4. Capping solution: 5% acetic anhydride and 6% 2,6-lutidine in DMF (v/v).
5. Deprotection solution: 20% piperidine in DMF (v/v).
6. Cleavage mixture: 80% TFA/5% triisopropylsilane in 1,2-dichloroethane.
7. Matrix solution for quality control via matrix-assisted laser desorption/ionization time-of-flight (MALDI-TOF MS): 0.7 M 3-hydroxypicolinic acid, 70 mM ammonium citrate in 50% aq. acetonitrile.

Protected from air and moisture, Fmoc monomer solutions can be stored in aliquots at –20°C for at least half a year. Before use, they should be mixed thoroughly because sometimes solubility problems arise, especially for the C monomer. The matrix solution can also be kept in aliquots at –20°C, whereas all other solutions are usually prepared freshly before synthesis.

2.2. Production of PNA Microarrays

2.2.1. Instrumentation

Contact printing device: SDDC-2 DNA Micro-Arrayer (Engineering Services Inc., Toronto, Canada) equipped with SMP3 pins (TeleChem International Inc., Sunnyvale, CA).

2.2.2. Support Media

1. Presliced, thermally oxidized silicon wafers of 2 × 2 cm (GeSiM, Rostock, Germany).
2. Nonderivatized microscope glass slides (Menzel-Gläser, Braunschweig, Germany).

2.2.3. Solutions for Surface Derivatization and Spotting Procedure

1. 10% NaOH (w/w).
2. Silanization mixture: 10 mL (3-aminopropyl)triethoxysilane in 200 mL of 95% aq. ethanol.
3. Activation mixture: 1.5 g of N-N'-disuccinimidyl carbonate and 5 mL of DIEA in 145 mL of dried acetone.
4. Betaine spotting buffer: 1 M betaine in water, pH adjusted to 7.5 with NaOH.
5. Deactivation mixture: 50 mM succinic anhydride, 150 mM 1-methylimidazole in 1,2-dichloroethane.
6. Washing buffer, heated to 90°C: 5 mM sodium phosphate, 0.1% sodium dodecyl sulfate.

2.3. DNA Hybridization

2.3.1. Instrumentation

Slide incubation was performed in hybridization chambers of TeleChem (Sunnyvale, CA) or Hauser Präzisionstechnik (Vöhringen, Germany).

2.3.2. Hybridization Solution

0.1X SSarc buffer: 60 mM sodium chloride, 6 mM sodium citrate, 0.72% (v/v) N-lauroylsarcosine sodium salt solution.

3. Methods

3.1. Parallel PNA Synthesis in Microwell Plates

PNA synthesis is conducted in filter-bottom microwell plates by means of an AutoSpot pipetting robot. All reactions proceed at room temperature. Synthesis runs fully automatically overnight and starts with the Fmoc removal at the rink resin. First, the AutoSpot software is used to calculate the volumes that needed of HATU, the base mix solutions and each of the four monomers (A, G, C, T). The respective reagents are aspired by the dispenser needle and mixed thoroughly in prearranged tubes. After preactivation, the activated monomer solutions are pipetted, one at a time, into the respective wells. Because the robot was originally designed for peptide synthesis, the number of monomer solutions that can be worked with during a run is not limited to four. Therefore, all kinds of molecules, including amino acids, fluorescence labels, or linkers, can easily be introduced into the growing PNA chain using identical reaction conditions.

3.1.1. Synthesis Protocol

1. Prepare the respective solutions for the whole process and place them to the robotic system. This includes all monomer solutions required for the sequences of the desired oligomers, HATU, base mix, deprotection, and capping solutions, as well as DMF and 1,2-dichloroethane for the washing procedures (*see also* **Subheading 2.**).
2. Swell Fmoc-protected rink resin in DMF (20 mg/mL) for 1 h. Mix well and transfer 100 µL of this suspension into each well of the filter-bottom microtiter plate. A vacuum is applied to the plate for the removal of the reagents during synthesis.
3. Enter into the computer the reaction conditions of the synthesis run as well as for each well the desired sequence of the PNA oligomer. The conditions that were found to be optimal for a 0.4 µmol scale synthesis are detailed in **Table 1**.
4. Start the synthesis process.

Table 1
Optimized Reaction Conditions for PNA Synthesis in Microwells

Synthesis procedure	Reagents for each well	Time repetitions
Solid support	100 μL Well rink resin LS (20 mg/mL in DMF)	—
Deprotection	30 μL · 20% peperidine in DMF (v/v) with a DMF washing step in-between	Twice (1 and 5 min)
Washing	80 μL DMF	Five times
Coupling	8 μL of activated monomer solution consisting of 4 μL Fmoc monomer (0.3 M in NMP), 2 μL HATU (0.6 M in DMF), 2 μL base mix (0.6 M in diisopropylamine, 0.9 M 2,6-lutidine in DMF)	Preactivation time: 1 min; coupling time: 30 min; two subsequent couplings with a DMF washing step in-between
Washing	80 μL of DMF	Three times
Capping	30 μL Capping solution (5% acetic anhydride and 6% 2,6-lutidine in DMF)	Five times
Washing	80 μL DMF	Five times
Washing after final Fmoc removal	80 μL DMF 200 μL 1,2-dichloroethane	Five times Two times
Cleavage from the resin	150 μL 80% TFA/5% triisopopylsilane in 1,2-dichloroethane	1 h

5. Synthesis stops with the final removal of the Fmoc group after the last cycle and subsequent washing. Place another microwell plate below the synthesis plate. Cleave the products from the resin by adding to each well 150 µL of cleavage mixture. After incubation, elute the released and completely deprotected oligomers with another 150 µL of the cleavage mixture into the second microwell plate.
6. Precipitate with 1 mL of ice-cold diethyl ether twice and dissolve after complete evaporation of any trades of ether each PNA in 100 µL of water. Store this solution at 4°C until use (*see* also **Note 2**).

3.1.2. Quality Control

1. Calculate the amount of PNAs produced by UV measurement at 260 nm. This allows an early and rapid evaluation of the synthesis performance. A standard synthesis on 2 mg of resin should yield about 100 nmol of PNA product, although the yield varies with the length of the oligomer sequences, of course.
2. Mix 0.5 µL of the diluted PNA solution (1 µL of aliquot in 20 µL of water) with 0.5 µL of matrix directly on the MALDI target and analyse the dried mixture by MALDI-TOF mass spectrometry. Typical spectra obtained in the positive ion mode are shown in **Fig. 2**.

3.2. Production of PNA Microarrays

For the production of PNA microarrays, either glass or thermally oxidized silicone slides are used as solid supports (*see also* **Note 3**). Slides are first silanized, then activated and finally used for the spotting procedure. All reactions are conducted identically for both silicone and glass slides and are performed at room temperature. Reaction conditions given below are used for the simultaneous surface derivatization of 20 slides in a slide holder. Modified aminosilane slides are protected from moisture and stored at 4°C until use. Activation of slides is always done directly before the spotting process. Storage of spotted and deactivated PNA slides is performed at 4°C.

3.2.1. Silanization

1. Wash silicone slides with dimethyl sulfoxide, ethanol, and water.
2. Etch all slides in 10% NaOH (w/w) for 1 h, followed by sonification for 15 min.
3. Wash the slides in water (until the pH is neutral) and ethanol.
4. Immerse the slides in silanization mixture [10 mL of (3-aminopropyl)triethoxysilan in 200 mL 95% aq. ethanol] for 1 h on a shaker and another 15 min in an ultrasonic bath.
5. Finally, wash the slides twice with ethanol, once with water, dry them under a stream of nitrogen, and heat them to 110°C for 20 min.

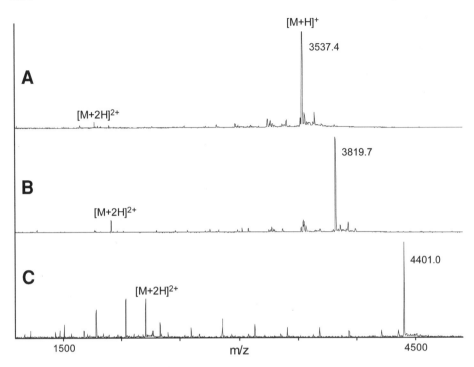

Fig. 2. Typical MALDI mass spectra of crude PNA products synthesized in microwell plates. (**A**) 13-mer (TTGAATCGCTCGA); (**B**) modified 13-mer with lysine and AEEA-OH spacer modification (Lys-AEEA-AGCTTACGGATCA); and (**C**) 13-mer with two AEEA-OH linker molecules (AEEA-AEEA-ACAAATTGCAGGATT).

3.2.2. Activation

1. Incubate aminosilane-derivatized slides with the activation mixture stirring gently for 2 h.
2. Wash twice with dried acetone and twice with 1,2-dichloroethane. After drying with nitrogen, use the succinimidyl ester activated slides directly for PNA spotting.

3.2.3. Spotting, On-Chip PNA Purification, and Deactivation of the Surface

1. Dilute crude PNA products to a concentration of 200 μM in betaine spotting buffer (*see* **Note 4**).
2. Spot the dilutions of the crude PNA products on the activated slides and let them incubate at room temperature overnight.
3. Immerse the slides in the deactivation mixture shaking gently for 2 h.
4. Remove unbound full-length product as well as all byproducts by extensive washing: twice with dichloroethane; twice with washing buffer (5 mM sodium phos-

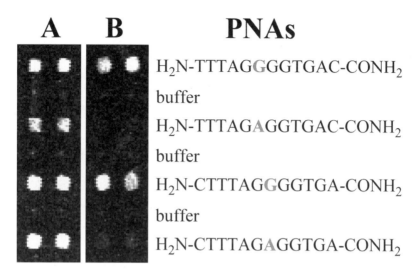

Fig. 3. Detection of a SNP. PNA 12-mers of identical sequence but for a G or A nucleotide at their sixth or seventh position (marked bold), respectively, were spotted in duplicate on succinimidyl activated glass slides, with one row of buffer spots placed in-between. Slide **A** was hybridized with a polymerase chain reaction product of a heterozygous sample and slide **B** with a polymerase chain reaction product of a homozygous sample.

phate, 0.1% sodium dodecyl sulfate) that has been heated to 90°C; once with hot water (90°C) for 10 min, briefly with 1 *M* NaCl in aq. TFA, and finally with pure water.
5. Dry the slides with nitrogen and store them at 4°C or use them directly for hybridization experiments.

3.3. DNA Hybridization

The exact hybridization conditions depend strongly on the sequence of the oligomer probes and the type and complexity of the sample that is being analyzed. Thus, only a set of conditions that is typical for the typing of single nucleotide polymorphisms is presented here. The result of such an analysis is shown in **Fig. 3**.

1. Add 18 µL of of the DNA onto an area of 2 × 2 cm and cover with a cover slip.
2. Incubate in a hybridization chamber at 38°C for 2 h.
3. Wash the slides twice with 0.1X SSarc of the same temperature, rinse with water, and dry with nitrogen.

4. Notes

1. Note that highly loaded resins are not suitable for PNA synthesis because the growing oligomers will aggregate, which is a major reason for synthesis failure. Therefore, resins with a loading capacity higher than the one used here should be avoided.
2. The ether precipitation can be avoided. The PNAs are eluted from the resin with 200 μL of water. After lyophilization, the oligomers are dissolved in 100 μL of water and stored at 4°C. Take care that the pH of the final spotting solution of the PNAs is 7.5. Although the cleaved Bhoc side chain groups are not removed by this procedure, purification takes place by the selective binding during the spotting procedure. Microarray performance is not affected by this simplification.
3. The combination of glass slides and fluorescence detection reflects the current standard procedure. Silicone slides can be used for fluorescence detection also, although being slightly inferior to glass slides. However, they are superior for the detection of DNA binding by mass spectrometry *(6)*. Here, glass slides exhibit charge effects.
4. To obtain homogeneous binding across the entire spot size and therefore homogeneous signal intensity, it is absolute essential to reduce the speed of evaporation of the tiny droplets that are applied to the array surface, thus permitting a longer reaction time across the spot. This can be achieved by using spotting buffers with a high content of salts or addition of reagents that prevent evaporation, such as betaine *(8)*, or both. However, because of the limited solubility of some PNA sequences, especially purine-rich sequences, buffers with high salt content proved to be inadequate, even though they are quite suitable for smaller sequences. Best results for sequences up to 20-mers are obtained with the spotting buffer used here.

Acknowledgments

This work was funded by the German Federal Ministry of Education and Research (BMBF).

References

1. Weiler, J., Gausepohl, H., Hauser, N., Jensen, O. N., and Hoheisel, J. D. (1997) Hybridisation based DNA screening on peptide nucleic acid (PNA) oligonucleotide arrays. *Nucleic Acids Res.* **25,** 2792–2799.
2. Nielsen, P. E., Egholm, M., Berg, R. H. and Buchardt, O. (1991) Sequence-selective recognition of DNA by strand displacement with a thymine-substituted polyamide. *Science* **254,** 1497–1500.
3. Egholm, M., Buchardt, O., Christensen, L., Behrens, C., Freier, S. M., Driver, D. A., et al. (1993) PNA hybridizes to complementary oligonucleotides obeying the Watson-Crick hydrogen-bonding rules. *Nature* **365,** 566–568.
4. Demidov, V. V., Potaman, V. N., Frank-Kamenetskii, M. D., Egholm, M., Buchard, O., Sonnichsen, S. H., et al. (1994) Stability of peptide nucleic acids in human serum and cellular extracts. *Biochem. Pharmacol.* **48,** 1310–1313.

5. Arlinghaus, H. F., Ostrop, M., Friedrichs, O., and Feldner, J. (2003) Genome Diagnostic with TOF-SIMS. *Appl. Surf. Sci.* **203/204,** 689–692.

6. Brandt, O., Feldner, J., Stephan, A., Schröder, M., Schnölzer, M. Arlinghaus, H.F., et al. (2003) PNA-microarrays for hybridisation of unlabelled DNA-samples. *Nucleic Acids Res.* **31,** e119.

7. Jacob, A., Brandt, O., Würtz, S., Stephan, A., Schnölzer, M., and Hoheisel, J.D. (2003) Production of PNA-arrays for nucleic acid detection, in *Peptide Nucleic Acids; Protocols and Applications* (Nielsen, P.E., ed.), Horizon Bioscience, Wymondham, Norfolk, UK, pp, 261–279.

8. Diehl, F., Grahlmann, S., Beier, M., and Hoheisel, J. D. (2001) Manufacturing DNA-microarrays of high spot homogeneity and reduced background signal. *Nucleic Acids Res.* **29,** e38.

22

Synthesis and Characterization of Deoxyribonucleic Acid-Conjugated Gold Nanoparticles

Pompi Hazarika, Tatiana Giorgi, Martina Reibner, Buelent Ceyhan, and Christof M. Niemeyer

Summary

Gold nanoparticles functionalized with thiol-modified single-stranded oligonucleotides are highly useful reagents for a variety of applications, ranging from materials science to bioanalytics. In this chapter, the preparation of citrate stabilized 15-nm Au nanoparticles is described. The nanoparticles are conjugated with 3'-thiol-modified deoxyribonucleic acid oligomers and the resulting conjugates are characterized by determining their shape, size, and surface coverage. The hybridization capabilities are quantified in a microplate assay.

Key words: Gold nanoparticles; thiol-modified DNA oligonucleotides; nucleic acid hybridization; chemisorption; electron microscopy; streptavidin; microplate analysis.

1. Introduction

Gold nanoparticles, functionalized with proteins have long been used as tools in biosciences *(1)*. The most prominent method for synthesis of colloidal gold, the "sodium citrate procedure" developed by Frens *(2)*, allows for the controlled reduction of an aqueous solution of tetrachloroauric acid using trisodium citrate. The citrate not only acts as a reducing agent but also functions as a protective group, forming a negative charged ligand shell surrounding the gold nanoparticle and thereby preventing its aggregation by electrostatic repulsion. Moreover, the citrate shell can be readily substituted by other more aurophilic molecules, such as thiols. For instance, antibody molecules adsorbed to 10- to 40-nm colloidal gold are routinely used in histology, allowing for the biospecific labeling of distinguished regions of tissue samples and subsequent electron microscopy analysis *(1)*.

Although protein-coated gold colloids have long been used in bioanalytical techniques, applications of deoxyribonucleic acid (DNA)-functionalized Au

From: *Methods in Molecular Biology, vol. 283: Bioconjugation Protocols: Strategies and Methods*
Edited by: C. M. Niemeyer © Humana Press Inc., Totowa, NJ

particles have only been introduced recently by Mirkin and co-workers *(3)*. Since their initial description, in numerous follow-up articles have been published that explore their basic properties *(4–6)* and describe how DNA-functionalized Au nanoparticles are used in a variety of applications, ranging from high-sensitivity DNA detection in homogeneous solution *(7)* and on chip-based substrates *(8–11)* to microarray-based protein detection *(12)* and supramolecular constructions on the nanometer length scale *(13–15)*.

In this chapter, the preparation and characterization of DNA-functionalized gold nanoparticles are described. First, ca. 15-nm Au nanoparticles are produced by the citrate method. Second, the particles are conjugated with thiol-modified DNA oligomers to the gold particles and, third, are characterized by determining their surface coverage, as well as by quantitating their hybridization capabilities in a microplate assay.

2. Materials

1. Tetrachloroauric acid trihydrate ($HAuCl_4 \cdot 3H_2O$; Acros).
2. Trisodiumcitrate dihydrate (Fluka).
3. Hellmanex solution (2%; Hellma).
4. Aqua regia (three parts HCl, one part HNO_3).
5. Au nanoparticles (stored at 4°C; ICN Biomedicals).
6. Thiol-, fluorescein-, and unmodified DNA oligonucleotides (Thermo Electron).
7. NAP 5 size exclusion chromatography columns (Amersham Biosciences).
8. 2-Mercaptoethanol.
9. KCN
10. $K_3Fe(CN)_6$.
11. Streptavidin (STV; stored at 4°C; Roche).
12. Microtiter plate (Nalge Nunc International).
13. TE buffer: 10 mM Tris-HCl, pH 7.5; 1 mM ethylenediamine tetraacetic acid.
14. TETBS buffer: 20 mM Tris-HCl, pH 7.35; 150 mM/300 mM NaCl; 5 mM ethylenediamine tetraacetic acid; 0.05% Tween-20.
15. TBS buffer: 20 mM Tris-HCl, pH 7.3; 150 mM NaCl,
16. PBS buffer: 20 mM phosphate buffer, pH 7; 200 mM NaCl.
17. 150 mM $NaNO_3$, pH 7.
18. Herring sperm DNA (HS-DNA; Roche).
19. Bovine serum albumin (BSA; Sigma-Aldrich).
20. Silver enhancing kit (ICN Biomedicals).
21. UV/vis spectrophotometer (Varian).
22. Atomic Absorption Spectroscope (AAS; Perkin-Elmer).
23. Transmission electron microscope (Philips).

3. Methods

The methods described below outline preparation of gold colloids (**Subheading 3.1.**), functionalization of gold nanoparticles with thiolated DNA

(**Subheading 3.2.**), determination of surface coverage of oligonucleotides by means of fluorescence and absorption spectroscopy (**Subheading 3.3.**), and quantification of their hybridization capabilities by DNA-directed immobilization in microplates (**Subheading 3.4.**).

3.1. Preparation of Colloidal Gold Particles

Citrate-protected gold colloids of average diameter 15 nm were synthesized. The synthesis involves following steps.

1. All glassware and stirring devices used in this preparation were cleaned properly to avoid any contamination during the experiment (*see* **Note 1**). The cleaning procedure involves the following steps: (1) all glassware was filled with aqua regia (three parts HCl, one part HNO_3) and left for 30 min; (2) the devices were rinsed thoroughly with ultrapure water; (3) all glassware was sonicated for 30 min in aqueous Hellmanex solution (2%) and rinsed seven to eight times with ultrapure water; and (4) finally, all glassware was dried at 50–70°C.
2. 50 mg of tetrachloroauric acid trihydrate ($HAuCl_4 \cdot 3H_2O$) was dissolved in ca. 5 mL of ultrapure water and added to 250 mL of boiling ultrapure water.
3. To the above solution, a solution of 125 mg of trisodiumcitrate dihydrate ($C_6H_5 Na_3O_7 \cdot 2H_2O$) in ca. 15 mL of ultrapure water was added under vigorous stirring (*see* **Note 2**).
4. After few seconds the solution became faintly blue and then blue color suddenly changed into dark red, indicating the formation of monodisperse spherical particles.
5. The solution was boiled for further 10 min and then cooled slowly to room temperature.

The resulting solution of colloidal particles was characterized by UV-vis spectrophotometry revealing an absorption maximum at 520 nm. The physical characterization of gold nanoparticles' shape and size was performed by transmission electron microscopy (TEM), which indicated a particle size of 14.0 nm ± 1.3 nm (**Fig. 1**).

The concentration of gold nanoparticles was calculated by using the following equation, which assumes the presence of ideal spherical particles *(16)*.

$$n = 3m / 4\pi r^3 s$$

Where n is the amount of gold particles per milliliter, m is the molar mass of gold in substance [g/mL], r is the particle radius [cm], and s is the specific gravity of colloidal gold [19.3 g/cm^3].

The value for m was determined by AAS, and r was determined by TEM. This formula gives the number of gold particles per milliliter. This concentration was then converted into number of gold particles per liter and divided by Avogadro's number (6.023×10^{23}) to get the final molar concentration of gold nanoparticles.

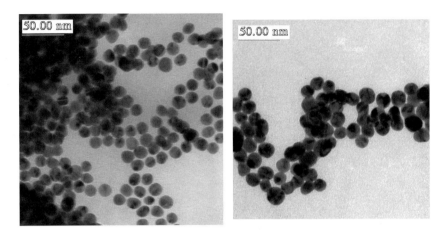

Fig. 1. TEM images of 15-nm Au colloid.

3.2. Functionalization of Gold Nanoparticles With Thiolated DNA

Gold nanoparticles were chemically modified with thiolated oligonucleotides (*see* **Note 3**). The oligonucleotide-modified nanoparticles exhibit an extraordinary high stability in solution containing elevated salt concentrations (300 mM NaCl; *see* **Note 4**), whereas unmodified nanoparticles are not stable in high salt buffers. The functionalization involves the following steps.

1. A solution of the oligonucleotide (e.g., 5'-AAG ACC ATC CTG-thiol-3', 100 μM solution in TE buffer) was previously purified by buffer exchange using a NAP 5 column. The oligonucleotide solution was added to an aqueous nanoparticle solution to a final oligonucleotide concentration of 4 μM.
2. Then to the above mixture, one volume of TETBS buffer containing 300 mM NaCl was added (final concentration of NaCl 150 mM), and the mixture was incubated for 24 h at room temperature.
3. Excess oligonucleotides were removed by centrifugation for 20 min at 20,800g.
4. After removal of supernatant, the red precipitate was washed twice with TETBS buffer containing 300 mM NaCl by repeated centrifugation and resuspension.
5. Finally, the precipitate was redispersed in TETBS buffer containing 300 mM NaCl.

The concentration of DNA-modified Au nanoparticles was determined by UV-visible spectroscopy using an extinction coefficient of $1.47 \times 10^9\ M^{-1}\text{cm}^{-1}$ for 23-nm Au nanoparticles.

3.3. Determination of Surface Coverage

Surface coverage of gold nanoparticles with oligonucleotides was determined by means of two alternative methods based on UV/vis- and fluorescence spectroscopy (*4*).

3.3.1. Determination of Surface Coverage by Fluorescence Spectroscopy

1. For the quantification of the surface coverage of DNA-modified gold nanoparticles, the particles were modified with oligonucleotides containing a fluorescein-modification at the 5'- and an alkanethiol-group at the 3'-end (e.g., fluorescein-5'- AAG ACC ATC CTG-thiol-3') by using the same procedure as described in **Subheading 3.2.** The concentration of DNA-modified gold nanoparticles in this stock solution was calculated by UV/vis spectroscopy using an extinction coefficient of 1.47×10^9 $M^{-1}cm^{-1}$ for 23 nm Au nanoparticles as discussed in **Subheading 3.2.**

2. To the gold nanoparticles coated with the fluorescein-labeled oligonucleotides, resuspended in TETBS buffer containing 300 mM NaCl, mercaptoethanol was added to a final concentration of 1.3 M. Then, the solution was incubated for 24 h at room temperature. Mercaptoethanol displaces the oligonucleotides from the gold nanoparticle's surface by an exchange reaction. The solution containing the displaced oligonucleotides was then separated from gold nanoparticles by centrifugation for 20 min (20,800g).

3. The fluorescence of the supernatant containing the displaced fluorescein modified oligonucleotides was measured. Fluorescence emission was used to calculate the number of oligonucleotides linked to the gold nanoparticle surface, and thus, the surface coverage of the particles with oligonucleotides. To this end, fluorescence units were converted to molar concentrations of the fluorescein-modified oligonucleotides by interpolation from a standard linear calibration curve. The standard calibration curve was prepared from known concentrations of fluorescein-modified oligonucleotides. Owing to sensitivity of optical properties of the fluorescein to different pH, ionic strength etc. identical experimental conditions, i.e., the same buffer, salt- and mercaptoethanol concentrations were used.

4. Then molar concentrations of the oligonucleotides measured were divided by the concentration of the oligonucleotide-modified gold nanoparticles of the stock solution, originally determined by UV/vis spectrophotometry (*see* **Subheading 3.3.1., step 1**) to calculate the average number of oligonucleotides linked to each nanoparticle. Finally, the measured oligonucleotides number was divided by particle surface area (determined by TEM), in the nanoparticle solution to get the normalized surface coverage value (*see* **Note 5**).

3.3.2. Determination of Surface Coverage by UV/vis Spectrophotometry

Surface coverage of gold nanoparticles with oligonucleotides was also determined using UV/vis spectrophotometry. For these studies fluorescein-labeled oligonucleotides modified gold nanoparticles also were used. A mixture of KCN/K_3Fe(CN)$_6$ was used to dissolve the gold nanoparticles containing the fluorescein-labeled oligonucleotide, by adding solutions of KCN (0.2 M) and K_3Fe(CN)$_6$ (2 mM) in water to the particles, present in TETBS buffer containing 300 mM NaCl. The final concentration of KCN was 0.08 M and K_3Fe(CN)$_6$ 0.8 mM. After about 30 min, absorbance measurements of the resulting pale yellow solution showed absorption bands at 494 nm (because of

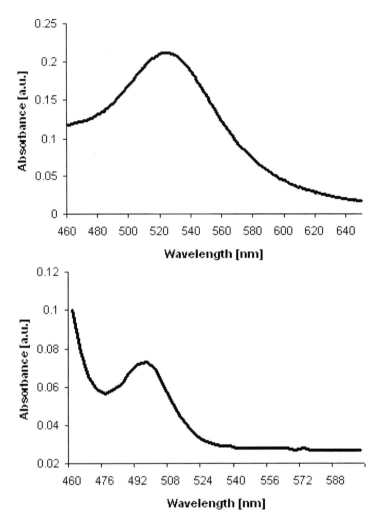

Fig. 2. UV/vis absorbance spectra of gold nanoparticles containing fluorescein-labeled oligonucleotides before (top) and after (bottom) the addition of KCN/ $K_3Fe(CN)_6$.

fluorescein oligonucleotides) and 420 nm (because of excess ferricyanide), whereas the nanoparticles surface plasmon resonance band at 526 nm had completely disappeared (**Fig. 2**). Standard linear absorption vs concentration calibration curve was prepared by using known concentrations of fluorescein-labeled oligonucleotides, under the same experimental conditions. Then, normalized surface coverage values were calculated as described in **Subheading 3.3.1.**

3.4. DNA-Directed Immobilization (DDI)

DDI allows the attachment of molecular components to surfaces in a site-selective and reversible manner *(15)*. Site selectivity is the result of the unique specificity of Watson–Crick base pairing of complementary nucleic acids. In this section, STV-coated microplates are functionalized with biotinylated capture oligonucleotides and are subsequently used for specific immobilization of the DNA-nanoparticle conjugates (*see* **Note 6**). The DDI assay involves the following steps.

3.4.1. Functionalization of Wells of Microtiter Plate With STV

STV is a 60,000-Dalton tetrameric protein composed of four identical subunits, each of which has a binding site for biotin. Coating of microplates with STV involves two steps.

1. 50 µL of STV solution in phosphate-buffered saline (final concentration 0.19 pmol/µL) was poured in each well of microtiter plate and then centrifuged the plate for 1 min (2250*g*) to better fill the wells with STV solution (*see* **Note 7**). The plate was then covered with aluminum foil and incubated for 2 d at 4°C.
2. Wells of the plate were washed with TBS three times' with each time taking ca. 1 min and, after that, 150 µL blocking buffer, HS-DNA/BSA solution (0.1 mg/mL) was added to each well. Then plate was covered with aluminum foil and incubated at 4°C for 1 d (*see* **Note 8**).

3.4.2. Immobilization of Capture Oligonucleotide

Biotinylated single-stranded DNA oligonucleotides were used as capture oligonucleotide, which tightly bind to the wells of STV-coated microtiter plates. It involves following steps.

1. Wells of the microtiter plate were washed with TETBS buffer containing 150 m*M* NaCl four times (twice for 30 s and twice for 5 min).
2. 50 µL of the solution of the biotinylated oligonucleotide (e.g., biotin-5'-GGT GAA GAG ATC-3') in TETBS buffer containing 150 m*M* NaCl (final concentration 0.24 µ*M*) were added to each well of the microplate and incubated for 30 min with continuous shaking at slow speed.

3.4.3. Addition of Linker Oligonucleotides

An unmodified DNA oligonucleotide (e.g., 5'-CAG GAT GGT CTT GAT CTC TTC ACC-3') containing two sequence stretches that are complementary to the immobilized capture strand as well as to the nanoparticle-bound DNA was used as linker. Excess of linker oligonucleotides was used to saturate the microplate-bound capture oligonucleotides:

1. The plate was washed four times as described above.

2. 50 µL of a solution of the linker oligonucleotide in TETBS buffer containing 300 mM NaCl (final concentration 0.2 µM) was added to the wells and incubated for 30 min with continuously shaking at slow speed.

3.4.4. Hybridization of DNA–Gold Nanoparticles

DNA-modified gold nanoparticles containing a DNA sequence complementary to the single-stranded region of the linker sequence were added to assemble the final adsorbate (*see* **Note 9**).

1. Wells of the microplate were washed four times as described previously.
2. 50 µL of blocking buffer, HS-DNA/BSA solution (final concentration 0.01 mg/mL) was added to each well.
3. 50 µL of the DNA-modified gold nanoparticles in TETBS buffer containing 300 mM NaCl was added to each well and incubated for 1 h under continuous slow shaking.

3.4.5. Silver Development

In this final step, signals of nanoparticle hybridization were generated by reduction of silver ions using hydroquinone. The reduction preferentially takes part at the surface of the gold nanoparticles. Though nanoparticles initiate the silver deposition, the silver precipitate also catalyzes further silver reduction (*see* **Note 10**). This process involves the following steps.

1. The wells of the plate were washed four times as discussed previously.
2. The wells were rinsed with 150 mM NaNO$_3$ three times, each time for about 1 min, to remove chloride ions.
3. 50 µL of the silver amplification solution was added to each well to get enhanced signals.

The growth of particles and the precipitation of silver increase with the increasing time of exposure to silver amplification solution, up to about 30 min. After this period, the process stops, and no further growth of particles and silver precipitate is observed. Absorbance signals were obtained by measuring at 490 nm by Synergy HT after regular time intervals.

4. Notes

1. During preparation of colloidal gold particles, all experimental devices should be cleaned properly because traces of metals or salts immediately cause aggregation of particles. Colloidal gold solutions are light sensitive and should be stored in dark bottles.
2. Different-sized gold colloids can be prepared by the same procedure by changing only the amount of citrate solution added (*2*).
3. During functionalization of nanoparticles with oligonucleotides, a high DNA surface density on the nanoparticle is advantageous in terms of particle stabilization, especially at elevated salt concentration (*6*).

4. DNA-modified gold nanoparticles are very stable, even at room temperature, for about 2–3 mo.
5. For the determination of the surface coverage of oligonucleotide-modified gold nanoparticles, one may also use the concentration of precipitated gold nanoparticles, by the addition of mercaptoethanol. This concentration can be readily determined by AAS. Using this method, particle concentrations that are slightly lower than the actual concentrations are obtained because of a partial loss of particles during centrifugation.
6. Similar DDI processes can as well be carried out using other solid surfaces, such as glass. This process can be used to form supramolecular multilayered nanostructures *(10,17)*.
7. During functionalization of the microplate wells with STV, care should be taken to fill the entire well with the STV solution to obtain a uniform STV coating.
8. The STV-coated microtiter plates should be stored in blocking buffer until further use.
9. Although for our studies we used Au nanoparticles, one may as well use other nanoparticle compositions, provided they are heavily functionalized with oligonucleotides.
10. The silver development step should be conducted in dark because of the light sensitivity of silver.

References

1. Kreuter, J. (1992) Introduction and overview, in *Microcapsules and Nanoparticles in Medicine and Pharmacy* (Donbrow, M., ed.), CRC Press, Boca Raton, FL.
2. Frens, G. (1973) Controlled nucleation for the regulation of the particle size in monodisperse gold suspensions. *Nat. Phys. Sci.* **241**, 20–22.
3. Mirkin, C. A., Letsinger, R. L., Mucic, R. C., and Storhoff, J. J. (1996) A DNA-based method for rationally assembling nanoparticles into macroscopic materials. *Nature* **382**, 607–609.
4. Demers, L. M., Mirkin, C. A., Mucic, R. C., et al. (2000) A fluorescence-based method for determining the surface coverage and hybridization efficiency of thiol-capped oligonucleotides bound to gold thin films and nanoparticles. *Anal. Chem.* **72**, 5535–5541.
5. Storhoff, J. J., Lazarides, A. A., Mucic, R. C., et al. (2000) What controls the optical properties of DNA-linked gold nanoparticle assemblies? *J. Am. Chem. Soc.* **122**, 4640–4650.
6. Jin, R., Wu, G., Li, Z., et al. (2003) What controls the melting properties of DNA-linked gold nanoparticle assemblies? *J. Am. Chem. Soc.* **125**, 1643–1654.
7. Reynolds, R. A., Mirkin, C. A., and Letsinger, R. L. (2000) Homogeneous, nanoparticle-based quantitative colorimetric detection of oligonucleotides. *J. Am. Chem. Soc.* **122**, 3795–3796.
8. Taton, T. A., Mirkin, C. A., and Letsinger, R. L. (2000) Scanometric DNA array detection with nanoparticle probes. *Science* **289**, 1757–1760.

9. Taton, T. A., Lu, G., and Mirkin, C. A. (2001) Two-color labeling of oligonucleotide arrays via size-selective scattering of nanoparticle probes. *J. Am. Chem. Soc.* **123**, 5164–5165.

10. Niemeyer, C. M., Ceyhan, B., Gao, S., et al. (2001) Site-selective immobilization of gold nanoparticles functionalized with DNA oligomers. *Colloid Polym. Sci.* **279**, 68–72.

11. Park, S. J., Taton, T. A., and Mirkin, C. A. (2002) Array-based electrical detection of DNA with nanoparticle probes. *Science* **295**, 1503–1506.

12. Niemeyer, C. M. and Ceyhan, B. (2001) DNA-directed functionalization of colloidal gold with proteins. *Angew. Chem. Int. Ed.* **40**, 3685–3688.

13. Mucic, R. C., Storhoff, J. J., Mirkin, C. A., and Letsinger, R. L. (1998) DNA-directed synthesis of binary nanoparticle network materials. *J. Am. Chem. Soc.* **120**, 12674–12675.

14. Mirkin, C. A. (2000) programming the assembly of two- and three-dimensional architectures with DNA and nanoscale inorganic building blocks. *Inorg. Chem.* **39**, 2258–2272.

15. Niemeyer, C. M. (2001) Nanoparticles, proteins, and nucleic acids: biotechnology meets materials science. *Angew. Chem. Int. Ed.* **40**, 4128–4158.

16. Ackerman, G. A., Yang, J., and Wolken, K. W. (1983) Differential surface labeling and internalization of glucagon by peripheral leukocytes. *J. Histochem. Cytochem.* **31**, 433–440.

17. Taton, T. A., Mucic, R. C., Mirkin, C. A., and Letsinger, R. L. (2000) The DNA-mediated formation of supramolecular mono- and multilayered nanoparticle structures. *J. Am. Chem. Soc.* **122**, 6305–6306.

23

Biofunctionalization of Carbon Nanotubes for Atomic Force Microscopy Imaging

Adam T. Woolley

Summary

The study of biological processes relies increasingly on methods for probing structure and function of biochemical machinery (proteins, nucleic acids, and so on) with submolecular resolution. Atomic force microscopy (AFM) has recently emerged as a promising approach for imaging biological structures with resolution approaching the nanometer scale. Two important limitations of AFM in biological imaging are (1) resolution is constrained by probe tip dimensions, and (2) typical probe tips lack chemical specificity to differentiate between functional groups in biological structures. Single-walled carbon nanotubes (SWNTs) offer an intriguing possibility for providing both high resolution and chemical selectivity in AFM imaging, thus overcoming the enumerated limitations. Procedures for generating SWNT tips for AFM will be described. Carboxylic acid functional groups at the SWNT ends can be functionalized using covalent coupling chemistry to attach biological moieties via primary amine groups. Herein, the focus will be on describing methods for attaching biotin to SWNT tips and probing streptavidin on surfaces; importantly, this same coupling chemistry can also be applied to other biomolecules possessing primary amine groups. Underivatized SWNT tips can also provide high-resolution AFM images of DNA. Biofunctionalization of SWNT AFM tips offers great potential to enable high-resolution, chemically selective imaging of biological structures.

Key Words: Scanning probe microscopy; chemical force microscopy; chemically sensitive imaging; single-walled carbon nanotubes; nanotube tips; biotin; streptavidin; carbodiimide coupling; DNA.

1. Introduction

Studying and understanding biological processes at the nanometer scale is an area of expanding significance in the postgenomic era. Two longstanding methods for structural determination, X-ray diffraction (1,2) and nuclear magnetic resonance (3,4) offer information about atomic positions in biological

From: *Methods in Molecular Biology, vol. 283: Bioconjugation Protocols: Strategies and Methods*
Edited by: C. M. Niemeyer © Humana Press Inc., Totowa, NJ

macromolecules. Although these approaches are highly effective in many circumstances, they also suffer from drawbacks, such as the need for a pure, crystalline sample in X-ray diffraction, and the inability to scale to very large (>100 kDa) molecules in nuclear magnetic resonance. Atomic force microscopy (AFM) (5,6) is an alternative strategy for probing biological structure, albeit with nanometer, as opposed to atomic resolution. AFM is a powerful tool for characterization of biomolecules (7,8), both under ambient conditions and in aqueous solutions that approximate an in vivo environment (9). Two chief drawbacks of AFM imaging, limited resolution and insufficient chemical selectivity, are both intrinsic to the probe tips themselves. The recent introduction of carbon nanotubes as probe tips for AFM (10) opened up a new avenue for improving both resolution (11) and chemical selectivity (12) in tips. In particular, single-walled carbon nanotubes (SWNTs), which have end radii in the subnanometer range and can be as small as 0.25 nm (13), enable high-resolution characterization of biomolecular structures (11,14–16). Moreover, because carbon nanotubes are macromolecules, they can be specifically functionalized using synthetic chemistry strategies. Indeed, oxidized ends of carbon nanotubes contain carboxyl groups, which can be derivatized readily with primary amine containing molecules, using carbodiimide coupling chemistry. The ability to control precisely the chemical functionality on nanotube tips allows high-resolution, chemically sensitive imaging in AFM (12,17–20) and offers significant potential for enhancing biomolecular characterization. The preparation, derivatization, and implementation of biotin functionalized SWNT AFM tips will be described as a general model for creating biofunctionalized nanotube probes and using them in chemically selective imaging. The application of underivatized SNWT probes to high-resolution AFM imaging of DNA will also be outlined.

2. Materials

1. AFM tips (FESP, Veeco, Sunnyvale, CA).
2. AFM instrumentation.
3. SWNTs (see **Note 1**).
4. 0.8-μm pore diameter filters (Millipore, Billerica, MA).
5. Combination micropositioning and optical microscopy system having a ×20 or ×40 long-working-distance objective.
6. Conductive carbon tape (Electron Microscopy Sciences, Fort Washington, PA).
7. UV-cure adhesive (Loctite 3105, Henkel-Loctite Corp., Rocky Hill, CT).
8. 0.0005-inch diameter tungsten wire (WireTronic, Inc., Pine Grove, CA).
9. Handheld UV lamp.
10. Sputtered Nb surface (Electron Microscopy Sciences).

11. Field emission scanning electron microscope (SEM).
12. 0.1 *M* 2-[*N*-morpholino]ethanesulfonic acid (MES) buffer, pH 6.0 (Sigma).
13. 1-ethyl-3-(3-dimethylaminopropyl) carbodiimide hydrochloride (EDC; Pierce, Rockford, IL).
14. 5-(biotinamido)pentylamine (5-BAPA; Pierce).
15. 0.1 *M* NaCl.
16. Phosphate-buffered saline (PBS), pH 5.6.
17. Biotinamidocaproyl-labeled bovine serum albumin (b-BSA; Sigma).
18. Streptavidin.
19. PBS, pH 7.0.
20. 0.2-μm pore diameter syringe filters (Millipore).
21. Mica.
22. 0.1% Poly-L-lysine (Ted Pella, Inc., Redding, CA).
23. Lambda DNA (New England Biolabs, Beverly, MA).
24. 10 m*M* Tris-HCl, 1 m*M* ethylenediamine tetraacetic acid (EDTA), pH 8.0 (TE; Fisher).
25. 1-inch diameter glass coverslips (Fisher).

3. Methods

In this section, techniques are outlined for fabrication of biotin functionalized SWNT tips for probing streptavidin with high resolution and chemical selectivity. In addition, methods for using unfunctionalized probes for high-resolution imaging of deoxyribonucleic acid (DNA) are detailed. The description is divided into the following sections: (1) SWNT attachment and reinforcement on AFM tips, (2) biotin functionalization of SWNT tips and streptavidin surface preparation, (3) specific probing of streptavidin with biotin modified SWNT tips, and (4) high-resolution AFM imaging of biomolecules using underivatized SWNT probes.

3.1. Nanotube Tip Preparation

The construction of AFM probe tips with protruding SWNTs is discussed in **Subheadings 3.1.1.–3.1.3.** and is shown schematically in **Fig. 1**. The procedure for making SWNT tips involves (1) attachment of a small SWNT bundle to a Si tip (*see* **Note 2**), (2) reinforcement of the SWNT bundle–tip pyramid junction, and (3) optimization of SWNT bundle length.

3.1.1. SWNT Bundle Attachment to AFM Tips

1. Purify SWNTs by using ultrasonication to suspend them at a concentration of approx 1 mg/mL, then filter the suspension through a 0.8-μm pore diameter membrane and allow the solvent to evaporate (*see* **Note 3**).
2. Place a small quantity (<1 mm³) of dry, purified SWNTs on a three-axis micromanipulator system on an optical microscope (**Fig. 1A**).

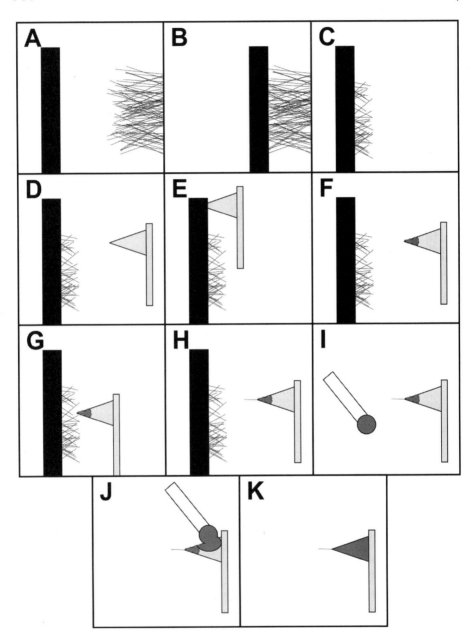

Fig. 1. Schematic of SWNT tip preparation. (**A**) Carbon tape with a sharply folded edge (left) and purified SWNT material (right) are mounted on micromanipulators in an optical microscope. (**B**) The carbon tape and SWNT material are brought into contact. (**C**) Some bundles of SWNTs remain affixed to the carbon tape after the SWNT material is pulled away. (**D**) An AFM tip (right) is mounted on the one of the micropositioners.

3. Place a folded piece of conductive carbon tape on a separate three-axis micropositioning system on the same optical microscope (**Fig. 1A**). The carbon tape should be folded to have as abrupt an edge as possible.

4. Using dark field illumination through a ×20 or preferably ×40 long-working-distance objective in the microscope to view the process, move the SWNT material into contact with the folded edge of the carbon tape (**Fig. 1B**) and then separate them. Bundles of SWNTs should be protruding from the folded edge of the adhesive tape (**Fig. 1C**).

5. Remove the purified SWNT material and then place an AFM tip on the now-available micropositioning setup (*see* **Note 4** and **Fig. 1D**).

6. Under dark field illumination of the microscope for viewing, use the micromanipulator to move the AFM tip into contact with the adhesive tape in a region where few SWNT bundles are present (**Fig. 1E**) and then move the tip back a short distance (**Fig. 1F**). This step transfers adhesive from the carbon tape to the pyramid of the AFM tip (*see* **Note 5**).

7. While still observing under dark field illumination, use the micromanipulator to brush the AFM tip pyramid gently against a protruding SWNT bundle (**Fig. 1G**) until it becomes attached to the tip (**Fig. 1H**). Successful SWNT bundle attachment can be confirmed under dark field illumination in the microscope if a thin bright line (SWNT bundle) is observed extending from the end of the tip pyramid (**Fig. 2**).

3.1.2. Reinforcement of the SWNT Bundle to Tip Pyramid Junction

For results that best reflect in vivo structure and functionality, chemically sensitive imaging should be performed in fluid. Many AFM instruments oscillate the fluid contents of the liquid cell to vibrate the AFM tip for intermittent contact (tapping mode) imaging. Because this oscillation can separate weakly attached SWNT bundles from the tip pyramid, the junction between the SWNT bundle and tip pyramid should be reinforced for fluid imaging, as outlined in this section.

1. Remove from the micromanipulator the carbon tape with SWNT bundles protruding from the folded edge, and replace it with a 0.0005-inch diameter tungsten

---

Fig. 1. (*continued*) (**E**) The AFM tip is brought into contact with the carbon tape edge in an area where few SWNTs are present. (**F**) After the tip is pulled away from the carbon tape, some adhesive remains on the pyramid. (**G**) The AFM tip is brushed against a SWNT bundle on the carbon tape. (**H**) A SWNT bundle sticks to the adhesive on the pyramid and is pulled away from the carbon tape. (**I**) A tungsten wire with a droplet of Loctite 3105 adhesive (left) is mounted on one of the micromanipulators. (**J**) The wire end is touched against the AFM tip, transferring adhesive to the pyramid. (**K**) After the adhesive is cured the SWNT bundle to AFM tip junction has been reinforced.

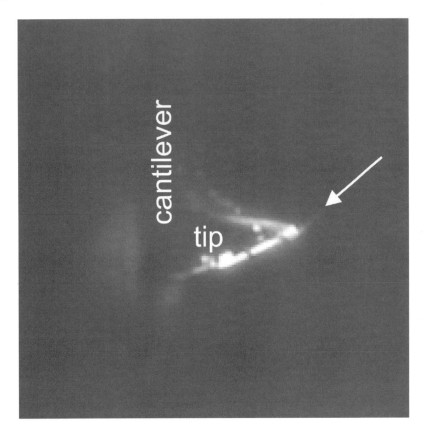

Fig. 2. Dark-field illumination optical micrograph of a SWNT bundle attached to a conventional AFM tip. The SWNT bundle is the light linear feature extending from the tip of the Si pyramid toward the white arrow.

wire having a droplet of Loctite 3105 UV-cure adhesive at the end (*see* **Note 6** and **Fig. 1I**).

2. While viewing under dark field illumination in the microscope, use the micropositioners to bring the wire end into contact with the AFM tip pyramid; be careful to avoid knocking off the attached SWNT bundle. Some adhesive should flow onto the tip pyramid (**Fig. 1J**) and help to strengthen the SWNT attachment to the AFM tip.

3. Cure the adhesive for 30 min under a handheld UV lamp (**Fig. 1K**).

4. Transfer the SWNT AFM tip to a field emission SEM, and focus down to a small (<1 μm²) area where the SWNT bundle to tip pyramid junction is visible. Repeatedly scan the same region for approx 5 min. This step deposits amorphous carbon on top of the SWNT bundle to tip pyramid junction to further strengthen attachment. Repeat this step in additional regions on the tip pyramid for greater stability.

3.1.3. Optimization of SWNT Tip Length

Attached SWNT bundles are often too long and flexible for high-resolution imaging, so SWNT tips are shortened to a protruding length of <1 μm, as described in this section.

1. Place an AFM tip with attached SWNT in the tip holder, engage the AFM system in intermittent contact (tapping) mode, and image a sputtered Nb substrate.
2. Switch to the instrument mode where the tip cycles between approaching toward and retracting from the surface. Set the z-scan range to approx 200 nm and adjust the z-position of the tip such that the tip only contacts the surface during the final 10–50 nm of each approach cycle (*see* **Note 7**).
3. Apply 5–20 V between the tip and surface; the tip-sample separation distance should increase, indicative of ablation of the terminal portion of the SWNT bundle.
4. Switch back to imaging the sputtered Nb surface; well-resolved Nb grains of approx 10 nm dimensions in the image indicate that the tip length is acceptable (**Fig. 3**).
5. Repeat **steps 2–4** as needed until the desired imaging resolution is achieved (*see* **Note 8**).
6. Tip functionalization (**Subheading 3.2.2.**) should be performed within a few days of SWNT length optimization. However, for high-resolution imaging with underivatized SWNT tips (**Subheading 3.4.**), the shortened probes can be stored for at least several weeks before use.

3.2. Biotin Functionalization of SWNT Tips and Streptavidin Surface Preparation

The procedures for functionalizing SWNT AFM tips with biotin and preparing a streptavidin surface are described below in **Subheadings 3.2.1.–3.2.3.** In summary, this process requires (1) preparing the necessary solutions, (2) carrying out the functionalization reactions on SWNT tips, and (3) preparing a streptavidin surface for study.

3.2.1. Solution Preparation

The MES and PBS buffer solutions listed in **Subheading 2.** are stable for a period of several months if sterilized.

1. Make a 10-mL solution containing 50 mM EDC and 5 mM 5-BAPA in MES buffer. This solution should be used as soon as it is prepared, and fresh solution should be made daily.
2. Prepare 5 mL of a solution with 30 μg/mL streptavidin in PBS, pH 7.0. Aliquot and store this solution either refrigerated (for as long as 1 mo) or frozen at −20°C (for several months). If the solutions are frozen, avoid multiple freeze-thaw cycles.

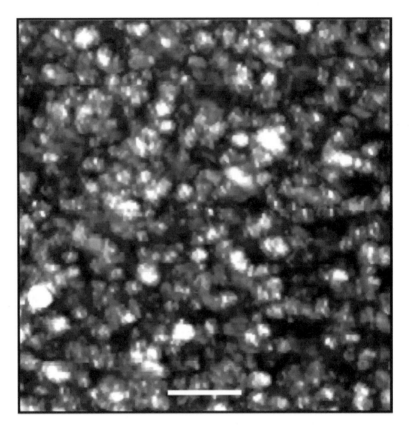

Fig. 3. AFM height image of a Nb surface taken with a SWNT tip after length optimization. Granular Nb features are clearly resolved, and the white bar represents 100 nm.

3. Prepare a solution with 250 µg/mL b-BSA in PBS pH 5.6; this solution should be refrigerated (for up to 1 mo).

3.2.2. SWNT Tip Functionalization

In this step carbodiimide coupling is used to form a covalent bridge between carboxyl groups on the SWNT tip and the amine group in 5-BAPA, generating a biotin functionalized probe.

1. Filter the solution with EDC and 5-BAPA in MES through a 0.2-µm syringe filter into a 1-inch diameter Petri dish.
2. Add 5 mL of deionized water to a 1-inch diameter Petri dish, 5 mL of MES buffer filtered through a 0.2-µm syringe filter to another 1-inch diameter Petri dish, and 5 mL of 0.1 M NaCl filtered through a 0.2-µm syringe filter to a different 1-inch diameter Petri dish.

3. Use tweezers to place one or more SWNT AFM tips in the bottom of the Petri dish containing the EDC and 5-BAPA in MES for 2 h (*see* **Notes 9** and **10**).
4. Rinse each SWNT probe by transferring the AFM tips successively into the Petri dishes containing MES buffer solution, 0.1 *M* NaCl, and deionized water. Each tip should be placed in each rinse solution for at least 2 min.
5. Allow the tips to dry after the water rinse step. The biotin-derivatized SWNT tips are now ready for use in probing streptavidin surfaces. Generally, functionalized tips should be used within a day of preparation, because their stability for longer time periods has not been studied exhaustively.

3.2.3. Surface Preparation

In this operation, freshly cleaved mica has a thin film of b-BSA adsorbed on the surface. Streptavidin is then complexed with biotin groups in the adsorbed b-BSA to create an upper layer of streptavidin molecules on the surface (*see* **Note 11; ref. *21***).

1. Place a freshly cleaved mica surface in a 1-inch diameter Petri dish containing 5 mL of b-BSA in PBS buffer, pH 5.6, for 2 h.
2. Rinse the mica in PBS, pH 7.0, buffer for at least 2 min.
3. Place the mica into a 1-inch diameter Petri dish having 5 mL of 30 µg/mL streptavidin in PBS buffer, pH 7.0, for 2 h.
4. Rinse the mica again in PBS buffer, pH 7.0, for at least 2 min. The streptavidin-coated mica surfaces should be stored in PBS buffer, pH 7.0, and used within a day of preparation.

3.3. Imaging and Measuring Forces With Biofunctionalized SWNT Probes

The methods used in obtaining AFM images and making force measurements with biotin modified SWNT tips are detailed in **Subheadings 3.3.1.–3.3.2.** Briefly, the important parts of this process are (1) setting up the AFM system and obtaining images, and (2) measuring the force necessary to pull apart biotin–streptavidin complexes.

3.3.1. Setup and AFM Imaging

Experiments to study the interaction between biofunctionalized SWNT tips and surfaces are typically performed in solution, so AFM instrumentation with a liquid cell is needed.

1. Place the biotin functionalized SWNT probe into the liquid cell tip holder, use the streptavidin surface as the substrate, and manually approach the tip toward the surface.
2. When the tip and surface are separated by less than several hundred microns, carefully fill the fluid cell with PBS buffer, pH 7.0, making sure that no bubbles are present in the liquid cell (*see* **Note 12**).

3. After the fluid cell is filled, select a scan size of zero area, and engage the AFM instrument in intermittent contact (tapping) mode. The zero area scan size prevents lateral forces (present during scanning) from removing the SWNT bundle from the pyramid if the tip approaches the surface too closely.

4. Once the AFM system is properly engaged, change the scan size to 1 μm × 1 μm and verify that the instrument is imaging well. Increase the scan area to the desired size if the 1 μm × 1 μm image is acceptable.

3.3.2. Probing Pull-Off Forces

In many cases it is desirable to measure tip-surface pull off forces, instead of obtaining surface images. In this case, the initial setup follows **Subheadings 3.3.1.1.–3.3.1.3.** Once the tip has reached the surface and imaging commences, the steps below should be taken.

1. Switch from imaging to the instrument mode where the tip cycles between approaching toward and retracting from the surface.

2. Lower the drive amplitude (or comparable parameter that controls tip oscillation in intermittent contact imaging) to zero, set the z-scan range to approx 200 nm, and adjust the z-position of the tip such that the tip only contacts the surface during the final 10–50 nm of each approach cycle (*see* **Note 7**).

3. Save the tip deflection vs z-distance data for as many cycles of contacting and then pulling the tip off the surface as are desired (*see* **Notes 13** and **14**).

3.4. High-Resolution Imaging of DNA Using Underivatized SWNT Probes

The methods for using unfunctionalized, length-optimized SWNT AFM tips in high-resolution DNA imaging are described in **Subheadings 3.4.1.–3.4.3.** In summary, this procedure requires (1) making of the necessary solutions, (2) preparing a DNA surface for study, and (3) AFM imaging.

3.4.1. Solution Preparation

The TE buffer solution listed in **Subheading 2.** is stable for a period of several months if sterilized.

1. Dilute the lambda DNA solution in TE buffer to a concentration of approx 1 ng/μL. DNA solutions should be aliquoted and stored at –20°C (for several months); multiple freeze-thaw cycles should be avoided.

2. Serially perform three 10-fold dilutions of the 0.1% poly-L-lysine solution in water to achieve a final concentration of 1 part per million (ppm) poly-L-lysine. Be sure to mix each diluted solution thoroughly before performing the next dilution. The diluted poly-L-lysine solutions should be prepared from the concentrated stock solution at least weekly.

3.4.2. Surface Preparation

In this step, DNA is aligned onto freshly cleaved mica treated with poly-L-lysine. The positively charged poly-L-lysine is bound electrostatically to the mica substrate, and the negatively charged DNA is held on the surface through interactions with the affixed poly-L-lysine *(22,23)*.

1. Place approx 25 μL of 1 ppm poly-L-lysine on a freshly cleaved mica substrate for 5 min; remove the solution, rinse with water, and then dry the mica surface under a gentle stream of compressed air.
2. Set the poly-L-lysine treated mica surface on a three-axis micromanipulation stage having a 1-inch diameter glass coverslip positioned orthogonally and above the mica surface.
3. Pipet a 1 μL droplet of 1 ng/μL lambda DNA solution onto the mica surface and use the micromanipulator to lower the cover slip edge sufficiently close to the mica surface (<1 mm) to hold the droplet in place by surface tension.
4. Translate the mica surface relative to the cover slip at a linear speed of approx 2 mm/s. The moving droplet should leave a 1–4-mm long liquid trail, which evaporates in <3 s. This step deposits DNA aligned on the surface in the direction of droplet motion.
5. Rinse the surface with water and dry the mica under a gentle stream of compressed air. Mica substrates with aligned DNA can be stored for months in a low humidity (<20%) environment, and minimal degradation of the surface DNA is observed.

3.4.3. Setting Up and AFM Imaging

High-resolution imaging of DNA on surfaces using SWNT tips can typically be performed under ambient conditions, instead of requiring use of the fluid cell tip holder.

1. Place the unfunctionalized SWNT probe into the tip holder, with the aligned DNA surface as the substrate, and engage the AFM equipment in intermittent contact (tapping) mode.
2. Once the AFM system is properly engaged, select the desired scan range and image the surface (*see* **Fig. 4** and **Notes 15** and **16**).

4. Notes

1. SWNTs can either be made in house by laser vaporization *(24)* or purchased from a range of commercial sources *(25)*. Although nanotubes from these companies have not been tested exhaustively for efficacy in making biofunctionalized probes, it is likely that most commercially available SWNTs are suitable for mounting on AFM tips and subsequent biofunctionalization.
2. Multiwalled carbon nanotubes attached to conventional AFM tips can be purchased from a commercial source (nPoint, Madison, WI). Several other approaches for attachment of SWNTs to AFM tips have also been demonstrated, including chemical

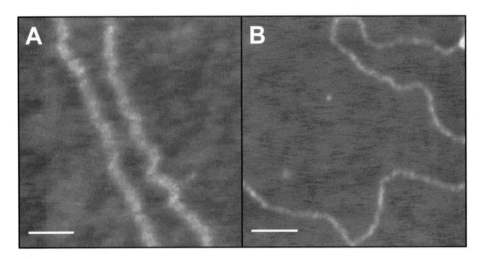

Fig. 4. AFM height images of lambda DNA aligned on mica. (**A**) Image taken with a conventional Si tip. (**B**) Image taken with a SWNT tip. The white bar represents 50 nm in each image.

vapor deposition growth directly on tips *(16,26,27)*, and "picking up" a SWNT on a tip by imaging a surface having protruding nanotubes *(23,28)*.

3. Commercially available SWNTs that have already been treated with acid or filtered may be of sufficient purity to make the filtration step unnecessary.

4. It is best to prepare a batch of 5–10 SWNT tips in a group because of the setup time in getting SWNT bundles on the carbon tape.

5. Directly coating the tip pyramid with Loctite 3105 UV-cure adhesive is an alternate approach to transferring adhesive from the carbon tape *(28)*.

6. For finer precision in applying UV-cure adhesive to the pyramid, the end of the tungsten wire can be etched to submicron dimensions *(29)*; larger diameter tungsten wires may be used if the tip is etched in this fashion.

7. Use caution as the z-position of the tip is moved toward the surface; typically 50 nm or shorter incremental increases toward the surface are best. If the z-position is adjusted too rapidly, such that the tip is pressed toward the surface more than 100–200 nm after initial contact, the SWNT bundle can be knocked off of the pyramid tip.

8. If the attached SWNT bundle will not shorten or the resolution does not become acceptable, it may be necessary to remove the SWNT bundle and make a new tip, starting from the beginning.

9. Use considerable caution in placing tips into solutions and make sure the tips are completely submerged and on the bottom of the Petri dish. AFM tips are small enough that surface tension allows them to float on top of aqueous solutions; floating tips often drift into the edge of the Petri dish, where the resulting impact can break off the cantilever tip with attached SWNT.

10. Functionalization of SWNT tips with other molecules having primary amine groups can be accomplished by substituting 5 mM of the desired amine for 5 mM 5-BAPA in the EDC and MES buffer solution. For example, benzylamine and ethylenediamine have been used to generate phenyl and amine terminated nanotube tips, respectively *(12,17)*.

11. A variety of other surface functionalization schemes can be also used. Two other common approaches for attaching molecules to surfaces are formation of self-assembled thiol monolayers on gold-coated surfaces *(30)* and the use of silane coupling to covalently functionalize Si, glass, or mica *(31)*.

12. The fluid cell should be filled very carefully and slowly; forcing liquid at a high flow rate through the cell can cause the SWNT bundle to separate from the AFM tip.

13. It is possible to exchange solutions when the tip and surface are still in the liquid cell. Typically, the tip should be separated from the surface by a distance of >100 µm before replacing solutions; also, use caution to avoid introducing bubbles into the fluid cell.

14. Best results are typically obtained when a new biofunctionalized SWNT tip is used with each new substrate that is to be characterized. Biofunctionalized SWNT tips can sometimes be taken out from the liquid cell and then used again on a different surface, but it cannot be predicted in advance whether the biofunctionalized SWNT tip will work effectively after having been used in previous experiments.

15. If the AFM probe resolution degrades during imaging, the terminal portion of the SWNT tip can be removed as described in **Subheading 3.1.3.** to provide a new SWNT tip.

16. The resolution in SWNT probe AFM images can vary somewhat from one tip to another. Images with very high resolution (**Fig. 4B**) can be achieved often with SWNT tips; generally the resolution with a SWNT tip is as good as can be obtained with a new Si tip, and typically the resolution is significantly better than what is obtained from a Si tip that has taken several images (**Fig. 4A**).

Acknowledgments

I thank Professor Charles M. Lieber of Harvard University for my rewarding stay in his laboratory as a postdoctoral fellow. This work was supported in part by a postdoctoral fellowship from the Runyon-Winchell Foundation, and in part by the Army Research Laboratory and the U.S. Army Research Office under grant number DAAD19-02-1-0353.

References

1. Glusker, J. P. (1994) X-ray crystallography of proteins. *Meth. Biochem. Anal.* **37,** 1–72.
2. Harris, K. D. M., Tremayne, M., and Kariuki, B. M. (2001) Contemporary advances in the use of powder x-ray diffraction for structure determination. *Angew. Chem. Int. Ed.* **40,** 1626–1651.

3. Wagner, G., Hyberts, S. G., and Havel, T. F. (1992) NMR structure determination in solution: a critique and comparison with x-ray crystallography. *Annu. Rev. Biophys. Biomol. Struct.* **21,** 167–198.
4. Staunton, D., Owen, J., and Campbell, I. D. (2003) NMR and structural genomics. *Acc. Chem. Res.* **36,** 207–214.
5. Binnig, G., Quate, C. F., and Gerber, C. (1986) Atomic force microscope. *Phys. Rev. Lett.* **56,** 930–933.
6. Yazdani, A. and Lieber, C. M. (1999) Up close and personal to atoms. *Nature* **401,** 227–230.
7. Hansma, H. G. and Pietrasanta, L. (1998) Atomic force microscopy and other scanning probe microscopies. *Curr. Opin. Chem. Biol.* **2,** 579–584.
8. Czajkowsky, D. M., Iwamoto, H., and Shao, Z. F. (2000) Atomic force microscopy in structural biology: from the subcellular to the submolecular. *J. Electron Microsc.* **49,** 395–406.
9. Bustamante, C., Rivetti, C., and Keller, D. J. (1997) Scanning force microscopy under aqueous solutions. *Curr. Opin. Struct. Biol.* **7,** 709–716.
10. Dai, H., Hafner, J. H., Rinzler, A. G., Colbert, D. T., and Smalley, R. E. (1996) Nanotubes as nanoprobes in scanning probe microscopy. *Nature* **384,** 147–150.
11. Wong, S. S., Harper, J. D., Lansbury, P. T., Jr., and Lieber, C. M. (1998) Carbon nanotube tips: high resolution probes for imaging biological systems. *J. Am. Chem. Soc.* **120,** 603–604.
12. Wong, S. S., Joselevich, E., Woolley, A. T., Cheung, C. L., and Lieber, C. M. (1998) Covalently functionalized nanotubes as nanometre-sized probes in chemistry and biology. *Nature* **394,** 52–55.
13. Sun, L. F., Xie, S. S., Liu, W., Zhou, W. Y., Liu, Z. Q., Tang, D. S., et al. (2000) Creating the narrowest carbon nanotubes. *Nature* **403,** 384.
14. Wong, S. S., Woolley, A. T., Odom, T. W., Huang, J.-L., Kim, P., Vezenov, D. V., et al. (1998) Single-walled carbon nanotube probes for high-resolution nanostructure imaging. *Appl. Phys. Lett.* **73,** 3465–3467.
15. Woolley, A. T., Guillemette, C., Cheung, C. L., Housman, D. E., and Lieber, C. M. (2000) Direct haplotyping of kilobase-size DNA using carbon nanotube probes. *Nat. Biotechnol.* **18,** 760–763.
16. Cheung, C. L., Hafner, J. H., and Lieber, C. M. (2000) Carbon nanotube atomic force microscopy tips: direct growth by chemical vapor deposition and application to high resolution imaging. *Proc. Natl. Acad. Sci. USA* **97,** 3809–3813.
17. Wong, S. S., Woolley, A. T., Joselevich, E., Cheung, C. L., and Lieber, C. M. (1998) Covalently-functionalized single-walled carbon nanotube probe tips for chemical force microscopy. *J. Am. Chem. Soc.* **120,** 8557–8558.
18. Wong, S. S., Woolley, A. T., Joselevich, E., and Lieber, C. M. (1999) Functionalization of carbon nanotube AFM probes using tip-activated gases. *Chem. Phys. Lett.* **306,** 219–225.
19. Woolley, A. T., Cheung, C. L., Hafner, J. H., and Lieber, C. M. (2000) Structural biology with carbon nanotube AFM probes. *Chem. Biol.* **7,** R193–R204.

20. Hafner, J. H., Cheung, C. L., Woolley, A. T., and Lieber, C. M. (2001) Structural and functional imaging with carbon nanotube AFM probes. *Prog. Biophys. Mol. Biol.* **77,** 73–110.
21. Florin, E.-L., Moy, V. T., and Gaub, H. E. (1994) Adhesion forces between individual ligand-receptor pairs. *Science* **264,** 415–417.
22. Woolley, A. T. and Kelly, R. T. (2001) Deposition and characterization of extended single-stranded DNA molecules on surfaces. *Nano Lett.* **1,** 345–348.
23. Hughes, S. D. and Woolley, A. T. (2003) Detailed characterization of conditions for alignment of single-stranded and double-stranded DNA fragments on surfaces. *Biomed. Microdevices* **5,** 69–74.
24. Thess, A., Lee, R., Nikolaev, P., Dai, H., Petit, P., Robert, J., et al. (1996) Crystalline ropes of metallic carbon nanotubes. *Science* **273,** 483–487.
25. A website listing a number of commercial SWNT suppliers is the following Website: http://www.pa.msu.edu/cmp/csc/NTSite/nanotube-sources-com.html.
26. Hafner, J. H., Cheung, C. L., and Lieber, C. M. (1999) Growth of nanotubes for probe microscopy tips. *Nature* **398,** 761–762.
27. Hafner, J. H., Cheung, C. L., and Lieber, C. M. (1999) Direct growth of single-walled carbon nanotube scanning probe microscopy tips. *J. Am. Chem. Soc.* **121,** 9750–9751.
28. Hafner, J. H., Cheung, C. L., Oosterkamp, T. H., and Lieber, C. M. (2001) High-yield assembly of individual single-walled carbon nanotube tips for scanning probe microscopies. *J. Phys. Chem. B* **105,** 743–746.
29. Ekvall, I., Wahlström, E., Claesson, D., Olin, H., and Olsson, E. (1999) Preparation and characterization of electrochemically etched W tips for STM. *Meas. Sci. Technol.* **10,** 11–18.
30. Bain, C. D., Troughton, E. B., Tao, Y.-T., Evall, J., Whitesides, G. M., and Nuzzo, R. G. (1989) Formation of monolayer films by the spontaneous assembly of organic thiols from solution onto gold. *J. Am. Chem. Soc.* **111,** 321–335.
31. Sagiv, J. (1980) Organized monolayers by adsorption. I. Formation and structure of oleophobic mixed monolayers on solid surfaces. *J. Am. Chem. Soc.* **102,** 92–98.

Index